IMPROVING THE COLLECTION, MANAGEMENT, AND USE OF MARINE FISHERIES DATA

Ocean Studies Board

Commission on Geosciences, Environment, and Resources

National Research Council

NATIONAL ACADEMY PRESS
Washington, D.C.

NATIONAL ACADEMY PRESS • 2101 Constitution Ave., N.W. • Washington, DC 20418

NOTICE: The project that is the subject of this report was approved by the Governing Board of the National Research Council, whose members are drawn from the councils of the National Academy of Sciences, the National Academy of Engineering, and the Institute of Medicine. The members of the committee responsible for the report were chosen for their special competencies and with regard for appropriate balance.

This report was supported by a grant from the National Oceanic and Atmospheric Administration. The views expressed herein are those of the authors and do not necessarily reflect the views of the sponsor.

Library of Congress Catalog Number 00-107450
International Standard Book Number 0-309-07085-6

Additional copies of this report are available from:
National Academy Press
2101 Constitution Avenue, NW
Box 285
Washington, D.C. 20055
800-624-6242
202-334-3313 (in the Washington Metropolitan area)
http://www.nap.edu

Printed in the United States of America

THE NATIONAL ACADEMIES

National Academy of Sciences
National Academy of Engineering
Institute of Medicine
National Research Council

The **National Academy of Sciences** is a private, nonprofit, self-perpetuating society of distinguished scholars engaged in scientific and engineering research, dedicated to the furtherance of science and technology and to their use for the general welfare. Upon the authority of the charter granted to it by the Congress in 1863, the Academy has a mandate that requires it to advise the federal government on scientific and technical matters. Dr. Bruce M. Alberts is president of the National Academy of Sciences.

The **National Academy of Engineering** was established in 1964, under the charter of the National Academy of Sciences, as a parallel organization of outstanding engineers. It is autonomous in its administration and in the selection of its members, sharing with the National Academy of Sciences the responsibility for advising the federal government. The National Academy of Engineering also sponsors engineering programs aimed at meeting national needs, encourages education and research, and recognizes the superior achievements of engineers. Dr. William A. Wulf is president of the National Academy of Engineering.

The **Institute of Medicine** was established in 1970 by the National Academy of Sciences to secure the services of eminent members of appropriate professions in the examination of policy matters pertaining to the health of the public. The Institute acts under the responsibility given to the National Academy of Sciences by its congressional charter to be an adviser to the federal government and, upon its own initiative, to identify issues of medical care, research, and education. Dr. Kenneth I. Shine is president of the Institute of Medicine.

The **National Research Council** was organized by the National Academy of Sciences in 1916 to associate the broad community of science and technology with the Academy's purposes of furthering knowledge and advising the federal government. Functioning in accordance with general policies determined by the Academy, the Council has become the principal operating agency of both the National Academy of Sciences and the National Academy of Engineering in providing services to the government, the public, and the scientific and engineering communities. The Council is administered jointly by both Academies and the Institute of Medicine. Dr. Bruce M. Alberts and Dr. William A. Wulf are chairman and vice chairman, respectively, of the National Research Council.

COMMITTEE ON IMPROVING THE COLLECTION AND USE OF FISHERIES DATA

PATRICK SULLIVAN *(Chair),* Cornell University, Ithaca, New York
KENNETH ABLE, Rutgers University, New Brunswick, New Jersey
CYNTHIA JONES, Old Dominion University, Norfolk, Virginia
KAREN M. KAYE, U.S. Geological Survey, Reston, Virginia
BARBARA KNUTH, Cornell University, Ithaca, New York
BRENDA NORCROSS, University of Alaska, Fairbanks
ESTELLE RUSSEK-COHEN, University of Maryland, College Park
JOHN SIBERT, University of Hawaii, Manoa
STEPHEN SMITH, Bedford Institute of Oceanography, Dartmouth, Nova Scotia, Canada
STEVEN K. THOMPSON, Pennsylvania State University, University Park
RICHARD YOUNG, Commercial Fisherman, Crescent City, California

Consultant

JOHN G. POPE, NRC (Europe), Ltd.

Staff

EDWARD R. URBAN, JR., Study Director
ANN CARLISLE, Senior Project Assistant

Acknowledgments

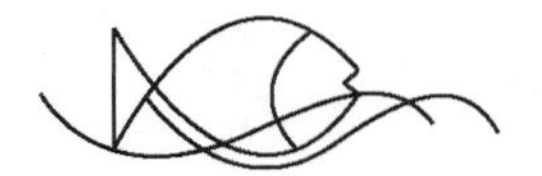

This report has been reviewed in draft form by individuals chosen for their diverse perspectives and technical expertise, in accordance with procedures approved by the National Research Council's (NRC) Report Review Committee. The purpose of this independent review is to provide candid and critical comments that will assist the institution in making the published report as sound as possible and to ensure that the report meets institutional standards for objectivity, evidence, and responsiveness to the study charge. The review comments and draft manuscript remain confidential to protect the integrity of the deliberative process. We wish to thank the following individuals for their participation in the review of this report: John Bailar (University of Chicago), Deb Southworth Green (U.S. Fish and Wildlife Service), David Hoel (Medical University of South Carolina), Pete Leipzig (Fishermen's Marketing Association), Douglas Lipton (University of Maryland), Bonnie McCay (Rutgers University), Kenneth Pollock (North Carolina State University), Terrance Quinn (University of Alaska, Fairbanks), Jake Rice (Canadian Department of Fisheries and Oceans), David Sampson (Oregon State University), Larry Six (Consultant), and Andrew Solow (Woods Hole Oceanographic Institution). While the individuals listed above provided constructive comments and suggestions, it must be emphasized that responsibility for the final content of this report rests entirely with the authoring committee and the institution.

Preface

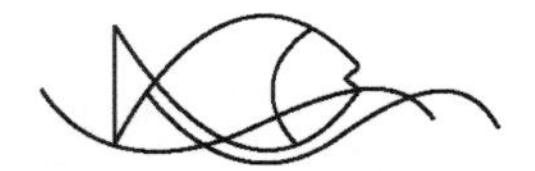

Congress has promoted fisheries science for over a century and its involvement in fisheries management took a great leap forward with passage of the Fisheries Conservation and Management Act of 1976. In the past decade, Congress has requested advice from the National Research Council (NRC) on both national issues (e.g., individual fishing quotas and community development quotas) and the assessments related to specific fisheries (Northeast groundfish). This report was produced, in part, in response to another congressional request, this time related to the assessments of the summer flounder stocks along the East Coast of the United States. Following the initial request, the NRC, National Marine Fisheries Service (NMFS), and congressional staff agreed to broaden the study into a more comprehensive review of marine fisheries data collection, management, and use.

National Research Council reviews of stock assessments result in unexpected tasks added to other responsibilities of stock assessment scientists and other NMFS personnel. The committee sent numerous questions to NMFS over the period of its study and NMFS was always responsive to the requests. Special thanks are due to David Sutherland, the committee's liaison at NMFS, and to Mark Terceiro, the stock assessment scientist responsible for summer flounder, who handled many queries from the committee. Mark also attended two of the committee's meetings to answer questions about the intricacies of summer flounder assessments and how the different data sources are used. Other NMFS personnel—particularly William Fox, Mark Holliday, and Maury Osborn—were also helpful in the committee's work and the committee greatly appreciates their efforts. The committee thanks Karl Haflinger (SeaState, Inc.) and Bill Karp (NMFS) for providing figures for the report. The committee was fortunate to be able to engage John Pope as a consultant to the project and to enlist the help of Bob Mohn of the Canadian Department of Fisheries and Oceans in assisting with the committee's analysis of summer flounder data. The committee could not have met its charge without substantial input from representatives of the commercial fishing industry, recreational fisheries sector, environmental advocacy groups, and others. Finally, the committee thanks Ed Urban and Ann Carlisle of the Ocean Studies Board staff for their support in carrying out this project.

Patrick Sullivan
Chair

Contents

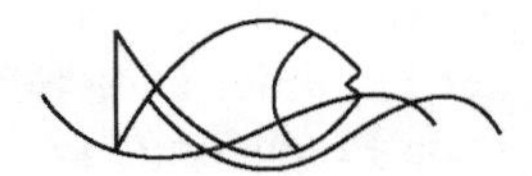

Executive Summary

Marine fish are important as a source of food, an item of commerce, the focus of recreational opportunity, and an element of cultural tradition in the United States and worldwide. Data from marine fisheries can contribute to our understanding of the marine environment and how humans relate to and use living marine resources. A comprehensive understanding of the problems currently challenging marine fisheries science and management requires consideration of both the biological and human dimensions of the fishery management process.

The dynamics of marine fish populations are affected directly by climate change, habitat availability, and water quality, but also are affected by human-influenced factors such as fishing and environmental degradation. In turn, human fishing practices are affected by the dynamics of the marine ecosystem and fluctuations in fish abundance. Thus, a complex relationship exists between fish and fishermen[1] that must be maintained to foster the existence of both. At the intersection of these complex interactions are fishery managers, who are required by the Magnuson-Stevens Fishery Conservation and Management Act of 1976 (the Magnuson-Stevens Act) to preserve both fish populations and the human harvesters who depend on these fish. Fisheries management requires high-quality observations and well-supported predictions about the status and dynamics of fish populations, and these will be influenced by and influence human activities. Stock assessment scientists, economists, and social scientists must work with managers to design appropriate methods to collect, manage, and use accurate and precise biological, economic, and social data to accomplish their management responsibilities.

Fisheries management responsibilities are shared among several partners in the United States. Most of the available information is collected and analyzed by the National Marine Fisheries Service (NMFS) in cooperation with state and interstate agencies. Some international fishery treaty organizations also collect and assess similar types of information. Eight regional fishery management councils formed pursuant to the Magnuson-Stevens Act work with interstate fish-

[1] The committee uses the terms "fisherman" and "fishermen" throughout the report because this is how practitioners of fishing (both male and female) tend to refer to themselves in the United States. Participants in recreational fishing will be referred to as "recreational fishermen" and "anglers."

ery commissions, treaty organizations, and states to implement fisheries management based on "best scientific information available" (required by National Standard 2 of the Magnuson-Stevens Act). Commercial and recreational fishermen participate to varying degrees in different fisheries by helping NMFS collect data. Participants in fisheries management often refer to fishery-independent data (collected by a resource agency, independent from fishing activities and using scientific sampling methods) and fishery-dependent data (measures of directed commercial and recreational fishing activity).

ORIGIN OF STUDY AND COMMITTEE APPROACH

This study reflects NMFS' desire to have the National Research Council assess methods for improving data for stock assessments and fisheries management, and a more specific interest by Congress to have the summer flounder stock assessments reviewed. The two objectives formed the two parts of this study.

Congress requested a one-time study of summer flounder stock assessments by the National Academy of Sciences as part of a conference report that accompanied the Departments of Commerce, Justice, and State, and the Judiciary and Related Appropriations Act of 1998 (PL 105-119). The following statement of task was developed to make the committee's work more useful nationally and to address the issue of data quality, which the NRC (1998a) identified as a major factor in the performance of stock assessment models:

> This study will evaluate the use of data in fish stock assessments and fishery management, including a variety of issues that range from those specific to summer flounder to more generic topics of data use for assessments of marine fish stocks. These issues will include methods of commercial data collection; accuracy and precision of fishery-independent surveys; institutional arrangements for data collection, analysis, and sharing among state and federal agencies; and appropriateness of data quality control procedures.

The congressional request to review the summer flounder assessments resulted from industry concern that NMFS had underestimated summer flounder stock size. Such a review also serves as an example of how fisheries data are used to provide assessment advice in general, how the quality of the data may affect the advice, and how public perceptions of data and assessments can affect their acceptance. Thus, the report first reviews the 1996 and 1999 assessments[2] of summer flounder and then uses the insights this review provides to help develop ideas on the appropriate collection of fisheries data more generally. Of course, not all stocks are distributed, exploited, or managed in the same way as summer flounder, nor are they all assessed with similar data sets. For example, groundfish in East Coast fisheries (including summer flounder) are sampled more frequently than fish in other regions, so it is important to recognize that summer flounder data are more extensive than those for many other species on the East Coast and fish species from other regions. Some species on the U.S. West Coast are surveyed only once every three years, whereas some U.S. fish stocks are not surveyed at all. Data requirements may be different for other species, but a practical example provided by the summer flounder fishery provides insight and force to the committee's broader recommendations later in the report.

Evaluation of the Summer Flounder Assessments

The summer flounder fishery was used as a case study for broader data issues of greater interest to NMFS. Summer flounder is a particularly appropriate focus of a case study because

[2] The committee first reviewed the 1996 assessment and used 1996 data for analyses in Chapter 2 because the 1999 assessment was not available until late in the project.

this species supports a fishery that spans the state waters from Maine to North Carolina together with the U.S. exclusive economic zone adjacent to the waters of these states. This area is targeted by both commercial and recreational fishermen, and has an abundance of data available for assessments. The committee fully reviewed the summer flounder stock assessments of 1996 and 1999 and held public meetings to learn the concerns of commercial fishermen, recreational fishermen, and environmental advocacy groups related to the summer flounder assessments. The committee evaluated the summer flounder assessments using different stock assessment models as a means to explore issues related to the summer flounder data.

Broader Data Collection and Analysis Issues

The broader issue of the need to improve the quality of data used in stock assessment was highlighted in the NRC report entitled *Improving Fish Stock Assessments* (NRC, 1998a), which showed that the quality of data used in five stock assessment models was more important than the particular model used. The present committee examined all forms of data available for stock assessments and fisheries management, including data from fishery-independent surveys, fishery-dependent data from commercial and recreational fishermen, and auxiliary data collected from a variety of other sources. The committee also examined traditional and new methods for collecting data, and discussed the current state of data management and several new developments in data collection and management.

FINDINGS AND RECOMMENDATIONS

Review of Summer Flounder Stock Assessments

Three analyses of summer flounder data by three individuals using three stock assessment models yielded the same general trends indicated by the 1999 NMFS assessment. These analyses showed that the spawning stock biomass had recovered substantially from a trough in the early 1990s, and that fishing mortality[3] dropped substantially in the same period. The committee believes that both changes are probably due to strict management measures implemented in 1992.

The models yielded some differences in predictions that are relevant to summer flounder management. These differences probably arose because there are assumptions in each modeling method that are not explicitly stated, but that affect model results. Thus, it is especially important to document all assumptions made within these complex models—whether explicit or implicit.

First, the model predictions of spawning stock biomass vary somewhat over time, especially in the most recent years. Some models indicate that the biomass has peaked and is falling again. This variability in the estimates of summer flounder biomass should be considered in managing the fishery, taking into account the range of possible spawning stock estimates from the models. All model biomass estimates are lower than the NMFS estimate.

Second, estimated fishing mortality rates varied greatly among the models. This result has implications for fishery managers in terms of how well they may be able to meet quantitative management targets, called biological reference points, based on fishing mortality rates. In the last year of the series, each method produced almost the same estimate of fishing mortality, all of which are above the fisheries management target level of fishing mortality (0.24 year^{-1}). The bottom-line conclusion of the committee's review of the summer flounder assessments is that the managers responsible for this species should be aware of the uncertainty that arises from the choice of model and should manage more cautiously (e.g., reducing fishing mor-

[3] Fishing mortality is a measure of the rate of removal of fish per unit of time from a population by fishing. Fishing mortality, expressed in a variety of ways, is a major fishery management indicator (see Box 1-2).

tality) in the presence of such uncertainty. This will generally require that they be somewhat more protective of fish stocks. Stock assessment scientists responsible for summer flounder should investigate how differences among model estimates arise and whether such differences indicate changes needed in the models or assumptions used.

These model runs were performed using most of the explicit assumptions used in the NMFS analyses, and assumed that NMFS and state surveys accurately portray the summer flounder population. But, what if the assumptions used by NMFS are incorrect? The committee addressed this possibility by analyzing the summer flounder data using the virtual population analysis method a number of times, changing single assumptions for different runs. Through these model runs, the committee came to the following conclusions:

• There is little direct evidence for the existence of a large number of unsurveyed large summer flounder, as claimed by some in the industry. Nevertheless, it is plausible that NMFS survey methods could miss such fish if larger fish are less susceptible than smaller fish to capture using NMFS trawl gear and methods. The committee demonstrated that this is an important issue in terms of the fishing mortality targets and total allowable catch, and suggested ways that NMFS could try to determine whether there are a significant number of unaccounted-for large summer flounder. If the NMFS surveys are less likely to catch larger fish than smaller fish, the previously documented size difference between female and male summer flounders will also need to be considered in assessments.

The age structure of the summer flounder population is important and its determination will require that NMFS and fishermen work cooperatively. The first stage should be joint trawling exercises by NMFS and commercial fishermen, using both traditional and adaptive sampling techniques, to test the effects of trawl gear and methods on catch. This would be the most direct approach. NMFS could also take additional, less direct steps, such as conducting egg surveys and tagging studies, but these actions could be more expensive and would take years to gather meaningful data. Observers could assist in tagging and recovering tags from fish.

• Another major need in stock assessments for summer flounder is for NMFS and industry to improve reporting of the discard rate. Because of the effect of discards on estimation of biomass and fishing mortality rates, industry and NMFS need to work together to devise means to encourage accurate reporting. One short-term solution would be to increase observer coverage to a high enough level to provide statistically meaningful estimates of discards. This could be accomplished by allowing fishermen on observed trips to catch more flounder to pay observer costs.

• Recreational fishing for summer flounder has been contentious during the past several years because of catch overruns in the recreational portion of the fishery. Recreational fisheries have been allotted 40 percent of the total allowable catch, but have taken more than 50 percent in the past two to three years. In part, this has occurred because the population of young fish, those more likely caught by inshore anglers, has rebounded in response to strict management controls. Because recreational data have a several-month lag time between collection and availability for management, overfishing can occur before fishery managers are aware of the problem. This problem is not unique to summer flounder; it is shared with most fisheries that have a significant recreational sector. Effective management cannot be achieved if only commercial fishermen are regulated strictly, while little control is exerted over recreational catches. Anglers, states, regional councils, and NMFS should work together to solve this particularly difficult problem. The ultimate goal should be data collection and management systems that allow in-season management of summer flounder fisheries, if this goal can be achieved cost effectively. In the absence of such a capability, the populations of summer flounder can be sustained only by more conservative management of recreational fisheries.

The actions recommended here may seem too extensive to be justified on the basis of the benefits of better data for the summer flounder fishery alone, particularly considering that NMFS must collect data for many other species. The premise of this study, however, is that lessons from the examination of the summer flounder data and assessments can be applied more broadly because summer flounder stocks, like those of many other groundfish species, are subject to both commercial and recreational fishing and cross legal boundaries of many states and the federal exclusive economic zone. The committee recommends that NMFS and the regional councils implement the recommendations of this report as a test of new ways of cooperating with commercial and recreational fishermen to improve both data quality and acceptance of stock assessment results.

Broader Data Collection, Use, and Management

The committee believes that all the participants in fisheries management should take actions to improve the collection, management, and use of fisheries data. The committee developed the following recommendations to Congress, NMFS, the regional councils, interstate commissions, and commercial and recreational fishermen with the objective of improving fisheries data and thereby fisheries management.

Recommendations to Congress

The U.S. Congress affects fisheries science and policy in two primary ways. First, Congress is the architect of the centerpiece of federal fisheries legislation, the Magnuson-Stevens Act. At present, Congress is formulating legislation to reauthorize the Magnuson-Stevens Act, whose funding authority expires on October 1, 2000. The committee recommends several ways in which the reauthorization could improve fisheries data collection, management, and use in the United States.

Second, Congress appropriates funding for NMFS, the regional councils, and interstate and international commissions to carry out their activities related to fisheries science and management. The committee highlights several items for which additional funding could improve fisheries data collection, management, and use and, consequently, fisheries management. Funding for more capable research vessels and for planning a Fisheries Information System are recent examples of positive congressional steps toward modernization of fisheries data.

Fisheries management is based on ad hoc methods of data collection developed over the past 25 years that may no longer lead to the best management. Congress should support and encourage NMFS to re-evaluate its systems of data collection, management, and use, and to conduct research to increase the effectiveness of these activities. Another important need is for a fishery-by-fishery analysis of the costs and benefits associated with data collection and fisheries management.

In the most recent reauthorization of the Magnuson-Stevens Act, Congress requested that NMFS develop a preliminary design for a Fisheries Information System. The committee believes that such a system could improve and standardize the management of U.S. marine fisheries data and thereby help managers understand regional trends and how they fit into the national context. The committee believes that the Fisheries Information System should be funded on an experimental basis for a fixed term, perhaps 10 years, with quantifiable and measurable objectives that can be evaluated at the end of that period.

Congress should continue to support the acquisition and calibration of new NMFS fishery research vessels that are more effective in data collection and handling than the vessels currently available in the aging NMFS fleet. The so-called "fish for research" programs used by NMFS and regional fishery management councils have proven to be a useful means of involving commercial fishermen in research and sampling. Congress should continue to support such programs, with the details of implementation left to the discretion of regional councils.

Congress should amend the Magnuson-Stevens Act to limit the confidentiality of commercial data. By providing better access to commercial data, such a step would help managers and scientists better understand the biology, sociology, and economics of fisheries. Sunset periods on confidentiality are logical outcomes of the public ownership of marine fish resources and the public trust responsibilities of NMFS and the regional councils in fisheries management. The proprietary periods may vary by data type (e.g., they may be shorter for fishing locations than for economic data) and by specific fishery, and these periods should be determined cooperatively between fishery managers and stakeholders. As part of the effort to gather and disseminate needed data, Congress should lift the prohibitions in the Magnuson-Stevens Act on collection of economic data (Sec. 303[b][7] and 402[a]).

Recommendations to NMFS

NMFS has many, in some cases conflicting, responsibilities. NMFS and the regional fishery management councils often suffer from a credibility problem and are more or less continuously engaged in conflicts with commercial and recreational fishermen and environmental advocates who disagree with fishery management plans or other aspects of fisheries management. These conflicts range from criticism voiced in regional council meetings and other public meetings to legal challenges to fishery management plans approved by the councils and NMFS. Some of these conflicts are probably unavoidable results of the dynamics of the regulator-regulated relationship between NMFS and fishermen and their different perceived objectives—such conflict is to be expected. Nevertheless, NMFS and fishermen do share a fundamental objective: the long-term sustainable use of marine living resources and the acquisition of whatever data are necessary to achieve this objective. NMFS and fishery stakeholders should work together to resolve their conflicts to achieve "win-win" solutions. Conflicts might be reduced by greater cooperation between NMFS and fishermen in data collection, so that NMFS develops trust in data from commercial and recreational fisheries and fishermen become confident that NMFS provides accurate data and assessments.

NMFS should continue to explore more cost-effective ways of obtaining the fisheries data it needs, including implementing new remote sensing techniques (e.g., hydroacoustics); implementing electronic logbooks and vessel monitoring systems; increasing observer coverage where needed; developing adaptive sampling in appropriate fisheries; and, especially, finding ways to improve commercial data to make it more useful for stock assessments and finding ways to estimate recreational catch more quickly to allow in-season management of recreational fisheries. NMFS also should consider creating mechanisms to obtain advice from commercial and recreational fishermen related to specific data collection policies and procedures. This could be accomplished through a combination of national meetings to discuss national-level policies and regional meetings to discuss data collection in specific fisheries, possibly through each regional council's scientific and statistical committee.

The Marine Recreational Fisheries Statistics Survey (MRFSS) should be fully funded and include all coastal states and territories that request inclusion. NMFS should invest in research related to MRFSS and investigate new ways to enlist recreational fishermen in data collection for routine monitoring and special studies, but only if the agency intends also to fund implementation of the results of the research. It appears that MRFSS funding and staffing levels are adequate only to maintain the existing survey and conduct a minimal amount of research, the results of which are not always implemented in a timely manner. Some recommendations have been implemented (e.g., changes in variance estimates), whereas others remain to be implemented (e.g., retention of previously contacted anglers in subsequent surveys).

NMFS should standardize the data sets and protocols included in the proposed Fisheries Information System, using the standards for spatial and other data established by the Federal Geographic Data Committee. The agency should consider moving away from proprietary data management software to software that is available from many vendors and for which data access and analysis routines can be written easily.

NMFS should evaluate the success of commercial data management firms in providing real-time value-added data products for specific operational purposes, and should determine ways to encourage such entrepreneurial activities. At the same time, NMFS should endeavor to obtain useful data from such sources.

The committee identified a number of data collection activities that merit special attention from fishery scientists within both NMFS and the academic community:

- Developing methods for evaluating the ecological benefits of fish stocks and their role in marine ecosystems.
- Determining how to minimize changes in the relationship of actual abundance to indices of abundance (e.g., survey, commercial, or recreational catch per unit effort) and misreporting when management systems are changed.
- Testing adaptive sampling for data collection for both NMFS and industry.
- Testing electronic logbooks and vessel monitoring systems that offer value-added features to fishermen.
- Linking environmental, economic, and social data, as well as climate forecasts, to stock assessments.
- Improving understanding of the functioning of the marine ecosystems affected by fishing activities by studying important non-target species to determine their feeding habits, their distribution, and their prey and predators.
- Gaining a greater understanding of the economic and social motivations of fishermen so that data from commercial and recreational fisheries can be interpreted correctly.
- Validating procedures for determining fish ages and identifying stocks.

Recommendations to Regional Fishery Management Councils

Regional councils should be more proactive and innovative in developing mechanisms within fishery management plans that encourage NMFS to work more effectively with commercial and recreational fishermen in data collection. Councils should play a major role in promoting greater use of data from commercial and recreational fisheries by including programs for collecting and using such data in fishery management plans, and working with NMFS to design appropriate mixtures of data collection approaches (e.g., vessel monitoring systems, observers, logbooks). The design and implementation of fishery management plans should include consideration of how data quality might be enhanced and whether data of the required accuracy and precision are available or could be collected in a cost-effective manner. If sufficient data quality is unlikely to be achievable at a reasonable cost for a particular type of management, councils should consider alternative, less data-intensive management systems. Councils should give serious consideration to new "fish for research" programs that could engage fishermen in data collection and research. Councils should obtain the data needed to conduct in-season management of recreational fisheries or, conversely, manage recreational fisheries conservatively enough so that in-season data are not necessary. They should work with NMFS to improve outreach to commercial and recreational fishermen, and should encourage independent review of data collection and stock assessments on a regular basis.

Recommendations to Interstate Commissions

Interstate commissions should find ways to increase the standardization of state survey data used in federal stock assessments, consistent with important state uses of the data. Commissions should work with NMFS and the states to create and maintain regional databases, and coordinate them through the proposed Fisheries Information System.

Recommendations to Commercial Fishermen

Commercial fishermen are a critical source of data about the fish stocks they depend on, and more generally, about marine ecosystems. Under most existing management systems, however, commercial fishermen have many incentives to misreport catch data and few incentives to provide accurate and complete data. Although the extent of misreporting is hard to quantify, evidence suggests that it does occur. Many improvements in fisheries management will require active participation of commercial fishermen in data collection, including more extensive cooperation in sampling and a reduction in misreporting of commercial data. Commercial fishermen should work with NMFS to obtain accurate and precise measures of the relative abundance of fish stocks, both through commercial data and research surveys. Commercial fishermen could help improve both and it would be to their benefit to do so—the fish stocks on which they depend are more likely to be sustained if both fishermen and managers share the same accurate view of the abundance of fish stocks.

Recommendations to Recreational Fishermen

Recreational fishermen presently play a relatively small and passive role in data collection, although the interest of anglers in participating in fish-tagging studies have been well demonstrated through the efforts of the American Littoral Society and others to tag sportfish. Angler organizations should increase their cooperation with NMFS and academic scientists to assist in routine data collection and scientifically designed, targeted studies, in order to improve the recreational catch data that are needed in stock assessments. Although scientifically designed tagging studies demand careful implementation, they are crucial to the accurate assessment of fish mortality and movement. Angler assistance is particularly important in fisheries that have a significant recreational component, such as the summer flounder fishery.

CONCLUSION

The future of fisheries management will be based on complementary data from fishery-independent surveys, commercial fishermen, and recreational fishermen. A particular need is to improve the quality of data from commercial and recreational fisheries, so that stock assessment scientists can be justifiably confident about using such data in their models. Commercial and recreational sources could provide large quantities of data important for stock assessments and for understanding the social and economic aspects of marine fisheries, but these data are not always useful in their present form. The sustainable use of marine fish resources, and concomitant protection of marine environments, will require new levels of commitment by the public and their representatives in Congress and federal and state governments to fund and carry out appropriate data collection and management.

1

Introduction

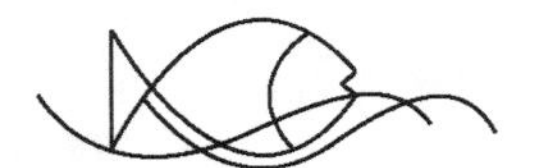

Some people fish to provide food for themselves and their families, others for commerce. Some fish for recreation, whereas others fish to carry on traditional ways of life and ceremonial customs. In the sense most relevant to management—and according to the Magnuson-Stevens Fishery Conservation and Management Act of 1976—fisheries include the humans who harvest fish. In addition to the direct benefits to humans, fisheries management must also consider ecosystem-level objectives such as preserving essential fish habitat and maintaining diversity of marine organisms. Because fishing by its nature causes some mortality to fish—even if the fish escape the gear or are caught and released—and because the summed activities of all individuals who fish may impact fish populations, it is critically important to understand and manage fisheries to ensure continuing benefits to all. The Magnuson-Stevens Act provides a set of 10 national standards for fishery conservation and management (Sec. 301 of the Act), to which the fishery management plans developed by regional fishery management councils[1] must adhere. Some of these standards can be formulated in quantifiable terms (although this is not always done). For example, National Standard 1 requires estimation of optimum yield and overfishing thresholds, National Standard 7 mandates minimizing costs of management, and National Standard 9 requires minimizing bycatch mortality (so these must be measured). Adherence to the national standards can be evaluated on the basis of data collected to characterize biological, economic, and social aspects of fisheries and the effects of management on these characteristics.

Even with the best data, fisheries and fisheries management are subject to uncertainty. We can never be completely sure of the current population abundance of a stock or how it will change. Environmental variables that affect the growth and reproduction of fish stocks are frequently unknown and are always difficult to both measure and predict. Even if the environment were

[1] The exclusive economic zone of the United States is divided among eight regions by the Magnuson-Stevens Act for the purposes of fisheries management. The fisheries of each region are managed by a separate fishery management council, as specified in Section 302 of the act. The regional councils are composed of designated state and federal fishery officials, as well as commercial fishermen, recreational fishermen, and environmental advocates nominated by state governors.

predictable, the effects of interactions of fish stocks with each other and with the environment are largely unknown. The expectations of fishermen, scientists, and managers can differ quite drastically from reality because of such uncertainties. The major goal of data collection is to support and enable biological, economic, and social analysis that will reduce these uncertainties, so that harvest can be sustained at the highest level that is commensurate with other management goals, such as maximizing long-term potential yield. NMFS (1999) estimates recent average yield from U.S. fisheries is 33% below the long-term potential yield.

Fisheries management includes three activities on an ongoing basis: (1) assessing the condition of a fish stock[2] in the context of its place in the ecosystem and in connection with the fishery it supports; (2) developing and implementing regulations to use and sustain the fish stock and the fishery; and (3) monitoring the biological, economic, and social effects of regulations. These activities may each require biological, economic, and social data. Because fish are publicly owned and publicly managed resources, the government has a stake in collecting biological, economic, and social data needed to encourage effective management. However, collection of private and proprietary social and economic data is sometimes viewed by the industry as being too intrusive (PFMC, 1998b). The legal precedents for the government's role in protecting resources for the good of the nation, the government's *public trust* responsibilities, are not as developed for marine resources as for terrestrial resources. A good summary of the legal precedents is given in NRC (1999b).

A fundamental goal of data collection and quantitative stock assessment processes is to estimate current and future stock abundance and the effects of fishing activities. Data management is an important link between the collection and assessment processes, because data of high quality must be available in a timely manner and accessible form to be useful for assessment scientists. High-quality biological, economic, and social data are essential for evaluating the effectiveness of regulations and, when necessary, designing new regulations.

A gulf commonly exists between the beliefs of fishermen and those of managers and scientists in terms of the current status of marine fisheries, the existence of problems in the fisheries, and how such problems can be solved. The committee will explore the nature of these beliefs in the following chapters. Differences in viewpoints related to fisheries problems frequently arise because[3]

- different fisheries stakeholders operate with different time horizons. Commercial fishermen often have to focus on cash flow within a given year to meet current expenses. Managers may evaluate the costs and benefits of different management options based on discount rates set by the government at 7% (OMB Circular A-94). Scientists and environmental groups focus on sustaining stocks indefinitely for both biological and economic reasons.
- fishermen take pride in their ability to catch fish and in their good working knowledge about fish behavior and distribution. They are frustrated by scientists who seem to be unwilling or unable

[2] "A fish stock can be defined as all fish belonging to a given species that live in a particular geographic area at a particular time, that is, all individuals actually capable of interbreeding. For practical management purposes, a stock is often further defined by political boundaries. That is, the management unit, often still called a stock, includes those members of a biological stock that are under management by a single governmental agency. Units so defined, however, do not necessarily reflect meaningful biological entities or the spatial heterogeneity of fish distributions." (NRC, 1998a, p. 8)

[3] These points are obviously generalizations and are based on statements made by participants at the committee's meetings, not on systematic sociological research on fishery stakeholders.

to utilize this knowledge and expertise. Scientists, on the other hand, frequently consider data from commercial and recreational fisheries to be biased (i.e., collected where fish density is high rather than representative of the full range of abundance); are skeptical of estimates of stock abundance that may be inflated due to technical innovation in fishing gear and methods; and believe that data from fisheries are qualitative (rather than quantitative) in nature. Scientists may have personal experience of receiving incorrect information from fishermen. Fishermen suspect that scientists present the most pessimistic results in assessments. Scientists are reluctant to include potentially biased data in their assessments because they have learned through their training and experience that such data may result in incorrect and misleading assessments. Fishermen are sometimes reluctant to share their information, including accurate logbook data, because they have learned through their experience that such data often are ignored or can be used against them.

- even when scientists use commercial data (e.g., catch, fishing effort, and catch-at-age data) these data are filtered through mathematical models in ways that make it difficult for non-scientists to understand the resulting output and its interpretation. Even in the cases in which industry generally understands the use to which their data are put, a knowledge of the likely errors in the data (e.g., suspected misreporting by other fishermen)—as well as the length of time between data gathering and the assessment—may make them suspicious of the assessment results.
- scientists treat an assessment as an accurate representation of a stock that can provide a basis for action; some fishermen appear to consider assessments as no more than an opening bid in negotiations to set total allowable catch.
- fishermen are made acutely aware daily of the variability of nature and usually ascribe stock changes to such variability, rather than to constant fishing pressure. Conversely, scientists interpret declining trends in fish populations as being exacerbated by fishing pressures.
- sustaining individual stocks and maintaining ecosystem structure and function may require that large standing stocks be left in the sea for reasons that may not be obvious to fishermen.
- different users may hold different goals for the fishery. For example, commercial fishermen may seek to maximize total biomass harvested (for the food market), whereas recreational fishermen may seek to maximize size of individual fish harvested (for trophies) and tend to want to maintain high standing stocks so they are more likely to catch fish on any particular fishing trip.

These differences in viewpoint among commercial fishermen, recreational fishermen, and scientists are often central to problems of stock management and conservation. Fishermen tend to be most vocal when management is most restrictive. Because of the uncertainty that always exists in stock assessments, fishermen may attack both the science and the stock assessments, arguing for risk-prone management decisions and total allowable catches (TACs) at the high end of assessment ranges (Sissenwine and Rosenberg, 1993). This can result in a downward spiral in the fishery, as increasingly dire warnings are met with increasingly strident objections until, ultimately, all doubt is removed by a fishery collapse. Some complaints from fishermen are valid and have a basis in fact, and in some cases fishermen are correct and NMFS could learn from them. In other cases, different opinions of fishermen and scientists arise from fishermen's misunderstanding of fish population dynamics that needs to be addressed if fishermen and managers are to become partners in sustainable fisheries management. However, even if all participants in fisheries have the same understanding of the problems, differing incentives will lead participants to argue for different solutions.

This report examines existing practice in fisheries data collection, management, and use in the United States and recommends how these important functions of fisheries management can be carried out more effectively.

DATA COLLECTION

Biological Data

A common problem in understanding and managing marine fisheries is that the fish populations are not directly observable. As a consequence, key pieces of information that determine the type of management, such as the size of the stock and the rate at which fish are being removed, rarely can be observed directly and often can only be determined in a relative sense. For example, a 10,000 metric-ton catch of a species might have been taken by removing 1 percent of a 1 million metric-ton stock or 50 percent of a 20,000 metric-ton stock. Fisheries assessment science concerns itself with determining which is most likely the case from the available data. It uses the resulting estimates to provide biological advice to fishery managers.

Many kinds of biological data are useful for fisheries management. An integrated age-based[4] assessment (see Box 1-1), for example, requires accurate estimates of the following biological information characterizing the population and the way the fishery interacts with it:

- Total fishing mortality = landed catch + discarded catch*percent discard mortality + mortality of unlanded fish caused by gear
- Proportion of catch comprising each age of fish
- Relationship of weight and age in the population
- Proportion of fish of each age in the population that are mature
- Recruitment indices[5]
- Indices of abundance such as catch per unit effort from commercial vessels or from a scientific survey
- Mortality rate from causes other than fishing (natural mortality)

[4] Age-0 fish are less than 12 months old; age-1 fish are 13-24 months old; and so on.

[5] Recruitment is the addition of new individuals to the population and is typically viewed as individuals becoming vulnerable to capture as they grow or change their behavior as they mature.

The effort required to catch a given number or weight of fish is critical to understanding how catch is related to fish population abundance. Data are collected directly from commercial and recreational fisheries (these provide fishery-dependent data)[6] and through various statistically designed surveys (these provide fishery-independent data) conducted by state and federal fishery agencies and some international commissions. Fishery-independent data are collected to provide measures of relative abundance that are not confounded by the commercial and recreational strategies of targeting areas where fish densities are highest. Collection procedures are designed according to sound sampling principles. Fishery-dependent and fishery-independent data collection methods will be discussed in greater detail later in the report.

Other auxiliary information about fish populations can be used in assessments. These include egg surveys, which count the numbers of fish eggs of a species from a specific area of the ocean and use these numbers to estimate the size of the population of spawning females. Acoustic

[6] Fishery-dependent data can be collected from anyone who harvests fish—recreational, commercial, ceremonial, and subsistence users. These groups have very different motivations, and if we are to make use of data collected in a fishery, we must understand and account for these motivations. More specifically, fishery-dependent data include catch (and appropriate descriptions of size, age, sex, and location of the catch), effort (and appropriate descriptions of the fleet or anglers applying the effort), catch per unit effort, fish prices, information on the costs of fishing, the technology employed in fishing, the number of jobs generated, the net economic value of the fishery, the distribution of these items by geographic area and time, and the effects of fishing on habitat. Fishery-dependent data can be obtained from onboard observers, vessel monitoring systems, logbooks, and port agents, as described in detail in Chapter 3.

surveys use echosounders or sonars to count numbers of fish or, more commonly, estimate the biomass. Both of these methods can be valuable because they estimate absolute rather than relative biomass or numbers of fish. Egg surveys can be expensive because they often require a number of scientific cruises to span the spawning season of a fish (often several months long) and work best for fish that spawn over a short time period (summer flounder spawn several times over a period of four to six months). Acoustic surveys tend to work best for species that give strong echos (associated with swim bladders), aggregate in single-species schools, and are found either at the surface or in mid-water. (Summer flounder, like other flatfish, have no swim bladder and being bottom-dwelling fish are poor subjects for acoustic surveys.) Another approach to estimating mortality and population size is to use tagging experiments. In practice, it is usually difficult to design such experiments to be wholly representative of a stock, but they can provide data to help test the assumptions used in stock assessment models.

Social and Economic Data

Traditionally, biological data, such as those described above, have been collected and used most commonly for fisheries management, although it has long been recognized that economic and social data related to fisheries, fishermen, and communities are also needed. However, economic and social data have not been routinely collected for fishery purposes. The success of fisheries management depends on knowing how management may affect individuals and communities as well as fish stocks. Management measures may intentionally or unintentionally change how fish are harvested.

Social and economic data are needed to understand catch, catch at age, and catch per unit effort because social and economic factors affect these variables. Also, some of the intended benefits of fisheries management are social and economic in nature. Several of the Magnuson-Stevens Act's national standards concern social and economic issues, so relevant data must be collected to assess the performance of fishery management plans in relation to these standards. Social and economic data are necessary for

1. understanding fishery-dependent information and comparing the predictions of stock assessment models with the observations of fishermen,
2. constructing regulatory impact review sections in fishery management plans,
3. evaluating the effects of management actions,
4. understanding multispecies fisheries in which fishermen switch among species based on both biological and economic factors, and
5. designing incentives and disincentives that are likely to result in compliance with regulations intended to encourage responsible fishing.

For example, economic data are needed to understand the age composition of the catch if certain age/size classes are being targeted for special markets (e.g., large fish being highly prized for the sushi market). The *West Coast Fisheries Economic Data Plan* produced by the Pacific Fishery Management Council (PFMC, 1998b) provides a list of the kinds of economic data that should be collected (Table 1-1). The next generation of stock assessment models may include social and economic factors that can affect fishing activities.

Fisheries assessment biologists are accustomed to observing a variation of 20% or more in their estimates. The magnitude of this observational variation depends on sample coverage, ecosystem variation, and the longevity of the species. Scientists have been frustrated in the past because even an increase in sample size did not guarantee an increase in the precision[7] of current

---

[7] Precision is a measure of the variability of data around its mean value, whereas accuracy is a measure of the closeness of a measured or computed value to its true value.

TABLE 1-1 Economic Data Needed for Fisheries Management

Harvesters	Processors	Charter Vessels	Recreational Anglers	Communities
Revenue and effort[a] data	Revenue data	Revenue data	Effort and catch[c] by target species	Tax revenues
Fixed and variable *cost data*	Fixed and variable *cost data*	Fixed and variable *cost data*	Cumulative per angler catch and effort	Fishery-related economic infrastructure
Cost of harvesting	Cost of processing	Cost of doing business		
Wages paid and jobs *provided*	Wages paid and jobs provided	Wages paid and jobs provided	Trip costs and angler demographics	Fishery-related income and employment
Capacity[b] information			Angler values and preferences	Geographic and physical characteristics

SOURCE: PFMC (1998b), p. ES-2, modified by committee (changes in italics).

[a] Effort is a measure of the resources devoted to catching fish, most often hours or days after fishing occurs.

[b] Capacity is a measure of the vessel and gear resources applied to a specific fishery, such as the net capacities, number of fixed gear units carried, and horsepower of the vessel.

[c] Catch is the number or weight of fish captured, including fish discarded.

biomass levels, due to the variation occurring in the driving forces of the environment and commerce. Now, however, there has been an information explosion and the rapid development of methods and technology for dealing with environmental and industrial variations. Managers are just beginning to acknowledge the uncertainty of scientists' predictions and the risks associated with overexploitation. Scientists will continue to refine their estimates and predictions, but managers and stakeholders must account for risk and uncertainty in their decisionmaking. Managers and stakeholders should be able to call on scientists and economists to help evaluate the risks and set reasonable objectives.

DATA MANAGEMENT

After data are collected, they must be processed to make them useful for stock assessments and fisheries management. Data processing often includes some degree of quality control. As computing and communication resources have advanced, attempts have been made to link data collection more closely to management systems. Presently, data are managed at state, regional, national, and international levels, not always with appropriate standardization and communication structures in place to allow sharing among organizations and levels. Incipient efforts have begun to standardize data collection among regions, and NMFS has produced a plan for a nationwide fisheries information system in response to a request of Congress. Important issues of fisheries data management relate to the appropriate level of data confidentiality, how fisheries data can be made compatible with other types of environmental data collected by the government, and the timeliness of data for decisionmaking.

DATA USE

Biological data are used primarily to understand the effects of fishing on fish populations and marine ecosystems. Social and economic data are used mainly to address the human components of fisheries and to develop and implement management methods that sustain fishermen and fishing communities and ensure future stock viability.

Assessments

Fisheries management, particularly when it is based on total allowable catch, relies heavily on assessments of stock abundance, which in turn rely on accurate and precise information. This information often comes in the form of data collected from commercial fisheries (e.g., logbooks, observers, and landing receipts), from the Marine Recreational Fishery Statistics Survey (e.g., intercept and telephone surveys), and from scientific field observations (e.g., annual state and federal trawl surveys). Clearly, the quality[8] and quantity of data directly affects the accuracy of assessments and resulting predictions. The National Research Council (NRC) (1998a) found that the quality of data and soundness of assumptions employed by stock assessment scientists are major factors influencing the performance of stock assessment models.

Scientists use a number of approaches to stock assessment, depending on the nature of a species' population dynamics, the value of the fishery, and the management requirements for advice. Thorough reviews of the theory of population dynamics and stock assessment are found in Hilborn and Walters (1992) and Quinn and Deriso (1999). Because most commercial fish species have life spans of several years and because there is often substantial variation in the numbers recruited each year, it is often appropriate to estimate the number of individuals in each year-class.[9] Such approaches are called age-based or age-structured assessments and are often developed within a broader assessment framework sometimes referred to as an integrated or fully analytical assessment (Box 1-1). Assessments for summer flounder are integrated assessments but other approaches are appropriate for other fisheries (see NRC, 1998a, for more details). Age-based methods require a finer level of detail in the data than required by other methods—for example, those based on stock production theory or those using equilibrium concepts—and thus tend to be more expensive than methods that do not distinguish among ages of individual fish.

Traditionally, an age-based assessment proceeds with the application of sequential population analysis to the catch-at-age data (together with an assumed rate of natural mortality). This approach provides estimates of the numbers of fish in the sea in earlier years. It also provides estimates of the removal rate (measured by assessment scientists as fishing mortality rate, F) for these earlier years. These estimates are possible due to the convergence of the estimates backward in time. Put simply, the life of every fish ends in either catch or natural death. At any given level of natural mortality, the total number of fish caught from a cohort through a cohort's lifetime must be the number of that cohort recruited, minus adjustments for deaths due to natural mortality. These techniques are often collectively referred to in the literature as cohort analysis or virtual population analysis (VPA), although both of these terms correspond to specific types of sequential population analyses (Megrey, 1989).

It was recognized early in the development of assessment methods based only on catch data

[8] The concept of data quality includes such attributes as bias, precision, accuracy, comparability, completeness, and representativeness (EPA, undated).

[9] Many fish species produce young at specific times of year and thus the population is formed of distinct age cohorts often called year-classes.

BOX 1-1
Age-based or Integrated Models

Age-based models use equations based on initial year-class size, natural mortality rate, and fishing mortality rates to determine the abundance of each year class. Because of the flexible nature of such general models, integration of a variety of data types is feasible, thus the models are often referred to as integrated. Estimation can be accomplished by maximum likelihood or least-squares procedures applied to age-specific indices of abundance, age-specific catch, and other types of auxiliary information. The main advantage of these models is that they make almost full use of available age-specific information. The primary disadvantage is that such models require many observations and include numerous parameters, thereby increasing the cost of using them. ADAPT, CAGEAN, and Stock Synthesis are all age-based models; the first two were used by the committee to run assessments on summer flounder data. ADAPT (Gavaris, 1988; Conser and Powers, 1989) is an age-structured assessment method based on least-squares comparison of observed catch rates (generally age-specific) and those predicted by a tunable[a] sequential population analysis. CAGEAN (Deriso et al., 1985) is an age-structured assessment method based on forward-recursion population equations, a least-squares objective function (although other objective functions are also described), and lognormal distributions for catch at age and fishing mortality. Stock Synthesis (Methot, 1989, 1990) is an age-structured assessment technique based on maximum likelihood methods, but with more flexibility to include auxiliary information and fitting criteria. The Stock Synthesis program allows inclusion of "environmental proxies" that might influence parameters such as natural mortality or growth. A length-based, age-structured model that has been used for several highly migratory pelagic species is MULTIFAN-CL (Fournier et al., 1998). Somewhat simpler, so-called ad hoc tuning methods, such as the Laurec-Shepherd technique (Laurec and Shepherd, 1983; Pope and Shepherd, 1982; Darby and Flatman, 1994), adjust the last age/last year assumptions of sequential population analysis to give the closest fit to auxiliary information. Because the Laurec-Shepherd method is relatively simple in its formulation and its behavior is well understood, the committee used it to analyze the potential problems and questions associated with the summer flounder assessments. It is significant that these models typically do not include information about environmental conditions and how fish populations respond to different conditions, although, in principle, there is no reason that such information could not be considered.

The assessment models applied in this review were chosen for their generality, ease of application, and the fact that they should be familiar to most in the fisheries community. These models represent a range of assumptions and are sensitive to different characteristics of a fishery. Such model attributes actually make these approaches useful for diagnosing problematic areas in an assessment, and thus they are useful tools to have on hand. Models better suited for a full assessment should deal with the complexities of these systems on a fishery-by-fishery basis. The software and technical methodology for dealing with these complexities are now available, making the need for good information more important.

---

[a] Tuning of a model involves adjusting parameter estimates to minimize differences between predicted population estimates and observations from indices of population (e.g., catch rate, survey index of abundance).

that such methods could not provide estimates of population abundance or mortality for the most recent years, because more recent cohorts have not yet been subject to the full extent of fishing or natural mortality. Abundance and mortality rates in the most recent years have to be related back to estimates for earlier years by using relative indices of abundance, such as catch per unit effort (CPUE)[10] of the fishery or relative trends from surveys.[11] In the case of summer flounder, only survey data are used to tune the assessments. Although some commercial catch, catch-at-age, and CPUE data are available for this stock, they are not used because concern has been expressed that CPUE indices from fishery logbooks are incomplete, inaccurate, or too variable. In part, this is because of changes in regulations for the summer flounder fishery over time.

More recent approaches use both catch-at-age data and the auxiliary data in the overall assessment. Such integrated approaches normally proceed by simultaneously fitting all data sources using least squares or maximum likelihood techniques. Traditional sequential population analysis techniques typically assume that catch-at-age data are precise and accurate and provide population estimates through a back-calculation technique. Integrated analyses assume that the catch observations may have error and that the parameters of the model should be estimated by balancing good fits of the model to the catch data and to auxiliary effort or survey abundance data.

Estimates of the numbers of fish in the most recent year, obtained from tuned sequential population analysis or integrated analyses combined with estimates of the recruitment, expected fishing mortality rate, natural mortality rate, and weight-at-age data, allow predictions to be made of the future catch. This is the basis for estimating TACs compatible with management plans for a sustained harvest.

Regulations and Management

Because fish typically are free-access common property resources, commercially valuable stocks tend to be harvested in an unsustainable or uneconomic fashion unless they are managed. Fisheries management uses stock assessment results and other data for decisionmaking. Managers can use only a few instruments to ensure a sustained harvest. Typically, managers employ rules that directly govern the actions of fishermen. Managers can attempt to control landings by

1. imposing catch quotas, bag limits, or trip limits;
2. limiting fishing effort (e.g., restrictions on the amount of time and gear that can be applied in a fishery);
3. closing key fishing areas;
4. closing seasons when a certain level of catch is achieved; and/or
5. restricting the kind or quantity of gear that may be used and the time and place it may be used or by limiting the number of users.

In principle, fiscal measures such as taxes on catches, fishing licenses, or fuel could also be used to manage fisheries. Frequently, a combi-

[10] Catch per unit effort (CPUE) is a measure of the average number or biomass of fish caught per unit of effort (e.g., fisherman-hours, length of tow, number of hooks). CPUE is often used as a measurement of relative abundance for particular fish stocks (e.g., Pacific halibut), although it can be misleading because fishermen are selective rather than random samplers of fish populations and because the nature and quality of the fishing gear and targeting practices may change over time, resulting in a CPUE that can be stable or increasing in the presence of an actual decline in a fish stock (Rose and Kulka, 1999).

[11] The major objective of fishery-independent surveys is to monitor temporal and spatial changes in the relative or absolute abundance of a target fish population or a particular component of that population (e.g., larvae, juveniles, spawning adults) in a manner that is not subject to the biases inherent in commercial or recreational fishery data, which are subject to changes in gear and targeting practices. Surveys usually use either fixed stations or randomly selected stations within the geographic range of the stock and often use gear that can be applied consistently year after year. More information is provided in Chapter 2 and Appendix C regarding the surveys used for summer flounder.

BOX 1-2
Biological Reference Points

Biological reference points are calculated quantities that describe a population's state and are used to evaluate objectively the consequences of management. They are used as targets for optimal fishing and for setting overfishing thresholds. (Councils must include measures in fishery management plans to prevent or end overfishing) (Magnuson-Stevens Act, Sec. 301[a][1], Sec. 303[a][10]). Biological reference points are calculated from the life-history characteristics of a given stock and are used to define harvest control rules. Biological reference points can be based on fishing mortality rate (F) (related to yield per recruit[a] or spawning biomass per recruit[b]) or biomass and often assume the existence of some optimal level for abundance.

$F_{0.1}$ and F_{max} are defined from the relationship between the yield per recruit and fishing mortality.

F_{max} is the fishing mortality rate that provides maximum yield per recruit; it is a threshold level. Levels of fishing mortality higher than F_{max} constitute growth overfishing because individual fish are harvested before they have had a chance to grow to a size that will maximize the yield per recruit. F_{max} is used as an overfishing definition for summer flounder.

$F_{0.1}$ is the fishing mortality rate beyond which increases in yield per recruit relative to increases in fishing effort are marginal. In technical terms, $F_{0.1}$ is defined as the rate of fishing mortality for which the increase in yield resulting from a small increase in fishing mortality is one-tenth the increase that would have resulted if the same small increase in fishing mortality had been applied to the unexploited stock (Gulland, 1969). $F_{0.1}$ is generally preferred over F_{max}, because $F_{0.1}$ is a target reference point that is lower than F_{max} and provides a buffer to avoid growth overfishing. It is widely used as a target F.

$F_{X\%}$ and F_{med} are based on the relationship between the spawning biomass per recruit and fishing mortality.

$F_{X\%}$ is the fishing mortality calculated to reduce the spawning biomass per recruit to X percent of its unfished level. Many of the overfishing definitions for U.S. fish stocks are based on this biological reference point (Rosenberg et al., 1994). Values from 20-30% have been used to define overfishing, and values from 35% to 50% have been used for target values.

F_{med} is the fishing mortality rate that allows the adult fish to be adequately replaced by new generations of fish. It theoretically gives the replacement spawning stock biomass per recruit that is equivalent to the median historic level. Sissenwine and Shepherd (1987) suggested F_{med} as an overfishing threshold because a stock harvested at this rate (or lower) should be able to replace itself. However, F_{med} is based on observed survival ratios, which depend on knowledge of the exploitation history of the stock, which is often unavailable.

---

[a] Yield per recruit is the total yield in weight harvested from a year-class of fish over its lifetime, divided by the number of fish recruited into the stock.

[b] Spawning biomass per recruit is "the ratio of the total weight of mature fish in a fish stock to the total weight that would exist if the stock were unfished" (Roberts et al., 1991).

nation of measures is employed to protect the long-term sustainability of the stock and long-term interests of users. However, most regulations ultimately control the current catch, and this may limit the income of fishermen in any given year.

Stock assessment scientists support the management process by advising on the nature of long-term sustainable harvest plans and the measures that are needed to accomplish such plans. They estimate the numbers of fish or biomass and a corresponding fishing mortality at each age in each year. Such estimates can be used to examine the rate of recruitment, current recruitment levels, and spawner-recruitment relationships and then to predict what might occur under various management scenarios. Biological reference points are prespecified management control objectives developed from observations and assumptions based on population characteristics (e.g., annual mortality rates, growth, maturity at age) that often prove to be useful statistics for comparing the output of stock assessment models (Box 1-2).

CONTENT OF REPORT

The committee summarizes its review of the 1996 and 1999 summer flounder assessments in Chapter 2 and provides additional detail about the assessments in Appendixes C and D. Chapter 3 provides background information for the more general data issues examined by the committee and Chapter 4 contains the committee's findings and recommendations.

2

Summer Flounder: Review and Insights

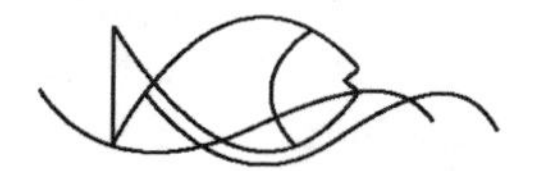

INTRODUCTION

The committee examined the summer flounder assessments with two objectives in mind. The first was to address specific issues raised by Congress and stakeholders regarding the quality and nature of these assessments and to gain a direct understanding of the significance these issues have for managing summer flounder. The second objective was to provide examples to illustrate the broader issues associated with the collection and use of fisheries data that affect the science and management of other stocks.

To focus these tasks, the committee elicited comments from representatives of the commercial and recreational fishing communities and environmental advocates concerning data collection and assessment methods applied to summer flounder. The committee reviewed the summer flounder data and the NMFS assessments and conducted its own independent assessments (Appendixes C and D) and an examination of modeling assumptions (this chapter) to generate its findings.

The committee provided opportunities at its two Washington, D.C. meetings for public input on summer flounder issues. Although more general issues were raised at the committee's third meeting in Seattle, Washington, some of the concerns raised there also had relevance to summer flounder. Commercial and recreational fishermen, scientists from the National Marine Fisheries Service (NMFS), representatives from the Mid-Atlantic Fishery Management Council, representatives from the Atlantic, Gulf, and Pacific States Marine Fisheries Commissions, and representatives from environmental groups made presentations to the committee at these meetings.

Commercial fishermen expressed a variety of views that indicated a general lack of trust in the management of summer flounder and other marine fisheries. Many recreational fishermen and their representatives were invited to the committee's meetings. Unfortunately, the committee was unable to elicit significant input from this sector regarding the summer flounder assessments or other data issues. Although the issues raised by stakeholders are not all resolved in this report, the report does explore a number of the more significant and persistent issues in the hope that NMFS and the regional councils will take steps to fix problems and correct misperceptions.

The committee's public sessions revealed

many perceptions of some commercial fishermen that affect their beliefs about fisheries data and stock assessments.

- Many fishermen believe that fisheries science and fisheries assessments are not objective scientific exercises; instead, they reflect political agendas.
- There are several steps between a stock assessment and the final management advice. Many fishermen believe that precaution is built into each step of the assessment process, so that the final results are too conservative. Fishermen advocate that scientists should provide completely objective advice and that managers should build in precaution only at the final management stage.
- Regulations change too frequently and before it has been determined whether management objectives have been achieved. This is especially problematic in cases that result in changes in gear regulations, because fishermen need to undertake the expense of replacing or modifying their gear with each change. The annual changes in the mesh size requirements for summer flounder were noted as an example.
- The fishing gear and survey techniques used by NMFS in its fishery-independent surveys have remained relatively constant over time. Fishermen no longer use the same kind of gear and consequently believe that NMFS' use of outdated fishing gear and fishing practices for surveys result in stock assessments that estimate smaller populations than the industry believes are realistic.
- Many fishermen question a number of assumptions about the biology and population dynamics used in specific stock assessments (e.g., in the case of summer flounder, this includes the assumed level of natural mortality and the assumption of constant selectivity for the older age groups).

Whether or not these concerns are valid from a scientific perspective, management of marine fisheries will be more difficult if these perceptions of fishermen are not addressed.

Because the National Research Council (NRC) was requested by Congress to focus on summer flounder, the committee examined the 1996 and 1999 summer flounder assessments and makes a number of recommendations in this chapter for improving assessments. It should be recognized, however, that NMFS cannot afford to conduct similar investigations for all the species it manages. Therefore, Chapter 4 highlights only those recommendations that should be applied broadly in fisheries data collection, management, and use in the United States. It is the responsibility of NMFS and the Mid-Atlantic and New England Fishery Management Councils to determine whether the steps recommended for summer flounder are a priority given the value of summer flounder compared to other fisheries. The lack of certainty in assumptions described below may hinder the summer flounder stock from recovering to its 1980s peak, although environmental factors may also be important deterrents to full stock recovery.

SUMMER FLOUNDER ASSESSMENT ISSUES

Here we examine the facts about summer flounder, determine the merits of the concerns, and recommend to Congress, NMFS, and the councils ways to correct problems and misperceptions. The committee investigated questions related to (1) the biology and population dynamics of summer flounder; (2) summer flounder sampling; and (3) the information content of the model and model assumptions currently in use (Box 2-1). In some cases, data were not available for the committee to investigate the issues.

Some of the concerns about the scientific assessment and management of summer flounder are specific and are dealt with in the following sections. Others result from the fishing industry's lack of trust in and an alienation from the management process. Such alienation suggests that there is a wider problem of cooperation and outreach, which will be addressed in Chapter 4.

BOX 2-1
Concerns About Summer Flounder Assessments Investigated by the Committee

1. **Questions related to the biology and population dynamics of summer flounder**
   a. Do the summer flounder in waters north of Cape Hatteras comprise a unit stock of fish?
   b. What natural mortality rate is appropriate to use in summer flounder assessment models?
   c. Are there differences between the growth and mortality of male and female summer flounder (sexual dimorphism) and, if so, how do the differences affect the assessments?

2. **Questions related to summer flounder sampling**
   a. What are the appropriate survey and commercial catchabilities for summer flounder?
   b. Do problems with determining the age of summer flounder discredit age-based assessments?
   c. Are effort data used appropriately and are the effects of effort changes incorporated properly?
   d. Is the observer program for summer flounder adequate?
   e. Can and should state surveys be standardized?
   f. Is the catch from recreational fishing estimated properly?
   g. Can precision of data be increased?

3. **Questions related to the information content of the model and model assumptions currently in use**
   What information does each model structure require and how do these requirements relate to the information in the data? Additional things to consider when answering this question are (1) the weight the model gives to the different sources of information that come into it, (2) how uncertainty is expressed in the assessment output provided to managers, and (3) the level of precision managers can expect with the data currently available.

QUESTIONS RELATED TO THE BIOLOGY AND POPULATION DYNAMICS OF SUMMER FLOUNDER

Summer flounder (*Paralichthys dentatus*; also called fluke) are found primarily from Cape Fear, North Carolina, to Cape Cod, Massachusetts (see Figure C-1) on a variety of seafloor substrates, including mud and sand (Able and Kaiser, 1994; Packer and Hoff, 1999). Summer flounder can reach a length of about 30 inches and a weight of some 20 pounds, and the maximum age is thought to be approximately 20 years. In a pattern typical for adults of many species of estuary-dependent fish, adult summer flounder move inshore in the spring and summer to feed in the bays and sounds and migrate offshore to spawn in fall and overwinter at the shelf break (Able and Fahay, 1998). Summer flounder appear to migrate northeast along the continental shelf as they grow larger. Considered lie-in-wait predators, they eat a large variety of prey, usually smaller fish, crabs, and shrimp.

Commercial catches of summer flounder along the Atlantic Coast have fluctuated between 2 and 18 million kg over the past 50 years. Recreational catch has been tracked only since 1981 and during this period, catch has fluctuated between

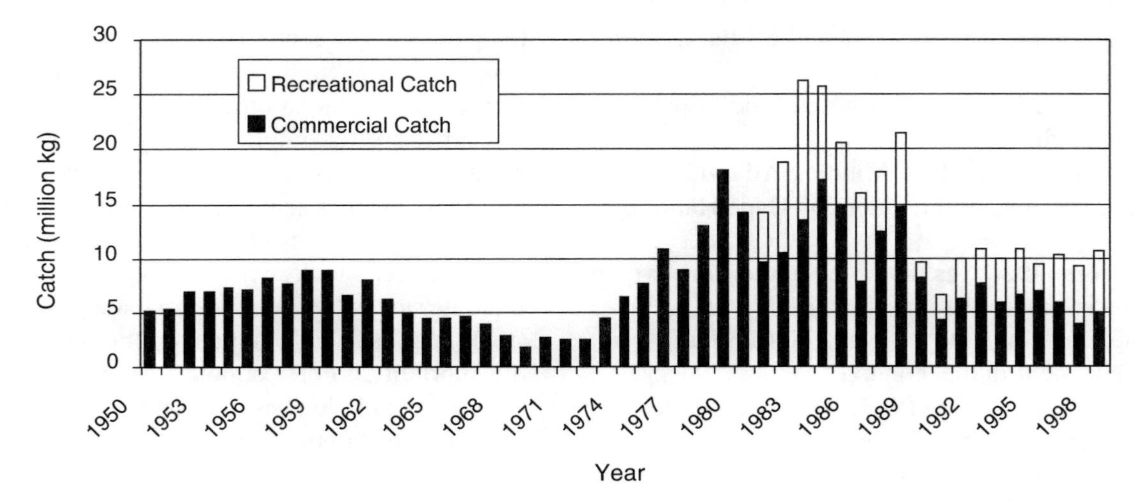

FIGURE 2-1 Commercial catch since 1950 and recreational catch since 1981 (earlier recreational data are unavailable) from Maine to North Carolina.

1.4 and 12.7 million kg annually. Amendment 2 to the Fishery Management Plan for summer flounder, approved in 1992, allocated 60 percent of the total allowable catch (TAC) to commercial fishermen and 40 percent to recreational fishermen. However, recreational catch has equaled commercial catches in several years and exceeded commercial catch in 1997 and 1998 (Figure 2-1), 69% over the recreational TAC. The proportion of fish in the population that are three or more years old has increased from 4 percent in 1993 to 43 percent in 1998 (Terceiro, 1999).

Do the summer flounder found in waters north of Cape Hatteras comprise a unit stock of fish?

National Standard 3 of the Magnuson-Stevens Act (Sec 301[a][3]) states "to the extent practicable, an individual stock of fish shall be managed as a unit throughout its range, and interrelated stocks of fish shall be managed as a unit or in close coordination." For ideal management, a species' distribution would be split into a set of discrete stocks, each of which spawns separately, has similar growth and mortality characteristics, and is fished by a different set of fishermen. This ideal is rarely achieved and the reality is that stock management areas have to be large, often comprising several separate, but usually overlapping, breeding populations and nursery areas. There are obvious administrative advantages of this approach and scientifically there may be advantages to combining sub-stocks rather than having to estimate migration rates among them. However, problems may result from combining a number of sub-stocks, potentially spanning a large gradient of latitude. The most important problem is that a combined stock may appear to be in reasonable shape even though some component is being fished to extinction.

Combining sub-stocks may also hinder stock assessments if growth, recruitment, natural mortality, or fishing mortality differ substantially among the sub-stocks of a combined stock. This may result in difficulties of estimating average weights at age or mortality rates. Thus, the adequacy of stock definition should be questioned in every assessment (see Appendix D of NRC, 1998a). In the case of summer flounder, Able and Kaiser (1994) published an extensive review of the species' life history. Three possible interpretations of the number and range of summer flounder stocks were presented:

1. Two major stocks: Middle Atlantic Bight and South Atlantic Bight
2. Two major stocks, both in the Middle Atlantic Bight
3. Three stocks: Middle Atlantic Bight, South Atlantic Bight, and trans-Cape Hatteras

The committee was not able to determine which of these interpretations is correct, but provides comments that may be useful for multi-stock management in this context.

Fogarty et al. (1983) used morphological characteristics and discriminant analysis to explore summer flounder stock structure. They found significant morphological differences among New York-New Jersey, Mid-Atlantic, and Cape Hatteras stocks, but were uncertain whether these differences had any biological significance. Although Jones and Quattro (1999) identified some genetic differences between summer flounder from Massachusetts and Rhode Island, it is unlikely that these are different stocks of fish that should be managed separately. Jones and Quattro pointed out that stock identification should be based not only on genetic data but should also consider life history attributes and morphological characteristics.

The existence of genetic, morphological, or other differences does not automatically mean that different populations should be treated as separate stocks for management purposes because (1) morphological or genetic differences may not translate into significant ecological differences and may simply reflect short-term or small-scale fluctuations in population structure and ultimate survival of a species; (2) the differences may not be related to factors that are important in stock management; and (3) the value of the fishery may not justify the cost of treating the stocks separately. Conversely, it is possible for populations to be genetically and morphologically identical throughout a vast geographic range and yet it may be preferable to manage them as separate stocks because of differences in growth, reproduction, and mortality. Such differences can be driven by differences in water temperature, productivity of phytoplankton and zooplankton, abundance of natural predators, fishing pressure, and other factors.

What are the implications for stock assessments and management if the unit stock assumption is wrong? If separate stocks are fished differently, the fishing mortality rate estimated for the combined stock will not be applicable to the separate stocks. Very intensive fishing of a small stock may have little impact on the overall fishing mortality. For example, the Arcto-Norwegian cod population has two components, one that feeds in the Barents Sea and one in the Bear Island-Svalbard area. Both components apparently spawn together on the Norwegian coast, mostly at Lofoten (it is not known whether they interbreed). Typically, the Barents Sea component is twice the size of the Svalbard component. Hence, if the former were not fished and the latter fished to extinction, the combined fishing mortality rate might be 0.3, which might seem reasonable, but the Svalbard stock component would have been eliminated. This is more or less how the North Sea herring collapsed, as separate units of a stock over its entire range were successively fished out. This scenario is more likely with schooling fish because fishermen can maintain high catch rates on declining, contracting populations rather than move to more abundant stock components. A similar situation exists with the various species of salmon on the west coast of North America. Species that are listed as endangered in California, Oregon, and Washington are abundant in Alaska. Thus, if the mortality of all salmon stocks is calculated together, the resulting number is not alarming. However, the number also does not reflect the true situation of the health of the salmon stocks. Northern cod provide another example (Rose et al., 2000).

If, on the other hand, a unit stock splits into two separately assessed stocks without accounting for movement between them, differential migration between the components will distort estimates of fishing mortality and population size. For example, some cod spawned at the southwest corner of Iceland drift as larvae to Greenland, where there is a separate fishery. They return as spawning adults to Iceland, where the extra numbers accruing to the Icelandic stock distort the fishing mortality rates downward if this factor is not taken into account (Shepherd and Pope, 1993).

Committee Investigations

The committee could not determine whether the summer flounder population should be considered multiple stocks. NMFS told the committee that "no meristic or morphometric data have been or are being regularly collected that would enable separation of the historical landings in potential stock components." However, any fishery that spans a large range might include multiple stocks.

Actions Needed

Testing the validity of the unit stock assumption requires an analysis of growth rates, fecundity, mortality, and other population factors among sub-areas. NMFS should conduct separate assessments of the stock if there are indications of different population factors (e.g., a difference in size-at-age between the northern and southern ends of the range), using whatever data are available specific to sub-areas of the range. Some relevant data may be available from the states, but a lack of standardization of their methods and differences in the timing and scope of their surveys may make such data inappropriate for reliable comparisons. The decision about implementing a split stock assessment depends on the benefit of doing so and the ability of the councils to use advice based on such assessments.

What natural mortality rate is appropriate to use in summer flounder assessment models?

Natural mortality rate (*M*) is the rate at which fish are removed from the population for reasons other than fishing activity, for example, predation, disease, and permanent emigration. In most cases, *M* is assumed to be the difference between total mortality and fishing mortality. For many temperate fish stocks, including summer flounder, stock assessment scientists assume that *M* is constant over time and at different ages of fish after they are large enough to be caught in the fishery and unaffected by fish population density.[1] In the case of summer flounder, *M* is set at a level of 0.2 per year on the fished ages of the stock; this level is commonly, though perhaps too pervasively, used for many fisheries worldwide. The meaning of this *M* value is that if a year-class of the population were not fished, its numbers would decline exponentially so that year-class number N(t) is given by the formula $N(t) = N(t_0)e^{(-0.2*(t-t_0))}$, where t is the current age in years and t_0 the initial age. For example, this formula predicts that after one year has elapsed, 18 percent of the year class would have died, after 5 years, 63 percent would have died, and that after 10 years, 86 percent would have died of causes other than fishing. For fished populations, fishing mortality rate (*F*) and natural mortality rate (*M*) are generally viewed as additive. Both *F* and *M* have the units of $year^{-1}$, but can also be expressed as instantaneous rates. In practice, fishing mortality is often the dominant cause of death at ages after fish have become large enough to be caught in the fishery.

Errors in the assumption about the level of natural mortality lead to some well-understood effects on the resulting assessment. For example, errors in the estimate of *M* lead to approximately equivalent and opposite errors in the estimates of *F*. Thus, if *M* is underestimated by 0.1, *F* will be overestimated by approximately 0.1, and *F* targets for future catches will be lower than they should be and harder to achieve. Thus, an incorrect choice of the natural mortality rate moves the estimates of current and target fishing levels in a direction opposite to the error in *M*. A lower natural mortality rate causes the estimated yield-

[1] Mortality of young-of-the-year fish, prior to their recruitment, is not included in the *M* used for management.

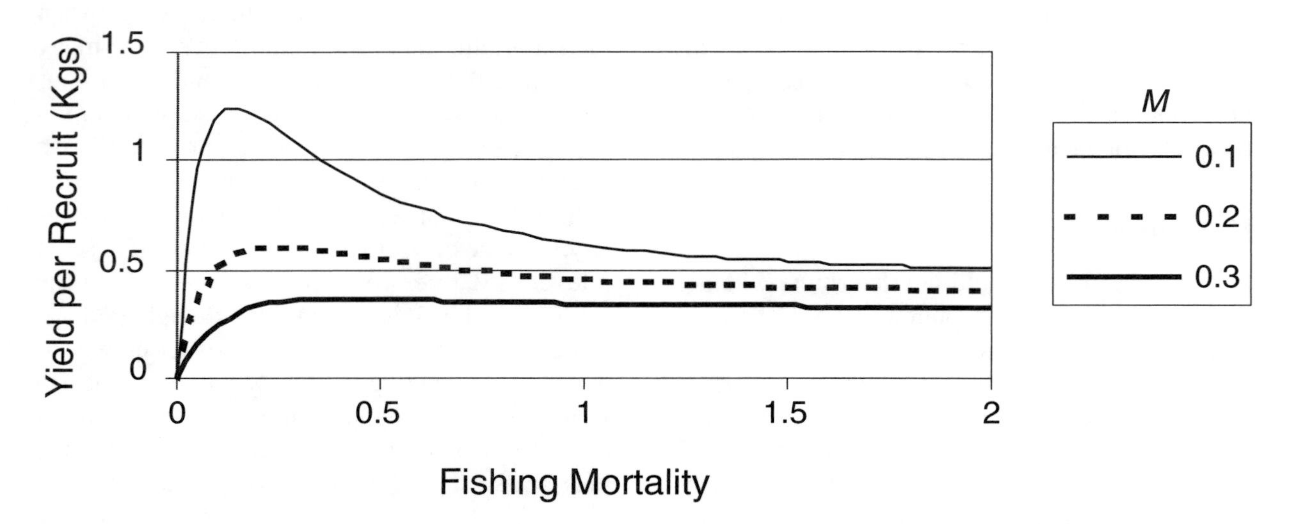

FIGURE 2-2 Summer flounder yield per recruit for three levels of natural mortality. F and M have the units of year^{-1}.

per-recruit curve[2] to steepen and peak at a lower level of fishing mortality rate and a higher M causes the curve to flatten and peak at higher levels of fishing mortality rate (Figure 2-2).

An error in M may cause an error in the opposite direction for calculated biological reference points such as F_{max} and $F_{0.1}$ (see Box 1-2 for definitions), which are often targets for management. However, under certain circumstances, an error in M may cause an error in the same direction for biological reference points, such as F_{med} (Sissenwine and Shepherd, 1987; Jakobsen, 1992). Thus, F_{med} may lead to management advice that moves F in the correct direction even if natural mortality is estimated incorrectly. Because it is managed to an F_{max} limit reference point, summer flounder management is affected by the choice of M.

The preceding discussion assumes that natural mortality rate is constant with fish age and through time. If M changes with age, the estimate of F also is affected differentially by age. If M changes through time for fish of specific size, perhaps as a result of changes in the abundance of or susceptibility to predators, the effects may be variable and subtle. One effect of a period of unaccounted-for increased natural mortality is to underestimate the number of young fish expected to have entered the fishery. This is particularly the case if natural mortality increases on fish of post-recruit ages. Such effects are sometimes investigated using multispecies assessment models by accounting for predation mortality. Research on fisheries in other areas of the world suggests that predation mortality may be quite high and variable on the younger and smaller sizes of fish (Sogard, 1997). However, predation mortality estimates for North Sea flounder species are generally low (ICES, 1997).

Multispecies models, in theory, allow estimation of natural mortality for prey species directly because they quantify M by observing predation deaths through stomach content data from predators. Multispecies models of the Georges Bank ecosystem have been constructed in the past, but have not been used to estimate predation

[2] Such curves describe the expected yield from an individual fish (a single recruit) over its lifetime.

mortality of summer flounder.[3] Although a number of methods exist for estimating natural mortality rate indirectly, few are reliable at this time because estimates of M are confounded with estimates of the proportion of fish vulnerable to fishing at each age.

Tagging experiments in which the initial mortality and tag loss rates on fish of catchable ages are well known and low provide another, more direct means of estimating M. The combination of knowing how many fish were released and the full time series of annual recaptures allows estimation of natural mortality rate (Hamre, 1980). Tagging data can also be used to estimate the dependence of M on fish age and size (Hampton, 1991). In fisheries characterized by a substantial mixed use (e.g., recreational and commercial harvest of summer flounder), each sector is likely to have a different reporting rate (typically recreational anglers are more likely to return tags) and each must be estimated properly. When tagging studies are combined with port sampling and intercept surveys wherein tags are obtained by the surveyor, both natural and fishing mortality can be estimated even in mixed-use fisheries (Brooks et al., 1998).

Committee Investigations

Because natural mortality rate is one of the more uncertain parameters used in a stock assessment, the committee investigated the effect of using a standard Laurec-Shepherd tuning of the virtual population analysis (VPA) of 1996 data (which gives results similar to the 25th Stock Assessment Workshop assessment for the same data [NEFSC, 1997]). The biological reference points and future TACs and spawning stock biomass (SSB) estimates were then calculated on the assumption that the 1997 catch was 7,162 metric tons. Calculations were also made under the assumption that fishing mortality remained at the status quo level (the 1996 level) in 1997. These factors were then recalculated assuming a natural mortality rate of 0.1 (instead of $M = 0.2$) at all steps of the calculation (Table 2-1).

Points to note from this analysis are that F_{max}, $F_{0.1}$, and $F_{20\%}$ are lower for $M = 0.1$ than for $M = 0.2$. However, F_{med} and the estimate of the current level of fishing mortality F_{1996} are estimated to be higher with $M = 0.1$ than $M = 0.2$. These specific results for summer flounder show how the various biological reference points change in response to a change in M, in line with the general theoretical expectations discussed earlier in this section. With respect to the predictions, the level of fishing mortality needed to take the TAC in 1997 was very similar in both cases, but represented a somewhat larger relative reduction in fishing in the $M = 0.1$ case (0.45 to 0.14), because the current level of fishing mortality appears higher. For $M = 0.1$, the equivalent catch in 1998 at F_{max} is much smaller. This reflects the need to reduce fishing mortality to a lower target level. Spawning stock biomass is similar in both runs in 1997 but is higher for $M = 0.1$ in 1998. This is due to the larger reduction in fishing mortality rate.

Catches and spawning stock biomass achieved in 1997, applying the same mortality rate as in 1996 (the status quo level), have similar results for both M values. Thus, these specific results for summer flounder also support the general conclusion that natural mortality estimates do not affect estimates of TACs much when M is held constant from one year to the next. It is clear that the correct value of natural mortality rate does influence the biological reference points and the likely target catch rates. A choice of a natural mortality of 0.1 would lead to more restrictive management decisions than one of 0.2, while one of 0.3 (not shown) would require lower reductions. M is a parameter that should be esti-

[3] The Northeast Stock Assessment Workshop in Fall 1995 (SAW 20) examined methods of Hoenig (1983), Pauly (1980), and $F_{max} = 3/M$ (e.g., Anthony, 1982) and chose to continue to use $M = 0.2$ for all ages for summer flounder. In stomach content analyses, young summer flounder do not seem to be preyed on heavily by other species, with the exception that young summer flounder are occasionally found in the stomachs of striped bass and dogfish (M. Terceiro, NMFS, personal communication, 1999).

TABLE 2-1 Effects of Changes in Assumed Natural Mortality Rate on Biological Reference Points, Estimates of Fishing Mortality, and Predicted Future Catch and Spawning Stock Biomass (SSB)

	$M = 0.2$	$M = 0.1$
Biological Reference Points (1996)		
$F_{0.1}$	0.15	0.09
F_{max}	0.24	0.14
$F_{20\%}$	0.28	0.17
F_{med}	1.48	1.77
Estimated fishing mortality in 1996[a]		
F_{1996}	0.95	1.03
Predictions for 1997 under status quo[b]		
F_{1997}	0.95	1.03
$Catch_{1997}$ (mt)	12,248	12,585
SSB_{1997} (mt)	12,263	11,403
Predictions for 1997 under TAC of 7,162 mt		
F_{1997}	0.44	0.45
$Catch_{1997}$ (mt)	7,162	7,162
SSB_{1997} (mt)	16,419	16,077
Predictions given 1997 TAC of 7,162 mt and $F=F_{max}$ in 1998		
F_{1998}	0.24	0.14
$Catch_{1998}$ (mt)	4,541	2,898
SSB_{1998} (mt)	32,677	37,204

NOTE: SSB = spawning stock biomass; TAC = total allowable catch. See Box 1-2 for other definitions. The assumed value of M not only affects estimates of fishing mortality, survivorship, and biomass, it also affects biological reference point estimates. Thus, the consequences of assuming an alternative M must be assessed in the context of all the estimates influenced by M and should be examined on a time scale appropriate to the life span of the fish.

[a] F_{1996} = the estimated fishing mortality for fully recruited summer flounder in 1996. This is the arithmetic mean of the fishing mortality over ages two, three, and four.

[b] The status quo condition assumes that fishing mortality stays at the 1996 level ($F = 0.95$).

mated correctly unless a management method less sensitive to its value (e.g., using F_{med}) is used.

What is the correct level of natural mortality for summer flounder? As with other species, the natural mortality rates of flounder species are difficult to estimate and often have been assigned values below 0.2. However, the M value for summer flounder has been estimated using a number of indirect approaches in previous stock assessment workshop (SAW) assessments (NEFSC, 1997) and the SAW concluded that $M = 0.2$ is a reasonable working assumption.

Actions Needed

Given that F_{max} is a basis for management plans, and that it depends on the value assumed for M, it is essential to obtain an independent estimate of M for summer flounder. A well designed and carefully executed tagging program is an excellent method for estimating M. Such a tagging program has the added advantage of also providing estimates of F and of movement rates between different parts of the range of the population. Tagging programs are time consuming and difficult to conduct and require full cooperation of all stakeholders in the fishery to be effective.

Emigration of fish from the surveyed area can significantly confound tagging studies, so estimates of emigration should be obtained. Relatively large-scale attempts to estimate emigration for summer flounder of the Chesapeake Bay region are described in the unpublished dissertation by Desfosse (1995). The benefits of such an experiment (unless results from a suitable previous experiment exist) would take a number of years to become available. In addition, the cost associated with a large-scale tagging study should be balanced with the benefit of obtaining information on this component of population dynamics relative to other, more pressing data needs, such as the vulnerability of older fish to fishing pressure across the management area and over time. Multispecies models that estimate predation mortality from stomach content analyses is an approach to estimating the importance of predation as a component of natural mortality. However, unless these kinds of data are already available, a new intensive (and expensive) sampling program would have to be implemented to sample the stomach contents of animals that prey on summer flounder. Such a sampling program may not be cost effective. It might also be possible to estimate M internally in the model, using Bayesian or other methods, as has been done in previous assessments for other species.

Are there differences between the growth and mortality of male and female summer flounder (sexual dimorphism) and, if so, how do the differences affect the assessment?

Fish size generally increases with age up to some upper biological limit for the species. This limit may vary with environmental conditions, differs from species to species, and may differ within a species, for example, between sexes.

For a number of flatfish species, males grow at a faster rate than females, but often to a smaller maximum size. For such species, the largest fish are almost always females; for example, female North Sea plaice typically grow to a greater length and weight than do males (Rijnsdorp and Ibelings, 1989). Spring and fall surveys show a 20 percent larger size-at-age among female summer flounder at five years (NOAA, 1992). Eldridge (1962), cited in NMFS (1981, p. 15) also notes a 20 percent greater size in females than males; this difference increased to 30 percent by age 8. Differences in growth between sexes may also lead to different mortality rates at age between sexes if mortality is related to size. This difference may make it difficult to ascertain the full impact on the stock from data for both sexes combined. The sex of summer flounder can be determined by port samplers for commercial catch and by survey scientists from survey catches, and otoliths and scales can be collected to determine the ages for the same fish. Incorporating sex differences in growth and selectivity at age may be possible. If information collected from surveys can be used to identify sex-specific age and size characteristics that can be applied to commercial and recreational landing data, the assessment may be able to account for differences resulting from the sexual dimorphism.

Committee Investigations

The committee did not have suitable data to calculate the effect of ignoring sexual differences in growth. However, mortality at age should be greater for females than for males, until both are equally well retained by trawl nets in commercial fisheries and hooks in recreational fisheries. In turn, this greater mortality for females would lower the estimated spawning potential of the stock, decrease the effective spawning stock biomass, and increase the effective fishing mortality.

Actions Needed

As discussed in the next section, differences in catchability at different ages can affect stock assessments and the resulting estimates of fishing mortality and biomass. This is analogous to the multi-stock problem discussed in the previous section, where one stock is male and the other female. If catchability differs with size (and thus between sexes), this will influence the evaluation of stock status. NMFS should determine whether there is enough difference in size at age and/or catchability at age between the sexes to warrant concern about the spawning potential of the stock. Currently, sex is not determined for catch samples from the summer flounder commercial and recreational fisheries. As a result, these data alone do not provide sufficient information to account for differences in growth between males and females in the assessment. It is possible to determine the sex of some fish species in catch samples when fish are gutted at sampling time, or when identifiable external morphological characteristics are available. Apparently this is not the case for summer flounder. It may be possible, however, to use survey information to distinguish gender-related differences since sex, age, and length are determined for survey samples. Use of this information could improve assessment results that are influenced by sexual dimorphism. If such data become available, NMFS should conduct catch-at-age or catch-at-length analyses that explicitly account for sex differences in size-at-age and selectivity. Such data could be applied retrospectively, assuming constant size-at-age by sex for years when observations are not available.

QUESTIONS RELATED TO SUMMER FLOUNDER SAMPLING

A common concern voiced by fishermen is that the gear used for sampling fish is antiquated and thus survey results are biased. This section addresses some of the concerns about the gear and methods used in the fishery-independent surveys.

What are the appropriate survey and commercial catchabilities of summer flounder?

Commercial fishermen believe that NEFSC surveys fail to fish appropriately in the survey strata areas and depth where the industry believes large summer flounder are found. Presentations to the committee by commercial fishermen suggested that large numbers of large summer flounder used to be taken on the portion of Georges Bank near the U.S.-Canada border (the Hague Line). This distribution of larger summer flounder is supported by some data from the American Littoral Society tagging program, the dissertation by Desfosse (1995), and Rountree (1994). Fishermen claim that even within the strata fished, the randomized stations tend to miss the sites where aggregations of large summer flounder are found. They also argue that the research gear (roller-rigged)[4] and the method for using it (including 30-minute tows) are unsuited to catching the larger, faster swimming summer flounder. Fishermen believe that the groundfish surveys, particularly the fall survey, do not adjust for flounder migrations, so the surveys do not adequately sample the population. Thus, fishermen conclude that the age distribution observed is unrepresentative of the population. The tow duration issue is important to resolve, particularly because NMFS has considered reducing tow durations to 15 minutes in an attempt to increase the number of tows and thereby the precision of estimates based on the surveys. NMFS has not tested the effect of tow durations on catchability-at-age of summer flounder and other flatfish. However, the catch of flounders larger than 45 cm in length, in both the commercial fishery and surveys, has recently reached its highest point since 1982, according to NMFS.

In considering these criticisms, it is important to understand how commercial and recreational catch-at-age data and survey catch per unit effort (CPUE)-at-age data are used in assessment models to estimate population size and structure. If both were used directly as estimates of population structure, rather than interpreted in the con-

[4] Roller gear is a type of footrope (the rope attached to the bottom front of a trawl net) made of round rubber or steel bobbins interlaced with rubber discs and rigged to roll as the net is pulled across the bottom.

BOX 2-2
Gear Selectivity or Catchability

Selection of a gear is best described by the catchability *(q)* of the gear (*g*) at age (*a*). This is the amount of fishing mortality that a unit measure of fishing activity (e.g., one day's fishing) would generate on age *a* by gear type *g*. Catchability has several components that relate to the availability of the fish to the gear (Are fish in the area fished by the gear?), its accessibility to the gear (Does the species' behavior put it in the path of the gear?), and its vulnerability to the gear (Does the gear catch fish if they are encountered?). In the case of trawls, mesh selectivity can be measured by selectivity experiments (e.g., by using trawls with codends covered with finer mesh to estimate the proportion retained by size). Other factors contribute to gear selectivity, such as footrope size, headrope height, tow speed, and tow duration; however, these factors relate to distribution and behavior of fish and are less easy to estimate directly.

Overall catchability can be measured using the results of VPA, provided adequate data are available. Estimates of catchability at age may change over time due to changes in availability or accessibility resulting from different gear uses or changing environmental conditions. When fishery or survey effort or catch rates are used to tune the VPA, it is usual to adopt a null hypothesis that they are constant from year to year. Since various choices of $q(a', y, f)$ for the oldest age a' can give equally plausible interpretations of the data, it is often the practice to set this value to the same value as the average of one or more of the intermediate ages. This assumption implies that the oldest fish have the same catchability as those of the intermediate age. Alternatively, a "dome-shaped" selectivity curve is assumed if younger and older fish are less susceptible to the fishing method than fish of intermediate ages.

text of the stock assessment model, the industry concerns would be valid. However, most models include assumptions about how selective the gear is for different age and size classes. For summer flounder, catch-at-age data are used as direct measures of the fishing-induced mortality at age. Because no measure of effort is associated with landings, no auxiliary data are developed from commercial catch rates. For summer flounder survey CPUE-at-age data, however, an age structure is implied based on the type of selectivity curve assumed. Thus, the important question becomes whether or not the selectivity curves are appropriate.

Models can correct for changes in the proportion of different ages caught through incorporation of an exploitation pattern (see Figure 2-3 for an example) for the fishery and for the survey gear. Exploitation pattern is a function of the gear used (Box 2-2) and of the geographic distribution of fishing effort. For gear such as gillnets, designed to catch a specific size of fish, the exploitation pattern initially increases with size and then declines as bigger fish cannot entrap themselves in the mesh of the net. With such gear types as trawls, which potentially can catch all fish greater than a certain size, the exploitation pattern initially rises and then may plateau. The exploitation pattern, however, may decline at larger sizes if larger fish are better able to escape the gear or the gear is used more intensely in areas where smaller fish live. Thus, if at least some of the older ages are caught and the exploitation pattern can be estimated, assessment models can correct for changes in the proportion of different ages caught.

In practice, the data sets typically collected for use in stock assessments provide little information on the shape of the exploitation pattern. Pope and Shepherd (1982) showed that in simple one-fleet situations the exploitation pattern could not be determined by catch-at-age data alone and their general results indicate that the exploitation pattern remains undetermined despite the existence of additional survey or fleet results. In practice, therefore, assumptions have to be made about the shape of the exploitation pattern. The normal default assumption, and the one that is used in the summer flounder assessments, is that the oldest ages of fish in the survey have the same chance of being caught as younger fish. These assumptions are applied to the consolidated commercial catch in some methods of analysis and to the survey catch rate results in others. Although this is a sensible default assumption, it may be wrong. Some studies of Irish Sea fish stocks have found that egg numbers gave larger estimates of spawning stock size than those obtained by standard assessment models using the usual default assumptions (Horwood, 1993; ICES, 1998). Estimates of spawning stock biomass based on egg surveys can be higher than those based on landings when there is underreporting. Generally, however, unexpectedly high spawning stock biomass results from older fish being less selected by the gear than younger fish, leading to an incorrect estimation of the proportion of fish of various ages in the population.

A more important question about catchability bias in surveys is not whether the gear is biased but whether it is possible to correct any bias. This requires an estimation of the exploitation pattern of gear. It is difficult to do this directly because data that bear on this point are often unavailable. The limited span of ages sampled by the fishery and various state and federal survey series also adds to this problem. If data on older fish were available, any trend in the catchability of these fish could be indicative of changes in catchability over some of the older ages. Although the use of default values of exploitation pattern for the older ages is perhaps inevitable, given the lack of data, it may lead to underestimation of the size of the spawning stock biomass and an inappropriate estimate of the age structure. Such default assumptions also may lead to overestimation of the status quo fishing mortality; it is apparent from the VPA results that catchability is assumed constant for ages 3, 4, and the 5+ group. Because recreational catch accounts for about half of the total catch, the catchability of the recreational fishery and its effect on population estimates also is important.

Committee Investigations

A concern raised by industry representatives was that both the NEFSC surveys and commercial catch data failed to represent the true age structure of the summer flounder population, by missing large fish offshore. Issues relevant to this potential problem include (1) whether there is evidence that older fish are missed in the assessments and (2) if older fish are missed, how this could affect the assessments.

Is there evidence for a large stock of older fish offshore? Commercial flounder fishermen cite past catches of large fish offshore. They claim that older fish do not enter the assessments because

1. the trip limits in the flounder fishery since 1992 make it less economical for commercial fishermen to go offshore where the larger fish are found in the winter;
2. large fish that *are* caught are missed by the port surveys because they quickly go, unsampled, to the sushi markets in Japan;
3. NEFSC spring and fall surveys use gear inappropriate to catch flatfish; and
4. all NEFSC surveys use tows that are too short and too slow to tire and catch larger, stronger flounders.

Commercial fishermen claim that there are older, larger fish at the edge of the continental shelf, an area seldom fished or sampled. Packer and Hoff (1999), on the other hand, indicate low catches of summer flounder at the shelf edge in their description of essential habitat for this species, although this analysis is based on NEFSC data that the industry views as suspect. Even if a refuge for large flounder is missed by both commercial and survey vessels, one would expect to see these fish in the spawning areas offshore at some point. If they do not migrate to the spawning areas, they are not part of the spawning stock biomass and thus their exclusion from the stock assessment is appropriate. What would constitute evidence to the contrary? The existence of a population of large summer flounder offshore should be reflected by

1. unexpectedly large numbers of flounder eggs in egg surveys;
2. catches of large flounder as they migrate seasonally inshore in the spring and back offshore to winter grounds;
3. logbook and export records of large flounders, even if not represented in port samples; and
4. large flounders being caught in general recreational fisheries and tournaments (if tournaments take place in areas where large flounders are expected to be caught).

The committee was not able to locate data on all these factors and was given little evidence that an otherwise unassessed population of large summer flounder exists offshore. The one possible exception is the recent rise in the number of citations in Virginia for summer flounder weighing 6 pounds or more (see Figure 2-7), although the fishing effort and locations related to these catches are not reported.

If large flounders are less catchable, how would this affect stock assessments? The proposed inability of the survey to find and catch large flounder can be simulated by altering the assumptions of the standard Laurec-Shepherd VPA tuning to halve the relative fishing mortality on age 4 and older fish relative to the average fishing mortality on age 2-3 fish (Table 2-2).

The main effects of this alternative assumption are:

- the estimate of current fishing mortality F_{1996} is reduced from 0.95 to 0.63 if larger flounder are less catchable (although part of this reduction results from the lower estimate of the fishing mortality on 4-year-olds).
- estimates of the biological reference points $F_{0.1}$, F_{max}, and $F_{20\%}$ all increase and thus would require less restrictive management under the alternative assumption.
- F_{med} is reduced in line with F_{1996} and hence advice based on achieving some specified proportion of F_{med} would be more stable. In other words, the adjustment needed in fishing intensity under alternative assumptions is less for F_{med} than it is for the other biological reference points.[5]
- the fishing mortality that would have been required to achieve the 1997 TAC is reduced to

[5] The main short-term negative effect on fishermen generated by management is usually associated with the proportion by which fishing mortality, and hence fishing effort and short-term catch, must be reduced. Thus, how critical a factor (such as the right exploitation curve) is to management might be measured by asking how it changes the ratio between the current and desired fishing mortality rates. For example, reducing fishing mortality to F_{max} under the standard assumptions requires a reduction from $F = 0.44$ to $F = 0.24$, a 45 percent reduction. If the "alternative F for 4+ is half of F for ages 1-3" hypothesis is correct, a reduction from $F = 0.31$ to $F = 0.30$, only 3 percent, is required. Therefore, assumptions about catchability affect target F values, which affect the difficulty of implementing management to control fishing mortality. By contrast, managing at F_{med} would apparently allow F to increase from 0.44 to 1.48 in the first case, a 236 percent increase, and from 0.31 to 1.17 in the second case, a 277 percent increase. This example illustrates that F_{med} seems to be more robust under uncertainty than F_{max} and $F_{0.1}$, though the committee regards F_{med} as a threshold reference point rather than a target reference point.

TABLE 2-2 Effects of Changes in Assumed Fishing Mortality Rate at Age on Biological Reference Points and Predictions of Fishing Mortality (*F*), Catch, and Spawning Stock Biomass (SSB)

	F Constant at All Ages	*F* for Ages 4+ is Half That for Ages 1-3
Biological Reference Points (1996)		
$F_{0.1}$	0.15	0.18
F_{max}	0.24	0.30
$F_{20\%}$	0.28	0.30
F_{med}	1.48	1.17
Estimated fishing mortality in 1996		
F_{1996}	0.95	0.63
Predictions for 1997 under status quo		
F_{1997}	0.95	0.63
$Catch_{1997}$ (mt)	12,248	12,445
SSB_{1997} (mt)	12,263	17,181
Predictions for 1997 under TAC of 7,162 mt		
F_{1997}	0.44	0.31
$Catch_{1997}$ (mt)	7,162	7,162
SSB_{1997} (mt)	16,419	21,425
Predictions given 1997 TAC of 7,162 mt and $F=F_{max}$ in 1998		
F_{1998}	0.24	0.30
$Catch_{1998}$ (mt)	4,541	6,950
SSB_{1998} (mt)	32,677	33,289

NOTE: SSB = spawning stock biomass; TAC = total allowable catch. See Box 1-2 for other definitions.

0.31. Although lower in absolute terms than the 0.44 level of the standard run, a proportionally similar reduction in fishing is still required to achieve half the current exploitation level (the 1997 TAC at status quo fishing mortality is similar for both runs).

- the 1997 spawning biomass is higher for the new assumption, both at the 1997 TAC level of fishing and at status quo.
- since the fishing mortality rate required to catch the 1997 TAC (0.31) under the new assumption is close to the new estimate of F_{max} (0.30), the catch in 1998 under F_{max} is substantially higher with the new assumption. However, since there is less reduction in fishing mortality, the biomass that results is similar in 1998 to the standard run.

In summary, adopting an assumption that old fish are less catchable than young fish leads to a more optimistic view of the stock. It is recognized that $F_{0.1}$, F_{max}, and $F_{20\%}$ are proportionally more sensitive to current *F* than is F_{med}. Adopting a different assumption about the exploitation rate of older fish could cause substantial changes to estimates of future TACs under an F_{max} target.

Figure 2-3 illustrates the change in the commercial selection pattern between the two runs. It also illustrates that the fishing mortality rate value adopted for age 4 (the last true age in the catch-at-age data) has to be applied to age 5 fish and older. Since age 4 fish are typically less than half the weight of the largest fish, this may be a doubtful assumption. This suggests that including more ages in the assessment could be helpful (see Figure 2-4).

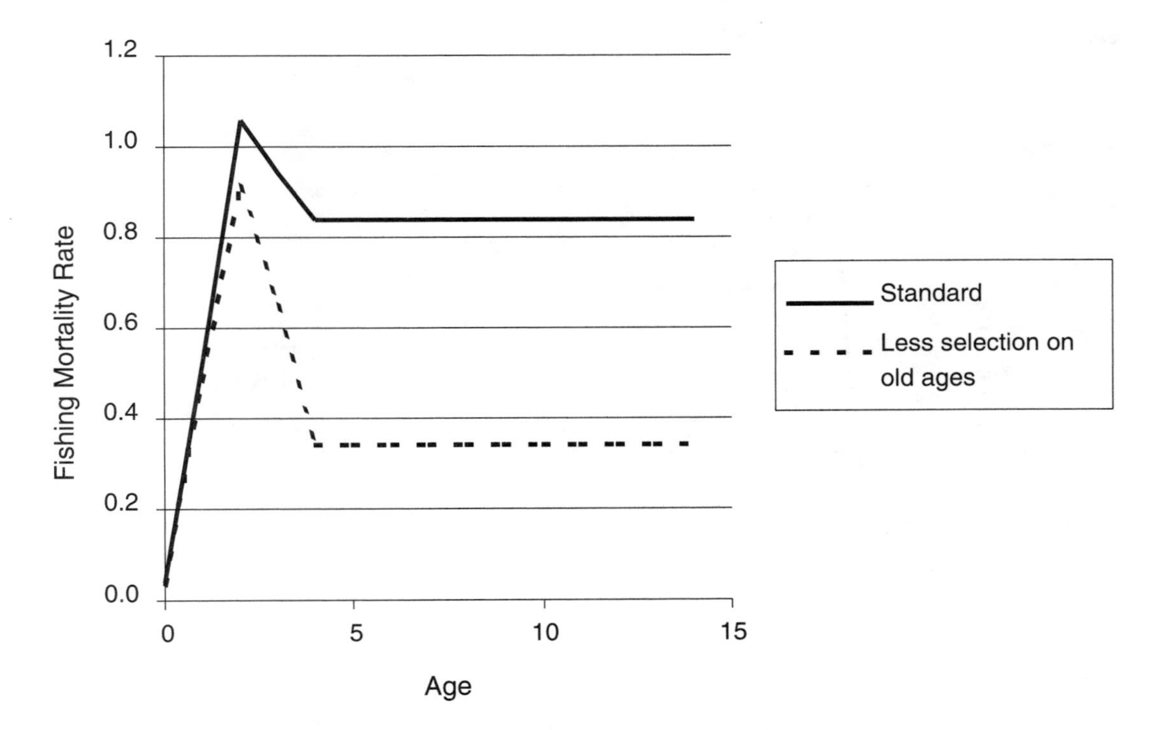

FIGURE 2-3 Assumptions about summer flounder fishing mortality on ages 0-14.

NOTE: The fishing mortality rate (not selection) of 0.8 on age-4 and older fish stems from the fact that the age-4 mortality is estimated as the average of the mortality rate on ages 1-3 and that on older ages set equal to the age-4 fish. This assumption does not provide a better or worse fit in the Laurec-Shepherd model than a curve with higher or lower mortality on age-4 fish. This is because the model fits are almost completely insensitive to the assumption of how domed the selection is (see Pope and Shepherd, 1982).

The SAW25 report (NEFSC, 1997) provides catch-at-age data that are relatively complete up to age 7. However, the NMFS VPA generally includes data only through age 4. Separable VPAs (Pope and Shepherd, 1982) conducted on these data suggest that selection has not changed through time and that regardless of whether the fishing mortality of 7-year-old fish is set at half, equal to, or 50 percent greater than the value of current fishing mortality on 2-year-olds, the exploitation pattern is relatively flat between ages 2 and 5. Runs of Laurec-Shepherd tuning were made on this data set with an assumption that fishing mortality on age 7 was either (1) equivalent to that on ages 3-6 or (2) was 50 percent of this level (Figure 2-4). These runs both provide estimates of biological reference points that are broadly similar to those when the Laurec-Shepherd is run with constant catchability on ages 0-4 (see Table 2-2). Both the runs, however, give a similar, more pessimistic view of the stock size than the equivalent Laurec-Shepherd runs on ages 0-4. The catch and SSB results from the two runs correspond quite closely, suggesting that adding the extra age classes to the analysis makes the decision about the level of exploitation of older fish less critical.

The more pessimistic view of the stock, given by both runs using the extended data set, occurred because the exploitation pattern remained high on ages 4 and 5 rather than decreasing as it did in the standard Laurec-Shepherd run conducted on ages 0-4 only (Figure 2-4). This result (based on actual catch data) does not support the view that

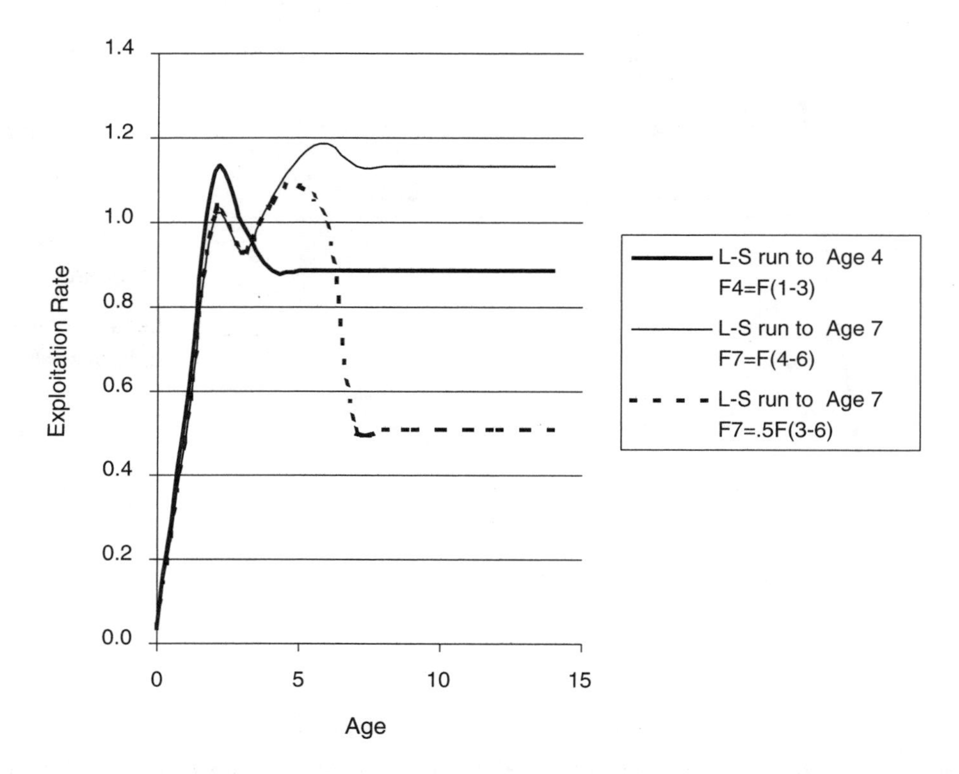

FIGURE 2-4 Comparison of exploitation pattern of summer flounder from runs including more ages in comparison to the standard runs. The middle line is the standard assessment with *F*(age 4) set as the average *F* on ages 1-3 (the standard line of Figure 2-3) and fishing mortality estimated on catch-at-age data up to age 7. The alternate lines show (1) where *F*(age 7) is set to the average of *F* on age as 4-6 and (2) where *F*(age 7) is set to 0.5 of the average *F* on ages 4-6.

exploitation is less on older ages. It seems that at least ages 4-6 years are exploited at much the same level as ages 2 and 3 years. Moreover, at age 6, the fish are a substantial proportion of their maximum size and weight. These fish may reasonably be expected to behave like even older fish and be equally catchable and may have the same fishing mortality rate. It should be noted that this interpretation relies on the older fish being properly sampled.

Actions Needed

Whether a population of large summer flounder is missed by surveys, recreational fisheries, and commercial fisheries—but contribute significantly to spawning stock biomass and recruitment—is one of the major points of contention between NMFS and commercial fishermen. If this population exists but is not surveyed, not fished, and does not contribute to stock spawning potential, it begs the question of whether it should be included in management considerations. Although the committee found no data to indicate reduced catchability of larger flounders or the presence of a large enough offshore population of large flounders to affect the assessments, NMFS should work with industry on this issue. The data normally used in population abundance assessments, commercial catch-at-age data and commercial and survey catch-rate-at-age data, do not provide information on the exploitation pattern of fish of the oldest ages. Since errors in the assumptions made to cover this deficiency may have quite important effects on management decisions, it is appropriate to consider what other sources of data could be useful to elucidate the population age structure and exploitation pattern. Tagging is one mechanism for collecting such information and managers will have to decide whether the cost of conducting an extended tagging experiment would be justified by the potential return. The potential return of tagging experiments would include improvements in estimation of exploitation rates at age and thus biological reference points, as shown in Table 2.2, decreasing the uncertainty of management based on biological reference points.

If older fish have a lower fishing mortality rate than is supposed in current assessments, they would have a higher average age than predicted. Thus, tagged fish would continue to be returned for a longer period than would have been predicted under the current assessment. A number of problems exist with tagging experiments, however, such as the need to accumulate the data over a number of years, the need for accurate estimates of tag loss and non-reporting (a common problem with earlier summer flounder tagging studies), and the need for complete mixing of the tagged fish in the population. If an extensive and expensive tagging study were undertaken, cooperation of the fishing community would be essential to make it successful. There are many ways this cooperation could be enlisted. Directly involving some fishermen with the tagging is one example. Commercial fishermen could be enlisted to assist with tagging large numbers of summer flounder offshore, either as paid participants or as an in-kind contribution to NMFS and industry cooperative efforts.

Alternatives to tagging could be used to answer the question about summer flounder catchability directly. One alternative that is faster, though inherently expensive (see Horwood, 1993), is to conduct egg surveys. These are used to estimate the number of eggs spawned and (with knowledge of average fecundity per gram of adult female) the biomass of the spawning stock. Summer flounder eggs are distinguishable from eggs of other species even at early stages, except for some difficulty in distinguishing between summer and windowpane flounder eggs (Berrien and Sibunka, 1999). Egg surveys with usable levels of precision can be difficult for summer flounder because this species' long spawning period would require multiple survey cruises to sample its eggs adequately. NMFS has not conducted egg surveys specifically for summer flounder, although egg and larval densities are available from the Marine Resources Monitoring, Assessment, and Prediction (MARMAP) program for 1977-1987.

A third (and possibly the least expensive) approach to deal with any catchability-at-age problem is to include more ages in the assessment. Survey data on age classes 5 and older typically are excluded from the analyses for all but the NEFSC winter survey (available only since 1992). This is due in part to the difficulty some assessment procedures have in dealing with sparse data, observed zero values, or missing data, which are more common for older age classes. Committee analyses determined that the assessments can be sensitive to the number of age classes included in assessments. Although this does not resolve the problem of the proportion of the older ages being caught, it may move the problem up to ages that are sufficiently scarce, thus reducing the problem to a negligible level. Therefore, future NMFS assessment activities for summer flounder should include more year classes (up to age 7) to tune the model and decrease the possibility of missing changes in fishing mortality on older ages. This will be particularly important as the summer flounder population recovers and includes a greater proportion of older fish.

Finally, NMFS should conduct studies of catchability at age as a function of tow times, using nets that include video cameras to study fish behavior and determine whether larger fish are more likely to outswim the nets than are smaller fish. Another means to test gear performance would be to compare survey data from cruises using spring roller rig trawl gear versus winter flatfish trawls; both surveys occur at approximately the same time when fish are not likely to be very mobile. Joint NMFS and industry survey exercises and sampling targeted at possible locations of large flounder should be undertaken. Because the debate about the population structure centers in the offshore area, this area should be targeted for special tagging and egg survey studies, as well as other observations designed to determine if larger, older summer flounder are more abundant offshore than expected.

The significant contribution of the recreational fishery to the fishing mortality rate makes the actual exploitation patterns of recreational anglers and model assumptions about these patterns more important in summer flounder stock assessments. NMFS should examine its assumptions about recreational exploitation patterns and determine whether these assumptions are valid.

Do problems with determining the age of summer flounder discredit age-based assessments?

The age of fish are most commonly determined by counting the number of rings on fish scales or ear bones (*otoliths*). These often show annual rings similar to those seen in trees. Interpreting the number of rings can be contentious and different readers may assign different ages to the same fish. Typically, scales are more appropriate for ageing younger fish, whereas otoliths are more appropriate for ageing older fish, because scales tend to erode as a fish ages. Operational assessments of summer flounder are based on ageing of scales, because ageing of scales is reliable when stocks are heavily fished and most fish are less than 3 or 4 years old. As the summer flounder population is rebuilt and age structure widens, scales will be increasingly problematic because they underestimate age in older fish and ageing may need to switch to otoliths. In recent years, significant effort has been devoted to developing accurate otolith ageing techniques.

A disagreement existed for several years between the age readers of NMFS' Northeast Fisheries Science Center and those at the North Carolina Division of Marine Fisheries in relation to fish ages based on scales and otoliths. (In 1996, landings in North Carolina accounted for 39 percent of the total commercial landings of summer flounder.) The disagreement on fish ages may have resulted from the problems described above. Groups of age readers from the two labs collaborated in several workshops in the 1990s and agreed in the latest one (in 1999) that summer flounder can be aged reliably using scales, if protocols agreed to by the two groups are followed (Bolz et al., in press). Participants in the 1999 workshop concluded that the majority of ageing disagreements between the two groups resulted from differing interpretations of "marginal scale increments due to highly variable timing of annulus formation and from the interpretation of first-year growth patterns and first annulus selection. It was agreed that the NEFSC and NCDMF [North Carolina Department of Marine Fisheries] age data used in the current assessment are valid for the respective components (NER [New England Region] and North Carolina waters) of the stock and fishery" (Terceiro, 1999). The level of agreement between age readers from NEFSC and the North Carolina Department of Marine Fisheries was 83% in this exercise. The committee is concerned that ages based on scales will not be adequate for older fish as the population recovers and believes that research to improve determinations of age from otoliths should continue. There is some evidence that measurements based on scales underestimate the ages of summer flounder older than five years (C. Jones, unpublished research).

In practice, a certain amount of error in ageing may not create serious differences to fishery assessments, particularly if the errors are consistent from year to year in magnitude and direction; the same ageing errors would occur in the catch-at-age, fecundity-at-age, and weight-at-age data, and the degree of mis-ageing should be similar throughout the range of the species. An effect that may be important is that any error in ageing tends to smear out a strong year-class over the adjacent year-classes. This may make the

TABLE 2-3 Effects of Changes in Assumed Fish Ageing Protocol on Biological Reference Points and Predictions of Fishing Mortality (*F*), Catch, and Spawning Stock Biomass (SSB)

	NEFSC and North Carolina Ages Used	NEFSC Ages Used Throughout
Biological Reference Points (1996)		
$F_{0.1}$	0.15	0.15
F_{max}	0.24	0.25
$F_{20\%}$	0.28	0.28
F_{med}	1.48	1.57
Estimated fishing mortality in 1996		
F_{1996}	0.95	1.33
Predictions for 1997 under status quo		
F_{1997}	0.95	1.33
$Catch_{1997}$ (mt)	12,248	11,415
SSB_{1997} (mt)	12,263	8,542
Predictions for 1997 under TAC of 7,162 mt		
F_{1997}	0.44	0.62
$Catch_{1997}$ (mt)	7,162	7,162
SSB_{1997} (mt)	12,132	16,419
Predictions given 1997 TAC of 7,162 mt and $F=F_{max}$ in 1998		
F_{1998}	0.24	0.25
$Catch_{1998}$ (mt)	4,541	3,199
SSB_{1998} (mt)	32,677	30,167

NOTE: SSB = spawning stock biomass; TAC = total allowable catch. See Box 1-2 for other definitions.

understanding of year-class variations more difficult. A useful discussion of the effects of misageing fish can be found in Gulland (1955).

Committee Investigations

To determine whether any age discrepancies between readers from North Carolina and NEFSC could affect the assessment, a Laurec-Shepherd analysis was conducted in which North Carolina-aged catches were replaced by an equivalent number of fish with the age distribution of the rest of the catch (as aged by NEFSC readers). As in previous subsections, the standard run of the Laurec-Shepherd model is compared with the modified run (Table 2-3).

The main points to note from this analysis are that

- the biological reference points were not very different under the two scenarios.
- when NEFSC ages are used exclusively, the current estimate of fishing mortality is increased by about a third and the fishing mortality required to catch the 1997 TAC is increased, presumably due to population estimates being decreased.
- spawning stock biomass estimates are also decreased for the same reason, as is the 1998 F_{max} TAC and the 1997 *status quo* TAC.
- the estimate of fishing mortality rate is higher using the NEFSC ages only. Since the

value of F_{max} is similar in both assessments, the reduction in effort required to meet F_{max} is 60 percent (NEFSC ages only) as opposed to 45 percent (NEFSC and North Carolina ages).

It appears that the potential results of the ageing discrepancies are significant in terms of the assessments, so it is fortunate that this issue has been resolved. This finding also points out the importance of ensuring that both groups continue to study and compare ages from both scales and otoliths, as necessary.

Actions Needed

The committee recommends that NEFSC and North Carolina continue efforts to ensure agreement on ages from scales and otoliths, perhaps using tags or oxytetracycline[6] marks, if necessary. Given that the disagreements relate to the early rings, an oxytetracycline experiment on young fish raised in culture could provide results in a short period of time. Work is ongoing and should continue until the issue is resolved. Other techniques, such as tagging and retrieving tetracycline-treated wild fish, and using other microchemical methods (Campana et al., 1990; Campana and Jones, 1998), could be pursued to resolve this issue.

Are effort data used appropriately and are the effects of effort changes incorporated properly?

In many fishery assessments, commercial fishing effort data and the associated commercial catch data are the primary means of establishing the trends in fishing mortality rate and population size and hence in establishing the current state of the stock and the current level of TACs. The summer flounder assessment is unusual in this respect in having a number of survey series to provide abundance data, and including no commercial fishing effort data in the VPA tuning. In a few fisheries, recreational effort is significant and is included through Marine Recreational Fisheries Statistics Survey (MRFSS) data. Commercial fishing effort data and the associated catch rates are both based on far more sampling than research vessel data and thus are likely to be less variable. However, because of changes in commercial practice and government regulations, commercial data may exhibit changes through time that, if not corrected, could bias fishery assessments. Namely, a mandatory logbook and trip limits were instituted in the commercial summer flounder fishery in 1992. NMFS believes that these changes invalidate the time series for commercial summer flounder data.

Committee Investigations

For fisheries in which a significant portion of the catch is taken by recreational anglers (e.g., the current summer flounder fishery), effort data can be obtained from MRFSS. Use of these effort data, however, is not straightforward. The data are obtained from telephone queries to angling households, in which the anglers provide information on the number of fishing trips they took, but not the target species or the number of hours they fished. Thus, the telephone survey does not provide directed effort, so directed effort must be estimated through a proportional conversion obtained from on-site interviews at access points, where anglers are asked which fish species they sought and what they caught. The ratio of directed effort to total effort can therefore be obtained from on-site interviews. Because this is a proportion in a given fishing mode and sampling period, the variance can be calculated as the variance of a proportion. Ultimately, the variance of directed effort is calculated as the product of the variance of the ratio and total effort for the fishing mode and sampling period. Total effort data for the Atlantic Coast for all species typically has a proportional standard error of less than 3 percent. When effort is requested for

[6] Oxytetracycline and other chemicals that mark calcium can be injected into a fish on a known date. Any subsequent age increments can be validated with time elapsed since the fish was injected.

specific regions (e.g., mid-Atlantic states), estimates are more variable, but have proportional standard errors still less than 10 percent. Additional increases in resolution increase variability. Effort data are not available on the MRFSS Web site, but are available to state and federal clients upon request.

The committee did not investigate the issue of including effort data in assessments. The committee believes, however, that inclusion of available effort data in stock assessments could help NMFS understand changes in the fishery, and fishermen could better relate what they observe to what the survey observes.

Actions Needed

The problems and benefits of including commercial catch and effort data in summer flounder assessments should be investigated. Comparisons of assessments with commercial catch rates is one way for NMFS to involve industry more closely and to monitor changes in catchability that may be occurring in the fishery. One approach to involve commercial fishermen in a manner more comparable to fishery-independent surveys is the sentinel survey, as used in the Atlantic Canada inshore cod fisheries. In these surveys, fishermen are hired by the government to fish using commercial gear in a systematic fashion to estimate cod abundance. Having the commercial CPUE used in a statistically designed survey to compare in time and space with the surveys is important. This approach provides results that appear valid to both scientists and fishermen.

One of the many examples of a sentinel survey (and probably one of the least contentious) is the 4VsW sentinel longline survey, which has been conducted on the east Scotia Shelf off Nova Scotia by commercial longliners in September-October each year since 1995. Locations are randomly selected from the strata used by the Canadian Department of Fisheries and Oceans' summer survey plus from three inshore strata. At present, the estimates from this survey are used to comment on general trends and compare with the departmental surveys (Mohn et al., 1998). Attempts will be made to include this survey along with the departmental surveys to fit an ADAPT model when five years of data are available in the time series.

Is the observer program for summer flounder adequate?

Observers on commercial fishing vessels can perform several different functions, including estimating bycatch and discard rates, estimating underreporting of landed catch, and assisting in tagging programs. In many fisheries, unwanted fish are discarded to the sea after being caught; they are very often killed by the fishing process, although discard mortality rates (the fraction of fish that die from being caught and discarded) vary by species, gear, and handling techniques. Discarding can occur for a number of reasons that fall into the categories of economic discards and regulatory discards. Economic discards occur when the fish are too small for the market, are damaged, or belong to a species for which there is no market or only a limited market. Regulatory discards occur when fish are required by law to be discarded, such as protected species, fish smaller than the legal minimum size, fish caught out of season, fish caught with the wrong kind of gear, and fish for which the fisherman holds no quota. Globally, the discarding rate for all species combined is thought to be approximately 27 million tons per year or about one-quarter of the world catch (Alverson et al., 1994).

An associated problem is landings that are undisclosed or misreported (i.e., catches that are not reported to the authorities or are misreported in terms of species or where caught) and thus cannot be used in assessments. For fishery assessments, it is important to know how many fish are actually removed from the population by fishing (harvested, discarded, and otherwise killed by gear) each year. This is necessary because undisclosed discards or catch may cause an underestimation of fishing mortality rates, and an underestimation of the benefits of fisheries management. In general, unreported or misreported discards or catches can affect fishery assessments in complex ways that are difficult to understand. However, some insights can be gained by considering the following simple cases.

- If half of all fish of all ages caught were discarded or landed, but unreported, estimates of the population size would be wrong by 50 percent but the estimates of fishing mortality and biological reference points would be correct. Catch would be underestimated by 50 percent, but if fishermen continued to catch twice as much as they report, this might not matter in a relative sense.
- If young fish are caught but not landed or reported as bycatch (due to a size-based limit, for example), no fishing mortality will be ascribed to that age. The benefits of reducing exploitation levels or of increasing the age of first capture (e.g., by increasing mesh size) might be underestimated in this case because some of the mortality that these measures would reduce exist but are not currently accounted for.
- If discarding or underreporting of all ages suddenly increases, this initially could be misinterpreted as a reduction in fishing mortality and vice versa. An increased tendency to discard or underreport old fish would give an impression of decreased fishing mortality on these ages.

The effects of more complex patterns of discarding, which could result from a restrictive quota system, are difficult to predict. One pattern could be more discards for years when the quota is particularly small or a large year-class of small fish enter the fishery. Another pattern could be an increasing level of discarding, particularly of smaller fish (i.e., as bycatch or discarding of smaller fish replaced by larger fish caught later), through time in response to more restrictive quotas. Such a pattern is of particular concern because, if not recorded, it can cause the stock to appear to be suffering increasing levels of fishing mortality and diminishing population size. This could lead to still tighter quotas being imposed and fishermen responding with even more increased discarding or undisclosed landings. Overall, undetected changes in discard rates and non-reporting rates can cause a downward spiraling negative feedback effect on assessments and fish populations.

In the case of the commercial summer flounder fishery, estimates of discards are available from onboard observers (see Table A10 in NEFSC, 1997) and from on-site interviews for anglers (see Table A14 in NEFSC, 1997). Estimates of commercial discards have been available since 1989 and recreational discards have been estimated since 1982. No estimates of the precision of discards by age are available. NMFS told the committee that "for summer flounder, the discard rates reported in the VTRs [vessel trip reports] look comparable to those observed by sea samplers [i.e., observers], but no in-depth study of possible biases has been conducted to date."

Committee Investigations

In assessing the importance of discards, the mortality of discards is the most important measure, because although discards may form a large part of the catch of the youngest age, it is possible that this discard mortality of young fish does not constitute a large mortality rate. In the case of summer flounder, 75% of age-0 fish caught are discarded and 38% of age-1 fish caught are discarded. In practice, it is this latter percentage that is of far greater importance because it represents a fishing mortality rate of up to about 0.4 (Figure 2-5); only a small number of age-0 fish are caught in the fishery, so the high discard rate is not as significant to the population. The discard mortality is assumed to be 80 percent for the commercial fishery (based on commercial advice to NMFS) and 10 percent for the recreational fishery.

As mentioned earlier, the observer coverage rate for summer flounder has been less than 1 percent for at least the past two years. Figure 3-2 shows that observer coverage below 10-25 per-

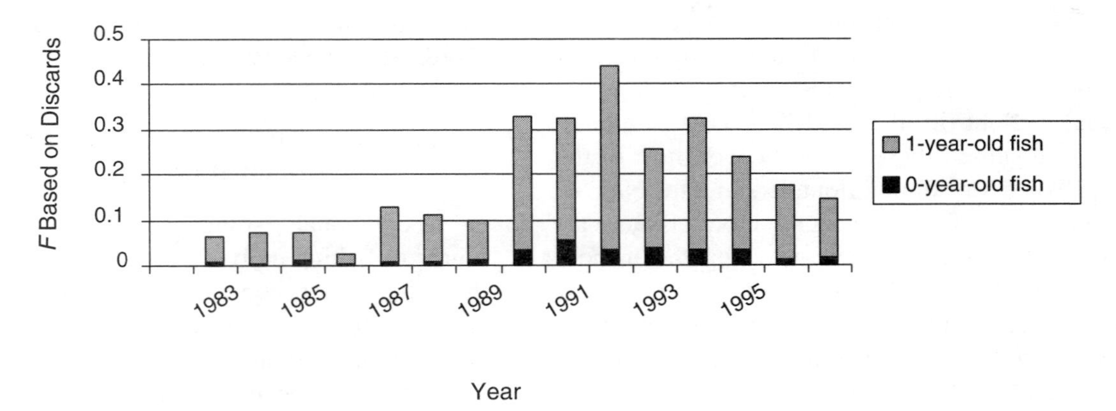

FIGURE 2-5 Discard fishing mortality for age-0 and age-1 fish.

cent can yield significant errors in estimates of catch and bycatch.

Actions Needed

Better data are needed regarding the rates of discarding and misreporting of summer flounder in both the directed fishery and other fisheries. At-sea sampling by observers is required to estimate discarding and may, if observers have the confidence of industry, also provide estimates of undisclosed landings. Observer programs are expensive, but can provide a possible supplement to shore-based sampling of landings. NMFS should investigate the scope and reasons for failures to report discards so that management methods decreasing such discards can be devised (e.g., full retention of catch or the use of temporary closed areas to restrict fishing where young fish are abundant, possibly based on industry information). Most groundfish fisheries in the U.S. Northeast region, including the summer flounder fishery, are observed to monitor interactions of the fisheries with marine mammals, although observer coverage is quite low. NMFS should continue to use the observer programs for marine mammals to observe bycatch and misreporting of catches and discards, while considering augmentation of observations specific to the summer flounder fishery. NMFS should consider increasing observer coverage to 25% or more of commercial summer flounder trips for several years to obtain a better estimate of bycatch discards and misreporting in the summer flounder fishery. Conversely, if the survey vessels use non-commercial trawls designed to capture small fish, the surveys can provide data on year-class strength before the fish have attained sizes susceptible to capture in the commercial fishery. This could provide early warning of potential future declines in the summer flounder population.

Can and should state surveys be standardized?

The summer flounder stock assessment is unusual in that a large number of fishery-independent surveys are available to assess the stock. In addition to the three annual NEFSC seasonal surveys, 9 state surveys were used in the 1996 assessment[7], and these contributed to 6 abundance indices for flounder ages 1 and older and 6 abundance indices for age-0 fish. The state surveys use different gear and survey design and cover different time periods (see Appendix C). Many of these surveys have undergone changes in gear,

[7] The Delaware state survey was added in the 1999 stock assessment.

survey design, and purpose over time, but all surveys used in the summer flounder stock assessments, including the NEFSC surveys, are given equal weight in the NMFS ADAPT model. Taken together, these surveys cover most of the area fished for summer flounder, with the NEFSC surveys generally covering the more offshore areas and the state surveys covering the inshore areas. The temporal coverage of the various surveys varies from sampling within one month to combining samples taken over a number of months (see Table C-4).

Committee Investigations

Trends from the individual surveys are not coincident (NEFSC, 1997). Results from the committee's analyses indicate that some of the surveys display greater lack of fit (residuals) than others, for example, in their catchability (see Figure D-6) and abundance (see Figure D-11).

The committee is not aware of any evaluation of the relationship between the timing of surveys and the movement of summer flounder from offshore to inshore and back. Such an evaluation would be important for interpreting trends of summer flounder in the individual survey series. Standardizing state surveys could increase the value of the data, advance survey methods, and result in increased technical support for data management and statistical analysis.

Actions Needed

The coastwide summer flounder stock assessment might be improved if the state surveys were standardized and coordinated. The committee recommends that NMFS investigate each federal and state survey of the summer flounder stock, including the temporal and spatial coverage of each survey and the nature and quality of the data produced, how the states use the data, and how standardization could affect state interests. However, the committee recognizes that the surveys may have purposes other than contributing to the coastwide stock assessment and that such purposes may make it difficult to standardize and coordinate state surveys.

Is the catch from recreational fishing estimated properly?

Recreational fishing used to be a minimal source of fishing mortality, but current estimates show that anglers contribute substantially to fishing mortality in several fisheries, including summer flounder. Summer flounder is one of several major flatfish species found along the Atlantic coast of the United States (others include winter flounder, American plaice, yellowtail flounder, and witch flounder); only a few are targeted by recreational anglers. Summer flounder are targeted by marine anglers along the U.S. Atlantic coast, with more than 80 percent of the catch usually being taken in the Mid-Atlantic region (Table 2-4). On average, recreational landings accounted for 37 percent of reported summer flounder landings between 1982 and 1996. Recreational landings have increased in their proportion of total landings since 1996, however, having approached and exceeded 50 percent of the total landings (see Figure 2-1 and Table 3-8). Therefore, accurate estimates of recreational effort and catch are important to stock assessments. Since 1979, NMFS has surveyed marine recreational fishing with the Marine Recreational Fisheries Statistics Survey (MRFSS). The survey is conducted in nearly all the coastal counties in the United States. The survey has undergone methodological and statistical scrutiny over the years and has been modified, as needed, to provide more accurate estimates of recreational catch and effort.

Because recreational fishing is geographically dispersed, it is expensive to monitor catches where and when they occur. Nonetheless, acceptable precision in catch and effort estimates can be obtained in some fisheries. Typically, precision of estimates of catch rates depends on the number of anglers sampled, the heterogeneity in the ability of anglers to catch fish, and the temporal and spatial heterogeneity of fish abun-

TABLE 2-4 Recreational Harvest of Summer Flounder (in kilograms) Categorized by Region

Year	North Atlantic	(PSE)	Mid-Atlantic	(PSE)	South Atlantic	(PSE)	Total	Total (PSE)
1981	176,503	(34.4)	4,182,492	(8.9)	221,541	(24.4)	4,580,535	(8.3)
1982	1,047,856	(16.2)	6,492,823	(31.7)	743,837	(23.4)	8,284,516	(25)
1983	599,673	(17.2)	11,838,757	(7.6)	265,938	(21.3)	12,704,368	(7.2)
1984	408,288	(15.7)	7,511,218	(8.2)	624,889	(29.4)	8,544,395	(7.6)
1985	2,667,214	(18.8)	4,906,649	(8.7)	533,753	(43.8)	8,112,385	(8.6)
1986	2,667,214	(18.8)	4,906,649	(8.7)	533,753	(43.8)	8,112,385	(8.6)
1987	606,703	(19.7)	4,823,630	(10.1)	116,437	(10.7)	5,546,771	(9.1)
1988	322,817	(11.6)	6,033,755	(4.5)	292,248	(9.2)	6,648,820	(4.1)
1989	143,213	(17.0)	1,160,805	(7.3)	136,504	(13.3)	1,440,522	(6.3)
1990	106,527	(21.3)	1,984,893	(4.6)	240,872	(10.6)	2,332,292	(4.2)
1991	161,371	(14.2)	3,343,168	(4.5)	195,748	(15.8)	3,700,287	(4.2)
1992	195,183	(12.5)	2,939,005	(4.6)	121,236	(8.1)	3,2476,425	(4.3)
1993	250,376	(11.3)	3,543,703	(4.8)	217,473	(7.1)	4,011,552	(4.3)
1994	444,675	(9.3)	3,576,988	(4.4)	218,234	(7.3)	4,239,898	(3.8)
1995	340,106	(10.1)	2,006,706	(5.7)	112,540	(24.0)	2,496,353	(5)
1996	542,131	(9.3)	3,737,110	(4.4)	193,783	(9.5)	4,473,024	(3.9)
1997	480,033	(13.2)	4,736,561	(4.7)	177,155	(8.8)	5,393,750	(4.3)
1998	911,821	(8.5)	4,529,118	(5.4)	239,410	(8.8)	5,680,349	(4.5)
1999	783,824	(10.9)	2,882,943	(5.1)	136,539	(12.6)	3,803,306	(28.6)

SOURCE: www.st.nmfs.gov/stl/recreational/database/quires/catch/time_series.html, accessed 06/12/00.

NOTE: Weights are expressed in kilograms for summer flounder that were landed whole and identifiable (MRFSS catch type A) and weight of fish caught and filleted, released, dead, or given away and not identifiable, but claimed by the angler to be summer flounder (MRFSS catch type B). These data do not include fish that were caught, released, and may have subsequently died. The North Atlantic region includes recreational landings in Maine, New Hampshire, Massachusetts, Rhode Island, and Connecticut. The Mid-Atlantic region includes landings in New York, New Jersey, Delaware, Maryland, and Virginia. The South Atlantic region includes landings in North Carolina, South Carolina, Georgia, and Florida. PSE = proportional standard error, which is the standard error of an estimate as a percentage of the estimate.

dance. MRFSS attempts to improve precision by increasing the number of anglers intercepted and interviewed. This approach works well in fisheries for which the fish is a popular target for anglers, fish distributions are consistent throughout the fishing season, and anglers predominantly use a few access sites. Summer flounder meets these criteria for acceptable precision in catch statistics for the majority of recreational fishing, with the exception of charter and party boats.

The broad habitat use, wide prey preferences, and seasonal migratory patterns of summer flounder make them vulnerable to many of the modes of recreational fishing, including all access modes surveyed by MRFSS (Table 2-5). The largest catch has come from anglers using private and rental boats, followed by party and charter vessels. Anglers using any of the access modes have equal probability of being contacted through the telephone survey of effort. The probability of intercept surveyors encountering anglers, the method of calculation of catch rate, and the precision of the catch rate differ substantially by access mode.

The private/rental access mode is readily sampled, yields interviews from completed trips, and produces catch estimates that are relatively precise; the proportional standard error of the private and rental access catch estimates is less than 10 percent. Hence, the bulk of the fishery should be sampled well.

The party/charter mode is more difficult to sample because access occurs at fewer sites, and at very specific times. The proportional standard errors of catch estimates from the charter and party component of catch commonly can be greater than 50 percent (Table 2-5). Moreover, the charter and party component is approximately 20 percent of the catch. The imprecision of this access mode contributes significantly to the imprecision in total catch. Additionally, all anglers on a given party/charter vessel experience the same fishing conditions, implying that data for individuals on a single angling trip may not be statistically independent. To the extent that fishermen on party and charter vessels fish farther offshore than fishermen using private and rental boats, this could affect the MRFSS assumptions. Aside from increasing the number of anglers interviewed, greater precision can be achieved by developing a specific sampling strategy that better suits the charter and party boat fleets. For example, a telephone survey that used a list frame of charter boat operators or charter boat logbooks would be more efficient than the current MRFSS random-digit dialing approach. This access mode can be surveyed with specifically designed sampling that takes advantage of the set schedules of the party/charter fleets. Occasionally, the interviewer rides on the party boat to obtain interviews. MRFSS acknowledges that charter and party boat catch estimates are a weakness of their sampling strategies and plans to address this issue in the future.

Committee Investigations

The committee did not examine the summer flounder data from the MRFSS, beyond that presented in Tables 2-4 and 2-5.

Actions Needed

The summer flounder fishery can be used to determine the extent to which other sampling approaches, such as logbooks, observers, and access-site-only surveys could increase the accuracy and precision of catch and catch-rate estimates for charter and party boats. Party and charter fleets are difficult to sample for most fisheries, and innovative approaches tested on summer flounder may be transferable to other fisheries. NMFS should proceed with such tests of the alternative sampling approaches mentioned above.

Another method for increasing precision of recreational data is to sample fishermen directly, rather than sampling the general population in coastal counties. Direct sampling approaches use sampling frames based on lists of fishermen who fish on charter and party boats and who possess a marine recreational fishing license, and by recontacting active anglers identified in earlier random-digit dialing (*longitudinal sampling*). Other ways for identifying the population of fishermen may increase sampling efficiency. NMFS should work with state agencies and recreational groups to better characterize this growing component of many fisheries. A related question, which the committee did not address, is whether the recreational catch is aged correctly. The typical practice in recreational fisheries that have important commercial components is to obtain length data from recreational access-site surveys and to estimate catch at age from age-length keys derived from the commercial catch. Hence, if there are problems with ageing the commercial catch for a species, the same problems will exist for the recreational catch.

Can the precision of data be improved?

Two main approaches can be used to investigate the precision of assessment data sets.

1. Make detailed investigations of the statistical characteristics (e.g., means and variability characteristics) of the samples taken in each sampling stratum.[8] These within-strata statistical characteristics are then combined to give the overall statistical characteristics of the data used in the assessment models.
2. Make models that explain separately each

[8] A stratum is a subarea of a sampling area delineated based on factors such as depth, habitat type, stock areas, and management areas.

TABLE 2-5 Recreational Harvest of Summer Flounder (in kilograms) Along the U.S. Coast, Categorized by MRFSS Access Mode

Year	Shore[a]	(PSE)	Charter[b]	(PSE)	Party/Charter[c]	(PSE)	Private/Rental[d]	(PSE)	Total	Total (PSE)
1981	1,071,014	(26.4)			863,184	(15.5)	2,646,337	(8.1)	4,580,535	(50.0)
1982	478,576	(16.9)			3,210,890	(63.6)	4,595,050	(7.4)	8,284,516	(87.9)
1983	1,454,854	(38.5)			2,294,449	(12.5)	8,955,065	(7.4)	12,704,368	(58.4)
1984	522,469	(15.3)			1,452,482	(16.2)	6,569,443	(9.1)	8,544,394	(40.6)
1985	376,621	(21.1)			631,049	(19.1)	4,657,614	(12.9)	5,665,284	(53.1)
1986	747,884	(32.4)	6,664	(89.1)	1,137,156	(18.6)	6,220,680	(9.9)	8,112,385	(150.0)
1987	195,216	(19.9)	6	(93.7)	1,101,880	(37.7)	4,249,669	(6.6)	5,546,771	(157.9)
1988	479,496	(9.8)	21	(102.1)	776,599	(9.7)	5,392,703	(4.8)	6,648,819	(129.4)
1989	113,347	(15.9)	79	(62.5)	124,870	(11.2)	1,202,226	(7.3)	1,440,522	(96.9)
1990	149,112	(16.6)	7	(63.3)	264,914	(8.4)	1,918,259	(4.7)	2,332,292	(93.2)
1991	343,043	(15.3)	122	(42.4)	370,028	(8.2)	2,987,103	(4.7)	3,700,287	(70.5)
1992	171,808	(15.1)	53	(47.2)	267,940	(10.2)	2,806,624	(4.7)	3,246,425	(77.2)
1993	183,799	(12.3)	30	(99.2)	626,812	(14.3)	3,200,911	(4.5)	4,011,552	(130.3)
1994	257,073	(8.0)	1,309	(17.4)	512,025	(8.4)	3,469,490	(4.5)	4,239,897	(38.3)
1995	152,180	(10.8)	488	(51.2)	184,529	(16.3)	2,122,155	(5.6)	259,352	(83.9)
1996	121,315	(12.1)	2,132	(42.1)	422,562	(8.8)	3,927,016	(4.3)	4,473,025	(67.3)
1997	163,432	(12.8)	1,034	(39.5)	729,456	(9.7)	4,449,827	(4.9)	5,393,749	(66.9)
1998	246,294	(11.2)	1,088	(29.9)	296,891	(11.8)	5,136,077	(4.9)	5,680,350	(57.8)
1999	178,486	(12.7)	94	92.9	304,421	(11.5)	3,320,303	(5.0)	3,803,304	(112.1)

SOURCE: http://www.st.nmfs.gov/stl/recreational/database/quires/catch/time_series.htm and http://www.st.nmfs.gov/stl/recreational/survey/glossary.html, accessed 06/07/2000.

NOTE: PSE = proportional standard error, which is the standard error of an estimate as a percentage of the estimate. Weights are expressed in kilograms, with the proportional standard error of the harvest values (MRFSS Catch Types A and B1) given in parentheses. These data do not include fish that were caught, released, and may have subsequently died. NMFS cautions that care should be exercised in using MRFSS weight data because weight estimates are minimums and may not reflect the actual total weight landed or harvested.

[a] Shore fishing includes all direct use of natural shorelines and artificial structures attached to the shore, such as piers, docks, jetties, and breakwaters.

[b] A charter boat is "a boat operating under charter for a price, time, etc. It is operated by a licensed captain and crew and the participants are part of a pre-formed group of anglers. Thus, charters are usually closed parties, as opposed to the open status of head boats." These data include catches from charter boats operating in the South Atlantic and Gulf of Mexico regions.

[c] The harvest values in this column refer to landings from charter boats and party boats in the North Atlantic and Mid-Atlantic regions. A party boat is "a boat on which fishing space and privileges are provided for a fee. The vessel is operated by a licensed captain and crew."

[d] Rental and private boats include rentals and private use in the North, Mid-, and South Atlantic regions. Rental boats are boats that are rented without crew and are operated by the renter.

of the assessment data sets. The models are constructed with restricted sets of parameters, and estimates of the statistical characteristics of the data are obtained from the deviations of the data from the models.

These approaches can be combined, with various degrees of reliance on data and models. The first approach requires access to the elements of the sampling data and a clear knowledge of how the samples are extrapolated to the overall figures. Although this is usually straightforward for survey data, sampling of commercial fisheries often contains gaps that have to be filled with default assumptions (e.g., the catch of a fleet in a given month was not sampled but was assumed to be like that of the same fleet in a different month). Such decisions often have to be based on professional judgement, although cross-validation techniques can be used to calibrate the likely accuracy of gap-filling decisions. These assumptions can be difficult to track in retrospect unless they are very well documented, although some information can be salvaged by interviewing the staff that handled the data.

An example of the second approach has been provided by Shepherd and Nicholson (1991), who developed a simple linear modeling approach that can be applied directly to survey catch rate-at-age data, weight-at-age data, and catch-at-age data. They noted that much of the systematic changes in the catch rate from a survey should be explained by the product of an age factor (to account for differential catchability across ages and the progressive decline of numbers with age due to fishing and natural mortality) and a year-class factor (to account for differences in the sizes of cohorts). This model is readily applied using an analysis of variance approach that fits age and year-class factors to a logarithmic transformation of the survey data. The lack of fit between the model and the data is ascribed to sampling error that is usually taken to have a log-normal distribution. However, a range of alternative sampling distributions can be fitted if a generalized linear model is used.

Weight-at-age data also can be investigated with similar models, again with age factors (to account for growth), year-class factors (to account for systematic changes in growth between year classes), and possibly year factors (to account for changes in growth between years) fitted to log-transformed data. Again, any lack of fit is ascribed to sampling error. The same approach also can be adopted for catch-at-age data using age and year-class factors to explain sources of systematic variation seen in the survey catch rate data and a year factor (to account for annual changes in fishing mortality rate as well as systematic changes through time).

The second approach is easier to apply in retrospect, but it suffers from the problem that the simple models adopted may not capture all the variation that results from systematic changes (e.g., changes in fishing patterns in catch-at-age data). Such variation may thus be ascribed incorrectly to sampling variation when it is a real signal in the data, inflating statistical variation and giving an impression of greater uncertainty than is the case. Most obviously, catch-at-age data may be affected systematically by changes in the selectivity of the fishing gear or the fishing practice of the commercial and recreational fishermen, or by the changing balance of catch between these users. Some of these changes can be examined by adding terms to the simple models described above, but others may be wrongly ascribed to variation in the data due to sampling variation. This can increase sampling variation and this possibility needs to be considered in the interpretation of the results. With this caveat the approach provides a simple means to examine the quality of the several data sets taken individually.

Committee Investigations

The analysis of design efficiency presented in Appendix C is an example of the detailed approach described above. In this analysis the stratified random designs used in the three NEFSC seasonal surveys in 1995 and the winter survey in 1996 were evaluated. In the first part of the analysis, the winter, spring, and fall surveys for 1995 were evaluated with respect to the application of the stratified design. A number of the strata in these surveys had only one station allocated to them and hence did not contribute to the variance estimates of the survey indices. In fact, sampling intensity is quite low for all of these surveys given the area covered, especially in the case of the 1995 fall survey, in which 48 out of a total of 56 strata had only one or two tows assigned to them.

Design efficiency is a measure of how much the survey design has contributed to increasing the precision of the survey estimates. In the case of the stratified random design, two components contribute: the strata boundaries and the number of tows allocated to each stratum. The efficiency of the design for the 1995 surveys could not be evaluated because of the occurrence of strata with only one tow. Instead, the winter survey in 1996 was evaluated for each of three major flounder species caught in the survey—summer flounder, winter flounder, and yellowtail flounder. The contribution of the strata component to the imprecision was substantial for all three species, indicating that the current strata boundaries are related to differences in the distribution of the flounder species. However, the allocation of tows to strata was not optimal with respect to precision and, in fact, worked against the advantages contributed by the strata component. Our investigations indicated that the precision of the survey estimates could be improved by basing the tow allocation scheme for the current year's survey on a combination of the previous year's survey results and the grouping of strata based on species distributions. The committee did not have the time to extend its analysis to the other years of the survey, but believes that such analyses would be very useful.

The analysis of variance model approach was used to examine suitable survey data sets (some only provided data for one age, which is not sufficient for fitting the model), the weight-at-age data, and the overall catch-at-age data used in the summer flounder assessments. Table 2-6 shows the estimate of the standard deviation of the log(e)-transformed data, by age and overall, for the various survey series and for the weight-at-age and catch-at-age data of each data set both by age and overall. The standard deviation of the log-transformed data approximates to the coefficient of variation of the untransformed data series when it has small values, so these figures (if converted to percentages) give a first approximation of the coefficients of variation of the various data sets.

Table 2-6 suggests that the individual surveys are rather variable. In practice they are used collectively to tune the virtual population analysis and indices used are all assigned the same weight.[9] Hence, it is some average of their values rather than their individual values that affects the assessment outcome. Table 2-6 shows an attempt to estimate their combined logarithmic standard deviation (calculated as the square root of the average variance from all the surveys). Note that individually, the NEFSC fall and winter surveys are more variable than the combined survey estimate. The weight-at-age data seem to have lower standard deviations of log-transformed data, which is adequate for most assessments.

Note, in particular, on a survey-by-survey basis, that the NEFSC fall and winter survey estimates are less variable than those of the state surveys. However, the combined survey estimates sometimes have half the standard deviations of those given for the NEFSC survey estimates alone. Clearly, some judgment has to be made about the precision, quality, and consistency of the survey estimates before they are averaged into a combined estimate for the assessment. By their nature, these data are susceptible to systematic variations that the simple model used here cannot capture. These variations are particularly likely to affect the catch numbers of

[9] Because of differing precision at different ages, some indices might be more informative at some ages than are other indices. Assessment methods are available to estimate unique weightings among surveys for each age group.

TABLE 2-6 Comparison of Federal and State Survey Precision in Terms of Standard Deviation of Log(e)-Transformed Data

Data Set	Age					
	0	1	2	3	4	Overall
Survey catch rate-at-age data						
NEFSC fall survey	n/a	n/a	0.42	0.54	0.46	0.46
MADMF fall survey	n/a	n/a	n/a	0.77	0.77	0.75
NEFSC winter survey	n/a	0.58	0.55	0.50	n/a	0.51
NJDF survey	n/a	0.82	0.82	n/a	n/a	0.79
CTDEP fall survey	n/a	n/a	0.85	0.60	0.60	0.68
CTDEP spring survey	n/a	n/a	0.67	0.58	0.53	0.58
NEFSC spring survey	n/a	0.73	0.62	0.64	0.65	0.65
MADMF spring survey	n/a	n/a	0.76	0.76	n/a	0.75
Combined survey estimate	n/a	0.41	0.26	0.24	0.27	0.23
Commercial data						
Weight-at-age data	0.19	0.11	0.11	0.13	0.14	0.14
Catch numbers-at-age data	0.40	0.30	0.24	0.29	0.39	0.32
Catch numbers-at-age data[a] (omitting age 0)	n/a	0.30	0.21	0.25	0.34	0.27

NOTE: Cells with "n/a" are for those ages with no available data. CTDEP = Connecticut Department of Environmental Protection; MADMF = Massachusetts Department of Marine Fisheries; NEFSC = Northeast Fisheries Science Center; NJDF = New Jersey Department of Fisheries.

[a] Omitting the age-0 catch numbers affects the values for older ages because of the way the ANOVA model used in this analysis fits factors for age and year-class to the data. Numbers at age 0 are highly variable and influence the overall estimates of variance for each combination of age and year-class.

the youngest ages. It is noticeable that the apparent coefficient of variation of these data improves if the age-0 fish are omitted from the analysis. It is thus possible that the estimates in the table present a pessimistic view of data quality. It would be worthwhile for NMFS to investigate further the quality of these data.

Actions Needed

The efficiency analysis conducted by the committee was based on only two years of surveys. A similar analysis over a longer time series needs to be done to understand the persistence of the distribution patterns for summer flounder, as well as other species routinely caught by the survey. This information could be used to develop compromise allocation schemes to produce near-optimal results similar to those presented in Appendix C. If spatial patterns are not predictably persistent over time, the application of other methods, such as adaptive allocation techniques (Thompson and Seber, 1996), need to be explored.

Simple tests of the quality of the assessment data suggest that the precision of the weight-at-age data are adequate but that the survey and to a somewhat lesser extent the catch-at-age data could be made more precise. As an initial step, the committee recommends that NMFS routinely calculate the variance associated with its routine catch-at-age data sampling.

The committee recommends that constructions of estimated variance for commercial samples be included in the NMFS computer programs used to estimate quantities from these samples and that assessments of the statistical quality of results be given with the estimates. This will help to ensure that sampling schemes are performing appropriately and that the sampling design is relevant and provides the best value for the sampling effort. Moreover, it produces a value that is helpful in monitoring data quality, and if necessary, defending appropriate sampling levels and methods. The problem of missing data is particularly acute in the VPA method, but some other methods are better able to deal with missing data, highlighting the usefulness of analyzing data with multiple models.

QUESTIONS RELATED TO THE INFORMATION CONTENT OF THE MODEL AND MODEL ASSUMPTIONS CURRENTLY IN USE

What information does each model structure require and how do these requirements relate to information in the data?

The structure and assumptions used in formulating a population model influence the modeling results in ways that are not always documented by stock assessment scientists and recognized by fishery managers and factored into their management decisions. The choice of a model and assumptions leads to some level of *modeling uncertainty*, as illustrated through the committee's re-analysis of summer flounder data. Much of the model structure and assumptions are derived from basic beliefs and principles, such as the notion that fish in a certain area are members of a closed population and that they die due to natural causes in proportion to the number present. However, some of the "information" or structure is imposed to provide estimates that cannot be derived from the data or to overcome deficiencies in knowledge about the system. Such imposed structure often includes assumptions that selectivity or catchability is constant through time, that natural mortality is constant and known, and that the same fishing mortality applies over the entire region. Although we know that such assumptions are unlikely to be completely valid, we cannot avoid making assumptions of some kind. We can, nevertheless, explore the sensitivity of our results to the choice of model and model assumptions by exploring the sensitivity of outputs to alternative models, alternative assumptions, and reasonable variations in the input data. Some of the assumptions have been examined earlier in this section. Here we consider how much difference in model formulations (with their typical assumptions) influence the outcome of assessment, including three related questions: (1) What weight does the model give to the different pieces of information that come into it? (2) How is uncertainty expressed in the assessment output provided to managers? and (3) Under what level of precision can managers expect to manage with data currently available?

Committee Investigations

Three alternative assessments of summer flounder—using VPA, ADAPT, and CAGEAN models—were carried out independently by different analysts and were compared with the NMFS 1999 assessment, which was done with an ADAPT model (see Appendix D). The same data were used in these alternative assessments, but the outputs from each of the models differed in response to the assumptions and structure of the model chosen (see Figures D-4 and D-5). The effects of a number of alternative assumptions were explored, including different weightings of the data components and whether the selectivity curve fell or remained constant for older ages.

The most significant difference among model results occurred as a function of the assumed consistency in fishing mortality across years. In the tuned VPA-type models, variations in F were controlled by the degree of shrinkage[10] of estimated Fs to the terminal F. In the CAGEAN-type model, the variation of fishing mortality was influenced by the assumption of constant selectivity. The ADAPT model apparently does not use

[10] Shrinkage is the term for a procedure that gives some weight to the assumption that fishing mortality rate (and in some models also population numbers) are unchanging through the recent past. Thus, the values at age for the most recent year may be estimated from the values at age of past years, which are known with more confidence due to convergent properties of the VPA.

shrinkage and could be emulated by the Laurec-Shepherd VPA model (by eliminating weighting towards recent years and by not using any shrinkage). However, the CAGEAN-type model in its pure form cannot reproduce the NMFS results, because of its different treatment of F, in that it interprets catch-at-age data in terms of F values that depend on an age-varying selectivity (constant through time) and a year-varying fishing intensity (constant over ages)—the so-called separability assumption. The VPA-based Laurec-Shepherd and ADAPT models, by contrast, allow fishing mortality rate to vary more flexibly but the price is that catch-at-age data have to be treated as though they are exact, which they are not (since F is based on a sampling process).

Model results also differ as a function of each model's assumptions about how the population estimates should be interpreted given the data. For data collected after 1994, a marked decline is seen in the commercial catch numbers of age-0 fish and a smaller decrease in the catch numbers of age-1 fish. In the survey data from years after 1994, the numbers of age-0 and age-1 fish are equal to or greater than the number in preceding years. The models may interpret these changes in one of two ways, depending on the model assumptions. In one sense, the relative number of young fish may be decreasing, as evidenced by the commercial catch-at-age data. This is the interpretation resulting when selectivity and fishing mortality at age are assumed to remain constant (as in the CAGEAN model). Conversely, there may be a shift to a lower selectivity in the fishery for the younger age classes. Such a shift to catching less young (small) fish might be expected from the increase in mesh size and minimum size regulations and seems to be corroborated by the survey's data regarding relative abundance data. The resulting estimates, if there is lower selectivity for young fish in the fishery, would be more optimistic and the overall population size could be interpreted as being larger than if fishing mortality and selectivity at age had not changed. Under the latter assumption, the population is estimated to have held steady or decreased. The actual situation probably falls somewhere between the two extremes, because neither assumption is entirely valid. The risks for the fishery are very different under the two hypotheses, so the consequences of both should be explored.

Some investigation is possible using the well-known properties of VPA estimates. Fishing mortality rates estimated by this approach for younger ages in earlier years can be regarded as being relatively unbiased by assumptions. Consequently, graphing the ratio of the fishing mortality at each age to the average of the fishing mortality rate on 2- and 3-year-olds can be used to track changes in age-specific catchability. Summer flounder mortality estimates seem to indicate that the catchability of younger fish has decreased since 1982 and to have dropped sharply since 1994 (Figure 2-6). The results for 1995 and 1996 are likely to be fairly robust to assumptions made in 1998, but the 1997-1998 results are still strongly influenced by the assumptions made in that year (as indicated in the discussion above). A second approach uses analysis of variance of catch-at-age data to look for evidence of changing catchability (extending the approach of Pope and Shepherd [1982]), but in the case of summer flounder this analysis was inconclusive.

A final suggested alternative would be to divide catch into the survey catch-per-unit-effort index to derive a measure of effort that should be linearly related to annual fishing mortality, selectivity, and survey catchability, and analyze the variance of these values. In all three approaches, departures from simple linear models can indicate trends that might not otherwise be noted in the nonlinear assessment model results.

A quick example of this method using the total catch-at-age data by year ($C[a,y]$) divided by the NEFSC spring catch-at-age data for ages 1-3 and their winter survey for the 0 groups by year ($U[a,y]$) is presented in Table 2-7.

In applying this last method to the summer flounder data, the total catch-at-age data from the fleet ($C(a,t)$) is divided by the NEFSC spring catch-at-age data ($U(a,t)$) for ages 1-3 and the winter survey for the age-0 groups. The quotient

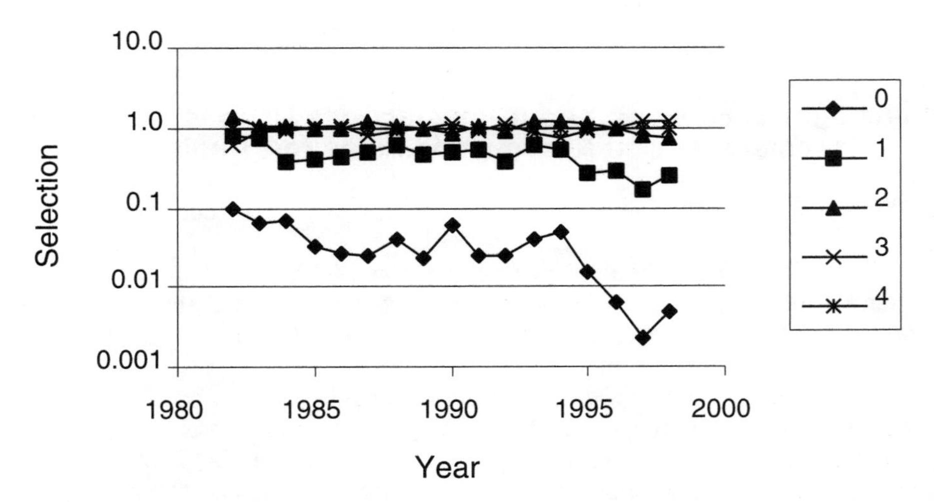

FIGURE 2-6 Summer flounder age-based selection relative to fishing mortality of age-2 and age-3 fish.

C(*a,t*)/U(*a,t*) is approximately *s(a)f(t)/q(a)*, where *s(a)* and *f(t)* are the selectivity at age *a* and full recruitment fishing mortality at year *t* for the fleet and *q(a)* is the catchability at age of the survey. Consequently, an ANOVA applied to the log ratio should be able to distinguish a year effect (log(*f(t)*)), and an age effect (log(*s(a)/q(a)*)). As a broad check for interactions, we divide the log ratio matrix into four quadrants (young ages-early years, old ages-early years, young ages-late years, old ages-late years) such that the sum of quadrant 1 and 4 minus the sum of quadrants 2 and 3 equal zero. From this an estimate is derived with one degree of freedom that indicates how selectivity at age in the commercial catch may be changing in earlier versus later years relative to the survey catchability at age, which we assume to be constant (Figure 2-6).

The significant quadrant effect shown in Table 2-7 indicates that there is a systematic shift in *s(a)/q(a)* through time. Assuming that *q(a)* in the survey is constant, we must conclude that *s(a)* is changing. This would indicate that the constant selection hypothesis must be rejected. If the constant selectivity assumption of the CAGEAN model is relaxed to allow selectivity to change progressively through time, the NMFS results are better approximated. This echoes what the ANOVA above indicates—the consistency of selectivity is a

TABLE 2-7 Analysis of Variance Table for Survey Data

Source of Variation	Degrees of Freedom	Sums of Squares	Mean Squares	F-statistic Value	Probability Value
Age (*a*)	3	59.30	19.77	35.20	0.0000
Year (*y*)	15	28.08	1.87	3.33	0.0010
Quadrant	1	7.94	7.94	14.14	0.0005
Residual	43	24.14	0.56		
Total	62	119.46	1.93		

BOX 2-3
Committee's Conclusions About Assumptions Used in Summer Flounder Assessments and Recommendations for Studies Needed

Using the virtual population analysis model runs, the committee concludes that:

- the common assumption that the natural mortality rate $M = 0.2$ is reasonable, given knowledge of natural mortality rates for other flounder species and data from the summer flounder stock. As long as M is not changed from year to year its effects on total allowable catch and spawning stock biomass are small.
- if older summer flounder are substantially less susceptible to trawl gear used in fishery-independent surveys than younger fish (NMFS assumes constant catchability over all ages), the fishing mortality rate would be substantially lower than estimated by NMFS (but would still not attain the target for summer flounder) and the total allowable catch and spawning stock biomass would be substantially higher.
- if the NMFS values of summer flounder ages are correct and the North Carolina values are incorrect, the spawning stock biomass and TAC would be lower than if both are assumed to be correct. Conversely, if the North Carolina values are correct and the NMFS values are incorrect, spawning stock biomass would be higher.
- the statistical precision of the winter flatfish surveys (including summer flounder) could be increased by allocating sampling effort in a more efficient manner.

Through other analysis and committee discussions, the committee concludes that:

- data available as part of the stock assessment were not adequate to determine whether the summer flounder population should be managed as a single unit.
- females are 20 percent larger than males at 5 years of age. This probably would not result in a major difference in the assessments now, because few flounder survive to age 5. However, if the fishery recovers to a broader age distribution, this sexual dimorphism may need to be considered in summer flounder stock assessments.
- the level of observer coverage in the fishery is probably too low to estimate discards accurately. Accurate estimation of discard levels is important because discard mortality is so high (assumed to be 80% for the commercial fishery) that a discarded fish has almost the same effect on the fishing mortality level as a fish that is landed.
- it is not apparent whether state surveys are timed and located to coincide with annual movements of summer flounder; it is therefore impossible to determine whether these surveys should be standardized.
- some components of recreational catch, particularly for the charter and party boat components, are relatively imprecise.

key assumption and certainly critical in determining biomass and fishing mortality estimates for the most recent years.

Box 2-3 contains a summary of the committee's findings and recommendations related to analyses of the summer flounder assessments and data.

POSSIBLE IMPROVEMENTS TO THE SUMMER FLOUNDER DATA SETS

Based on the answers to the questions in Box 2-1, a number of actions could be taken to improve data collection. Actions could include

The committee recommends actions in each case that NMFS could take to improve data and/or assumptions used. Priorities for changes in the summer flounder stock assessments and for research related to this species are discussed in the following points (not listed in priority order).

1. One of the major disagreements between NMFS and commercial fishermen regards whether the methods and locations of NEFSC trawl surveys undersample large summer flounder. Because this is a major area of contention and the committee's simulations show that incorrect assumptions could substantially affect the actual values for fishing mortality and spawning stock biomass, it is important that NMFS and industry work together to resolve this issue. The following steps should be considered:

a. Use data for fish up to 7 years of age, rather than 4 years, to constrain (tune) the model.

b. Use tagging studies to determine whether the average life expectancy of fish is greater than the average age of fish caught in surveys, indicating the existence of a greater number of older fish in the population than observed in the surveys.

c. Conduct sampling surveys of flounder eggs to determine whether the total biomass of spawning females is greater than assumed.

d. Study whether the ability of survey trawl gear to catch large summer flounder is overestimated, thus underestimating the biomass of these fish.

2. Conduct catch-at-age or catch-at-length analyses that explicitly account for sex differences in size at age and selectivity at age.

3. Find a way to include commercial effort data in the stock assessments and use commercial vessels to assist in conducting summer flounder surveys, as exemplified by the Canadian sentinel surveys and adaptive sampling exercises.

4. Increase observer coverage to improve the accuracy and precision of estimates of bycatch, discard rates, and landings of summer flounder, as well as to assist in tagging programs.

5. Determine how state surveys are designed, timed, and conducted in relation to summer flounder migrations. Use the findings of this exercise to determine how best to use the state survey data in the summer flounder assessment.

6. Increase the precision of recreational catch and effort estimates, perhaps by finding ways to better identify participants for inclusion in sampling surveys.

7. Maximize the precision of trawl surveys relative to sampling objectives, as demonstrated in Appendix C.

8. Work with fishermen and processors to ensure that larger (sushi market) flounder are properly sampled.

using tournament data, conducting social and economic studies of commercial and recreational fishermen, employing commercial fishermen as collectors, and improving observer program data. Any changes in the sampling intensity should be judged in the context of the contribution of specific data to the overall assessment and management of the species and the value of improving management of summer flounder in relation to the management of other species.

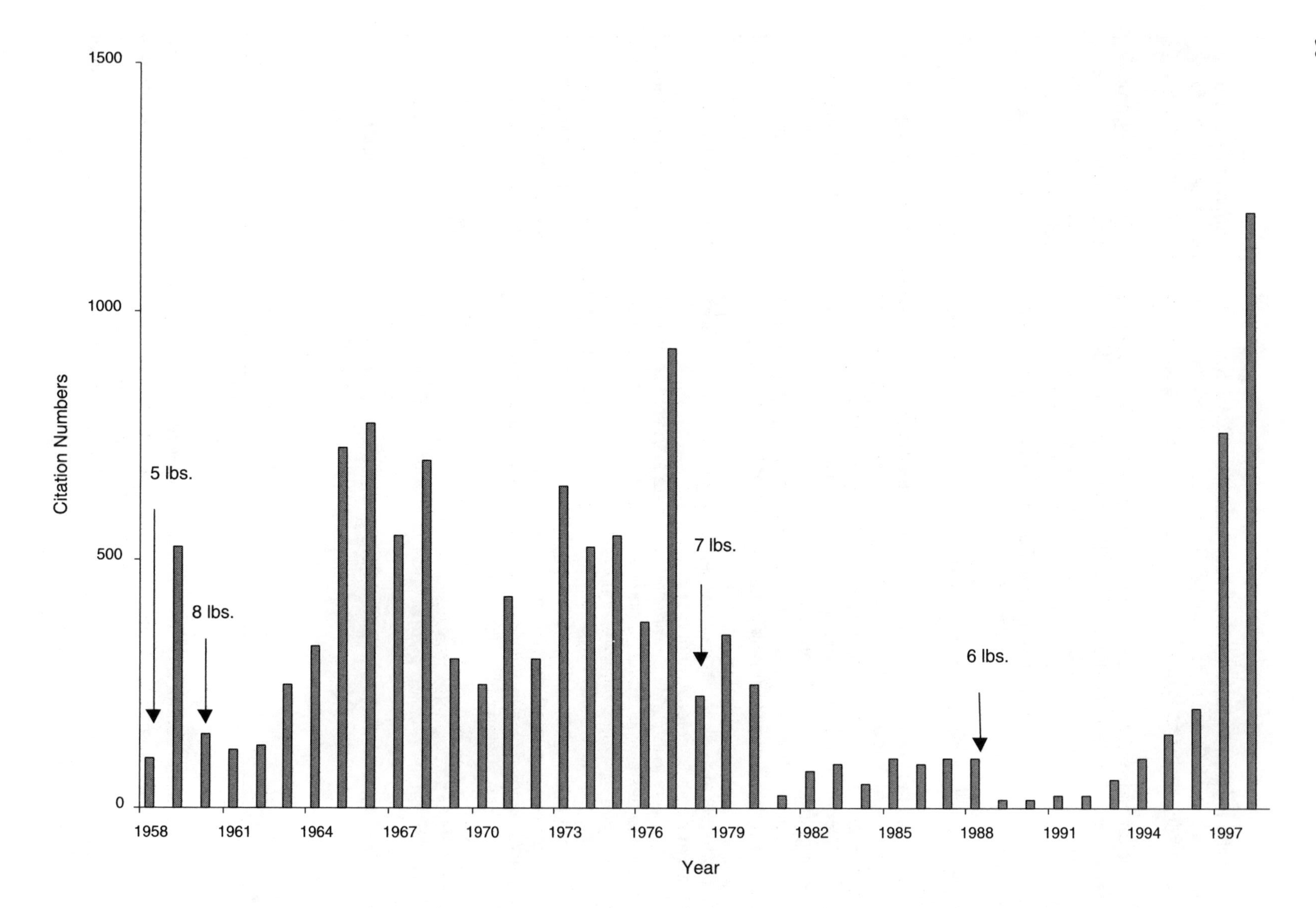

FIGURE 2-7 Virginia saltwater fishing tournament (summer flounder citations). Numbers indicate the minimum weight in pounds for a citation to be awarded.

Tracking and Management of Data Precision

In stock assessment, it is usually a mistake to measure one input (e.g., commercial or survey CPUE) with great precision, while only measuring other inputs approximately. The precision of all inputs, therefore, should be reviewed. Matching precision of inputs can be difficult because inputs often are sampled by different federal and state bodies drawing on different budgets and using different methods. Moreover, sampling for summer flounder is often carried out in conjunction with sampling for other species, with the result that costs are shared and optimization of sampling needs is viewed in terms of the wider program.

To achieve the greatest possible precision of outputs, we suggest that precision targets be set for all sampling schemes and that performance be monitored annually in terms of achieving these targets. If this is done at the same time that data are analyzed, the requirement need not increase the time needed to conduct assessments. Such tracking of precision would assist year-to-year adjustments of data sources to keep them in appropriate balance. A useful exercise would be to conduct a thorough sensitivity analysis to evaluate how a given percentage change in the precision of each of the different inputs is translated by the assessment model into changed precision in the assessment outputs. Coefficients of variation (CVs) between 10 percent and 20 percent for composite estimates (e.g., total catch at age over significant ages, or combined survey indices of abundance) would be appropriate for most important fish stocks, at present, though changing needs and problems could require greater precision. Precision of the subcomponents of such measures could also be set to achieve this target range of CVs. For example, in the case of components of catch-at-age data, this would require a CV that is approximately equal to the overall target CV/(the square root of the catch share) (Pope, 1983). For example, if the target were a 10 percent CV, a fleet sector that caught 25 percent of the catch should require a CV of 20 percent, that is, half the precision.

A Role for Commercial and Recreational Fishermen in Data Collection

The summer flounder fishery clearly demonstrates the need for better cooperation among recreational anglers, commercial fishermen, state agencies, and federal agencies in data collection so that data quality is improved and its credibility is enhanced. There is room for improvement from all sectors. Strengthening the active support of commercial fishermen is particularly important. Chapters 3 and 4 include discussions of and suggestions about different ways that commercial fishermen could be enlisted to help with surveys and improve the quality of commercial fishery-dependent data.

Summer flounder fishermen related several concerns about data collection methods currently in use. In other fisheries, mistrust between fishermen and scientists has been greatly reduced by their joint participation in data collection efforts: fishermen provide boats and advise on appropriate gear types and deployment procedures and scientists gain "indigenous" familiarity with the fishery and get to explain quality control limitations with which they must be concerned. Summer flounder fishery managers should examine these partnership approaches and test a working model in the summer flounder fishery.

New Sources of Recreational Fishery Data

Actions also could be taken to increase the participation of anglers in collection of summer flounder data. Anglers, who are particularly important in summer flounder population dynamics, participate in at least two types of programs that reward the catch of large fish and keep track of either exact fish weights (tournaments) or weights above a certain pre-set value (citations). Ancilliary data exist in state citation programs, and data from tournaments could provide qualitative trends in fish abundance. Many of the Atlantic Coast states sponsor citation programs in which anglers are awarded citations for fish that are at or above specific weights. The Virginia program is the best example of this type of data (Figure 2-7). The Virginia Saltwater Fishing Tournament records the number of citations awarded each year to anglers catching trophy-size fish of specified weights. Most other state programs are much smaller.

Citation data must be used with care because the effort of obtaining these trophy fish is not included and the weight that defines a trophy fish has changed over the years. Nonetheless, these data do reveal general trends in the abundance of large fish. For summer flounder there was a general increase in the number of trophy citations from 1958 to 1977, even with an increase in minimum citation weight. From 1978 to 1992, the number of citations remained low, despite a lowering of the minimum weight for a citation fish. Since the early 1990s, the number of citations awarded has increased dramatically. The trends in citations reflect, to some extent, the availability of large fish in the population. Although commercial fishermen expressed concern that older fish were potentially missed by NEFSC trawl surveys, some of these fish are apparently distributed near shore, where they are captured in recreational fisheries, primarily in the summer. A comparison of such data with commercial catch data from the same area could be used to cross-check data sources and assumptions. Tournaments provide another auxiliary source of recreational data, particularly snapshots of the abundance of legal-size fish at particular times and locations.

3

General Issues in the Collection, Management, and Use of Fisheries Data

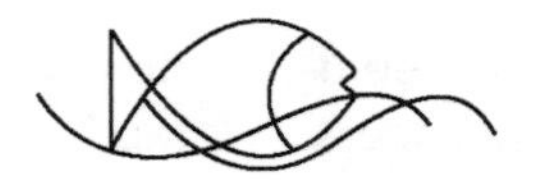

WHAT ARE FISHERIES DATA?

The phrase "fisheries data" is a general way of referring to data that may be of use in the management of a fishery as well as for commercial, recreational, cultural, and scientific purposes. Such data usually include biological information about the exploited fish and associated species, economic information about the fishermen and the markets for the catch, and information about the environmental conditions that affect the productivity of the species. This information is collected from many sources.

A primary source of information is the commercial and recreational fishermen themselves, so-called fishery-dependent data. Logbooks (also called trip tickets) are designed to collect data on the time and place of fishing, the effort expended, catch by species, and other information. In many jurisdictions, completion of logbooks is a condition of participation in the fishery. Often, information from logbooks is the most timely information on current fishery conditions; mechanisms for self-reporting are rare in recreational fisheries.

Catch sampling programs are another important source of information. Fish can be measured and weighed either at sea (by observers) or at landing sites (by port agents). Observers are placed on commercial fishing vessels to provide information on fishing activities that are not always reported in logbooks, such as effects of fishing activities on protected species and the extent and fate of bycatch and discarding. Samples can be obtained to determine the species composition, sex ratio, and age composition of the catch.

In some fisheries, scientific surveys are a vital component of the stock assessment process. Research vessels of the National Oceanic and Atmospheric Administration (NOAA) and commercial fishing vessels operating under charter agreements with NOAA are used to conduct surveys of fish abundance. These surveys are the primary source of fishery-independent data, including estimates of the age structure of fish populations and relative abundance of stocks. The National Research Council (NRC, 1998a) demonstrated the importance of accurate indices of abundance, which in many fisheries can be obtained only from fishery-independent surveys.

In fiscal year 1999, the National Marine Fisheries Service (NMFS) spent $28.8 million on ship time for surveys (not counting personnel and analyses), $3.9 million for recreational monitoring, $9.2 million on observer programs (with another $10 million provided by industry), and $2.8

million on vessel monitoring system (VMS) programs. The expenditure by NMFS for these data collection activities is thus on the order of $45 million. Additional expenditures were made by states and industry. The total fishery harvest in the United States (commercial and recreational) is valued at approximately $45.7 billion when the total economic effects are included (NMFS, 1995).

WHO USES FISHERIES DATA?

Fisheries data have many uses and many users—including stock assessment by scientists, strategic planning by industry, and fishery monitoring and allocation decisions by managers. Adequacy of data can be evaluated only in the context of the purposes for which they are used. Each use implies a set of users and a suite of requirements that the data must satisfy, including timeliness, level of detail, accuracy, accessibility to users, coverage or completeness, and credibility of the data collection process and the management process that uses the data.

Fisheries data are vital to strategic planning activities in coastal communities that rely on fisheries. Fishery management authorities are responsible to use fisheries data for creating policies for the orderly and sustainable development and management of fisheries. Civil authorities use fisheries data to site marinas, underwater pipes and cables, and other maritime facilities, and to develop infrastructure for the fishing industry. Bankers use fisheries data to plan economic development and loan packages to fishermen, fish processors, and ship suppliers. Fishermen themselves use fisheries data to plan future fishing activities, such as shifts to new fishing grounds, changes in fishing gear, and changes in species targeted. However, fishermen often use their own data sources, including their own logbooks and observations, and what they learn from other fishermen and buyers, instead of using government data. This may occur because of some fishermen's mistrust of government data, the frequent lag time in availability of such data (often too great to use government data in business planning), and the lack of data for the geographic area and type of fishery in which a specific fisherman is engaged.

Monitoring conditions in a fishery is the responsibility of regional fishery management councils and NMFS, and is the primary means of assessing compliance with and accomplishing enforcement of fishery regulations. Another major responsibility of the regional councils is allocation of harvest opportunities among different user groups. Environmental and other interest groups also have become increasingly involved in monitoring fishing activities. Monitoring often requires data with great detail in both time and space as well as frequent updates, often within a fishing season.

Stock assessment is a critical use of fisheries data and is often considered its primary use. The committee devoted a significant portion of its attention to the data used in stock assessments, using the summer flounder fishery as a case study. Scientists employed by state, interstate, national, and international fishery agencies are the primary users of data relevant to stock assessments; in addition, university and private sector scientists increasingly are becoming involved in stock assessments and related research. Current stock assessment practices use data aggregated over the entire fishing ground and over a fishing season. Although assessment methods may require a greater diversity of data, the resolution in space and time is usually rather coarse and may need updating only infrequently, such as annually or semi-annually.

The multiple users of fisheries data have different requirements in terms of resolution in time and space for each possible data element. Table 3-1 summarizes the requirements for data elements by various users (based on committee experience); specific details depend on the characteristics of individual fisheries. Data system designers, therefore, must consider that the demands will vary among users, and the system must be capable of accommodating users who require data at different spatial resolutions and different degrees of timeliness.

TABLE 3-1 Example of How Data Timeliness and Spatial Resolution Vary Among Users

	USE Users			
	MONITORING Councils NGOs	ASSESSMENT NMFS Scientists Academics Fishermen	ALLOCATION Councils	PLANNING Councils Local and State Governments Economists Bankers Fishermen
Update Frequency (timeliness)	Within season	Between seasons	Within and between seasons	1-5 years
Spatial Resolution	Detailed/mandated jurisdiction	Stock-wide	Stock-wide	Ad hoc
Required Elements				
Catch by species	✓	✓	✓	✓
Fishing effort	✓	✓	✓	✓
Catch at length	✓	✓		
Age composition		✓		
Sex ratio		✓		

Credibility is one of the major concerns surrounding current fisheries data collection activities. Many stakeholders believe that data collected by NMFS are neither accurate nor complete. These misgivings are exacerbated by problems of timeliness and accessibility and by perceived conflicts of interest; NMFS not only collects the data but also conducts stock assessments, makes policy recommendations to councils, enforces fishery regulations, and makes judgments about the policy recommendations and fishery management plans prepared by the regional councils. For many fishermen these multiple responsibilities of a single agency create some mistrust regarding the collection and use of fisheries data.

Two recent reports stress the importance of greater collaboration among scientists and stakeholders in data collection. First, the Consortium for Oceanographic Research and Education states

> Finally, collaborative data collection and research efforts should be encouraged among agency scientists, independent scientists, and representatives of industry and public interest groups. Not only would this build confidence among the different groups, but it would provide access to valuable, non-traditional sources of information (CORE, 2000).

Second, the General Accounting Office was asked by Congress to examine NMFS' compliance with several aspects of the Magnuson-Stevens Act, including use of the best available scientific information and consideration of economic effects of fisheries management on communities. The GAO recommended that NMFS:

- increase the involvement of the fishing industry, its expertise, and its vessels in fishery research activities in order to expand the frequency and scope of NMFS' data collection efforts,
- review data collection requirements placed on fishermen to limit requested informa-

tion to what is needed for conservation and management, regulation, and scientific purposes, and

• review data collection procedures for fisheries where the recreational sector constitutes a major portion of the fish caught to minimize the inconsistent treatment of commercial and recreational fishermen (GAO, 2000, p. 29).

It is clear from these activities, both initiated by Congress, that improving data collection is a priority for Congress.

Each region of the United States uses different methods to collect, manage, and use fisheries data. In part, such differences are based on differences in the biology and social aspects of the fisheries. Many differences may be due to tradition and familiarity with certain approaches and the accumulation of past actions, rather than rational choice. Other differences arise from state and federal legislation that requires or permits specific activities.

Data Needed for Different Management Methods

What biological, economic, and social data are most needed to provide assessments suited to five common management methods?

- current state of the fishery
- management goals and measures of their achievement of intended effects
- management actions needed to achieve management goals

Five common management methods include (Table 3-2):

- Total allowable catch (TAC)
- Effort management
- Gear restrictions and fish size limits
- Closed areas (see NRC, 2000)
- Closed seasons

Current State of the Fishery

Fishery status questions address not only the current status of the stock but also the fishery as a whole, including social and economic factors. Relevant questions include

- What is the current spawning stock biomass level?
- What is the current level of fishing mortality?
- Is recruitment being sustained?
- Is growth potential maximized?
- What is the effect of fishing, if any, on the ecosystem?
- What is the essential habitat for the species and what is the status of the habitat?
- What social and economic benefits are realized from this resource?
- What is the relation between current fishing capacity and the sustainable yield of the fishery?

All types of management have specific needs for answering the system status question. For TAC-based management, it is essential to know the current catch, and for effort management to know the current effort. For management based on gear restrictions or individual size limits, it is important to know about the sizes of fish currently being caught and the selectivity of the gear used, and is probably desirable to know the size of fish at maturity, if a goal of management is to allow fish a chance to spawn at least once before capture. For closed areas, it is important to know the distribution of fish relative to the extent of the closed areas and the rate at which fish move in and out of these areas. With closed seasons, it is important to know the seasonal distribution of fish and the timing of spawning.

In practice, data requirements may be simplified by substituting measurements of effort for fishing mortality, catch per unit effort (CPUE) for biomass, length distributions in catch and surveys for age, and recruitment survey CPUE for recruitment. Hence, data of these types should be

TABLE 3-2 Requirements for Biological, Social, and Economic Data For Five Common Management Methods

Management Method	Data Requirements
TAC-based	Catch and effort data
	Fishing mortality rate
	Annual TACs and estimation of recruitment
	Social and economic impacts of management
	Likelihood that regulations will foster misreporting of fishery-dependent data, including economic and regulatory discards[a]
	Economic contributions of recreational and commercial fisheries, including supporting industries
	Distribution of catch among gear types and between commercial and recreational fishermen
Effort management	Catch and effort data
	Fishing mortality rate
	Social and economic impacts of management
	Optimal harvesting and processing capacity
	Present participation of individuals in the fishery
	Dependence on the fishery
	How efficiency of effort has changed and how effort is allocated across different species and sizes of fish
	Likelihood that regulations will foster misreporting of fishery-dependent data, including economic and regulatory discards
	Economic contributions of recreational and commercial fisheries, including supporting industries
	Distribution of effort and likely impacts of capacity reduction approaches
Gear restrictions and fish size limits	Catch and effort data
	Size distribution of fish being caught
	Selectivity of gear
	Size at maturity
	Age at first capture
	Encounter rate and release mortality of undersized fish
	Compliance with size limits
	Social and economic impacts of management
	Economic contributions of recreational and commercial fisheries, including supporting industries
	Impact of regulations on fishing behavior, especially where fishery-dependent data are used for stock assessments
Closed areas	Distribution of fish (by size and maturity, within and outside the closed areas)
	When and at what rate fish move in and out of the area
	Catch and effort data outside the closed area
	Social and economic impacts of management
	Economic contributions of recreational and commercial fisheries, including supporting industries
	Distribution of catch among gear groups and between commercial and recreational fishermen

(continued)

TABLE 3-2 Continued

Management Method	Data Requirements
	Catch and effort data Likelihood of regulations to foster misreporting of fishery-dependent data and economic and regulatory discards Potential shifts in fishing areas and effort Knowledge of essential habitat
Closed seasons	Seasonal distribution and timing of spawning Catch and effort data Distribution of fish in closed and open seasons Social and economic impacts of management Economic contributions of recreational and commercial fisheries, including supporting industries Distribution of catch among gear groups and between commercial and recreational fishermen Likelihood of regulations to foster misreporting of fishery-dependent data and economic and regulatory discards Potential shifts in fishing times or areas

[a] Economic discards are fish discarded because they are unmarketable (because of their quality, size, species, or sex) or because a fisherman hopes to replace them with higher-value fish. Regulatory discards are fish discarded because they are prohibited by regulation from being landed, because of their size, species (prohibited species or species for which the seasonal quota has been filled), the gear used, or area fished.

desirable for almost any management system. However, for fisheries in which fishing mortality is a small proportion of total mortality, it might be argued that an intensive monitoring system is of marginal value relative to the low risk of overfishing.

Whatever method of management is chosen, managers need to know approximately what portion of a stock is being exploited and how exploitation must change to achieve management goals. A minimum requirement for such assessments would be some sort of general production model that includes at least catch and effort, hence the need for catch and effort data for all management types.

Management Goals and System Response

System response questions involve monitoring the changes in stock status in response to changes in the management control variable (e.g., catch, effort, gear, time, or area restrictions). In the case of catch and effort quotas, stock status usually is expressed by fishing mortality level or changes in relative abundance. In the case of mesh changes or size limits, system response is usually measurable in terms of average size of fish in the fishery. Closed areas or seasons are likely to require both catch and effort data, subdivided by area and time.

Management Actions

Answers to management implementation questions can help managers as they select actions to achieve management goals. Such questions require that biomass be estimated for TAC management and that recruitment also be monitored. However, for heavily exploited fisheries in which the spawning stock biomass has been substantially reduced from unfished levels, recruitment is often an important component that needs to be monitored, because only a few poor recruitment years are needed for the population to crash. Effort quotas require that changes in gear efficiency and targeting practices also be monitored. Mesh size or size limit management requires an understanding of gear selectivity. Closed areas and seasons require information about the distribution of fish in areas that might form part of extended closures. An important focus of both monitoring and research is how fisheries respond to regulations. This is an essential component of sustainable management, and implementation uncertainty is often one of the largest sources to total uncertainty about a stock-fishery system.

Data Quality Required

In an ideal world, all advice would be completely accurate and precise, but in practice data contain some level of bias and random variation, and incremental gains in precision and accuracy often require ever greater relative expenditures for sampling and analysis. Moreover, some sources of inaccuracy or imprecision may be impossible to eliminate, even with infinite sampling, because of the inherent randomness and chaos in natural systems. Pressures to maximize total allowable catch can lead to excess fishing that can harm a fishery before managers understand the dynamics of the target fish population(s). Fisheries can also be damaged when pressures to maximize total allowable catch cause managers to attempt to manage at a level of detail finer than available information will allow.

Every management system should be evaluated in light of the amount of inaccuracy and imprecision in management advice that can be tolerated and still allow the system to achieve its goals. The precision of data needed depends on the management regime and objectives chosen. For example, management with closed areas would lower the precision needed for data outside the closed area.[1] Taking a different management approach, with an objective of keeping catch (and employment) as high as possible, subject only to the fish being able to reproduce sustainably—the apparent goal of many U.S. fishery management plans—requires accurate and precise estimates of current stock status, minimum levels of spawning stock biomass, and fishing mortality. The higher the rate of exploitation, the more precision is needed to manage a stock adequately from a biological perspective. From economic and social views, a high level of precision may be necessary in order to avoid undue disruption in the industry.

Acceptable levels of imprecision and inaccuracy also depend on the extent of annual variations in management restrictions that will be tolerated by managers and fishermen and the ability of a stock to withstand inevitable over- or underexploitation caused by inaccurate or imprecise management. Fluctuations in total allowable catch due to imprecise data would require effort to move in or out of the fishery and would probably increase costs compared to a situation in which the TAC is lower but less variable. For a given level of precision, the amount of data required (though not necessarily its cost) is, as a first approximation, independent of the size of the stock. It would be wise to compare the management method (and data collection costs) to the potential benefits of management. Managers may not be prepared or able to pay for the levels of sampling that would provide an appropriately precise fisheries assessment for some low-value stocks.

If managers are not prepared to pay for greater precision, or if needed precision is not achievable at any price, managers may have to modify either their objectives or their control rules. One approach would be to select a lower level of exploitation so that stock abundances would change more slowly and fluctuations in numbers of young fish would be dampened by

[1] The amount of area that needs to be closed to avoid the need for high precision data is unknown in practice, but modeling studies have indicated that as much as 30-70 percent of total fishing area may need to be protected, if this is the only form of fishery management. See NRC (2000) for a summary of the state of knowledge regarding the use of marine protected areas for fisheries management.

the presence of more age classes in the fish stock. Another approach would be to adopt a form of management that needs less precision in estimates of the current stock status and estimates of next year's catch, for example, managing a fishery by limiting fishing effort rather than by limiting catch. Closed area management requires even less assessment information and is more robust to any lack of precision in assessment. If a sufficiently large part of the range of a fish stock (including spawning areas and nursery areas) were closed, it is unlikely that fishing could damage it. However, closed areas would probably not reduce the need for precise data for highly migratory stocks.

Timeliness of data is a final aspect of data quality. Sampling may be designed adequately to ensure that data are accurate and precise enough for management purposes, but data analysis may not be timely. Lack of timeliness can hinder good management, particularly in the case of heavily exploited fish populations with few year classes and for fisheries that depend on in-season management. State data are sometimes only available a year after their collection and recreational data often are not available until the following season. This lag in data availability results in management that responds to the situation that existed one year ago, a situation that may no longer exist. This may explain, in part, the finding of NRC (1998a) that assessment results tend to lag behind the actual situation by one or more years in detecting stock declines and rebuilding. Timeliness of data is also affected by the frequency of surveys, discussed elsewhere in this report.

METHODS OF DATA COLLECTION

Data are available from a number of sources, including from ceremonial and subsistence fisheries, from fishery-independent surveys conducted by the states and NMFS, and from commercial and recreational fisheries.

Data from Ceremonial and Subsistence Users

Many fisheries are exploited for ceremonial or subsistence uses. For example, halibut and salmon are prominent fish species used by Native Americans in the Pacific Northwest and Alaska for both ceremonies and subsistence. Non-natives in these areas are also subsistence users. Pacific Islanders use coastal fish species and tuna for similar purposes. Data related to ceremonial and subsistence users are collected for inland waters and Pacific coastal waters and used in stock assessments. Most Pacific coast ceremonial and subsistence data relate to salmon fisheries, but groundfish catches for these uses also are included in landings data that NMFS receives from states.

Subsistence use, although small in comparison to recreational and commercial use, is still significant in many U.S. fisheries. In some cases, subsistence use may be included in the recreational fishing category, accounting for individuals who regularly fish off the shore, piers, and other coastal access points to provide food for themselves and their families. Such individuals may be contacted by MRFSS intercept samplers, but they may be missed in telephone surveys because of language difficulties, mistrust of government agencies, or because they do not have telephones. Non-commercial catches may form a large percentage of the diet in some communities, but this has not been studied extensively. Another small component of subsistence use is the catch that commercial fishermen take for personal use. In any case, except for the examples given earlier, catch for ceremonial and subsistence fishing is a minor portion of the catch in most fisheries.

Data from Fishery-Independent Surveys

NMFS and individual states conduct a variety of surveys throughout the year in offshore and inshore waters. Some federal surveys are conducted as many as three times per year (East

Coast flatfish), whereas other species may be surveyed every three years (many West Coast and Gulf of Alaska fisheries) or never (Table 3-3). Appendix C illustrates and analyzes the variety of surveys conducted for summer flounder. In addition to surveys of fish abundance and population characteristics, the states, NMFS, other agencies, and academic scientists collect data related to other components of marine ecosystems and marine environmental conditions in an attempt to understand how fishing affects marine ecosystems and how marine environmental conditions affect fish populations.

In general, the purpose of stock assessment activities is to monitor changes in the abundance of fish populations over time in order to evaluate the effects of past and present fishing activities on fish population trends and to predict the consequences of future fishery management decisions. Stock assessments, together with monitoring of physical and biological variables, are also needed to evaluate the effects of the environment on fish populations.

Monitoring changes in abundance of fish stocks over time requires having at least one measure that reflects these changes without biases or with constant known biases. Catches from the commercial fishery may fluctuate from year to year due to causes unrelated to changes in absolute abundance. For example, changes in commercial catch can result from changes in the amount of fishing effort in any one year as a function of the price and abundance of alternate fish species, improvements in fishing technology (better nets, more precise acoustic detectors or navigational equipment), changes in management measures (closed areas, seasons, trip limits), or inaccessibility of the stock due to changes in the ranges of fish populations caused by environmental factors.

Year-to-year changes in the distribution of fishing effort should be considered when using fishery CPUE data to measure fish abundance. If a fishing fleet moves from fishing grounds where fish densities are low to grounds where fish densities are high, CPUE will increase even though the overall stock abundance remains constant or declines. Analyses of CPUE data must adequately consider the spatial aspects of fish population distributions and the fishing effort applied to catching fish. Often, however, CPUE data are simply combined over broad (and inappropriate) spatial scales.

In most fisheries, the best measure of relative fish abundance is obtained from fishery-independent surveys, in which the gear (and usually vessel), timing, survey design, and procedures are kept constant from year to year. As a result, annual changes in the abundance or biomass of a species are assumed to reflect actual changes in relative abundance. Surveys are intended to determine whether populations have changed relative to previous years; typically they are not designed to determine absolute abundance. In addition to tracking the relative abundance of fish stocks over time, fishery-independent surveys provide a means to gather information unattainable from landed catch (e.g., maturity indices, fishery indices for sublegal-sized fish).

Operationally, the general practice of this kind of survey is to use fishing gear of a type commonly used in the fishery. However, there have been cases (e.g., the crab fisheries in the Bering Sea and the Gulf of St. Lawrence) in which bottom trawls were used for the survey while crab pots or traps were the only gear used by the fishery. Even if the gear chosen for the survey resembles that in common use when the surveys were initiated, the fishing industry can continue to upgrade and improve its gear. This usually creates the perception in the fishing industry, many years after the survey series has started, that the survey gear is old-fashioned and sub-optimal. Such perceptions also lead to charges that the outmoded survey series is not useful because of the fishing gear used in the survey. Although it is true that there have been many improvements to fishing gear and practice over the past thirty or more years, criticism of a survey series should be based more on whether

TABLE 3-3 Research and Charter Vessel Surveys, NMFS Fiscal Year 2000

Type of Survey	Area	Species or Species Complex	Frequency or Seasonality of Survey	Number of Stations
		Atlantic and Gulf Surveys		
Autumn bottom trawl survey	Cape Hatteras to Nova Scotia, 4-200 fathoms	Fish and macro-invertebrates	Annual	355 trawl/CTD stations
Winter bottom trawl survey	Cape Hatteras to Georges Bank, 15-100 fathoms	Fish and macro-invertebrates	Annual	155 trawl/CTD stations
Spring bottom trawl survey	Cape Hatteras to Nova Scotia, 4-200 fathoms	Fish and macro-invertebrates	Annual	335 trawl/CTD stations
Northern shrimp bottom trawl survey	Gulf of Maine, 50-120 fathoms	Northern shrimp	Annual	65 trawl stations
Sea scallop survey	Cape Hatteras to Georges Bank, 15-60 fathoms	Sea scallop	Annual	600 dredge stations and 300 CTD profiles
Surf clam/ocean quahog survey	Cape Hatteras to Georges Bank, 4-40 fathoms	Surf clam/ocean quahog	Triennial	475 hydraulic dredge stations and CTD profiles
Apex predator survey	Key West to Delaware Bay, 5-40 fathoms	Shark	Triennial	100 longline stations and profiles
Atlantic herring hydroacoustic survey	Georges Bank and the Gulf of Maine, 10-200 fathoms	Atlantic herring	Annual	3,400 nautical miles of hydroacoustic trackline; ~70 pelagic trawl tows
Small pelagics hydroacoustic survey	Cape Hatteras to Nantucket Shoals, 10-100 fathoms	Atlantic mackerel, butterfish, loligo and illex squid, and Atlantic herring	Annual	3,400 nautical miles of hydroacoustic trackline; ~70 pelagic tows
Trawl survey standardization and technology development	Cape Hatteras to Nova Scotia, 5-100 fathoms	Gear efficiency study	Annual	Variable

Ecosystem monitoring survey—winter	Gulf of Maine, 15-200 fathoms Cape Hatteras to Georges Bank, 4-200 fathoms	Multi-species eggs and ichthyoplankton	Annual	30 bongo hauls 90 bongo hauls/CTD profiles
Ecosystem monitoring survey—early spring	Cape Hatteras to Nova Scotia, 4-200 fathoms	Multi-species eggs and ichthyoplankton	Annual: survey piggybacked	120 bongo hauls/CTD profiles
Ecosystem monitoring survey—late spring	Cape Hatteras to Nova Scotia, 4-200 fathoms	Multi-species eggs and ichthyoplankton	Annual	120 bongo hauls/CTD profiles
Ecosystem monitoring survey—summer	Cape Hatteras to Nova Scotia, 4-200 fathoms	Multi-species eggs and ichthyoplankton	Annual	120 bongo hauls/CTD profiles
Ecosystem monitoring survey—early autumn	Cape Hatteras to Nova Scotia, 4-200 fathoms	Multi-species eggs and ichthyoplankton	Annual: survey piggybacked with autumn trawl survey	120 bongo hauls/CTD profiles
Ecosystem monitoring survey—late autumn	Cape Hatteras to Nova Scotia, 4-200 fathoms	Multi-species eggs and ichthyoplankton	Annual	120 bongo hauls/CTD profiles
Northern right whale survey	Bay of Fundy to the Gulf of Maine, 15-200 fathoms	Whales	Annual	Visual line transect survey with 100 plankton and CTD stations
Harbor porpoise survey	Georges Bank and the Gulf of Maine, 15-200 fathoms	Harbor porpoise	Triennial	Visual line transect survey with 30 CTD profiles
Marine turtle survey	North Carolina to the Gulf of Maine, 4-200 fathoms	Turtles	Triennial	Visual line transect survey with 50 CTD profiles
Harbor porpoise and hydroacoustic survey	Gulf of Maine, 10-200 fathoms	Harbor porpoise	Annual	Visual line transect survey with 50 CTD profiles
Pelagic delphinid survey	North Carolina to the Gulf of Maine, 4 fathoms to EEZ boundary (abyssal depths)	Dolphins	Triennial	Visual line transect survey with ~10 plankton tows and 200 CTD profiles
SABRE/striped bass	North Carolina to Virginia, 5-100 fathoms	Striped bass and larval fish	Winter	80 striped bass stations 80-100 plankton tows

(continued)

TABLE 3-3 Continued

Type of Survey	Area	Species or Species Complex	Frequency or Seasonality of Survey	Number of Stations
West Florida shelf fishing reserve	Gulf of Mexico off Florida, 5-100 fathoms	Reef fish	Winter	130 to 260 stations
SEAMAP reef fish	Gulf of Mexico from Texas to Florida, 5-60 fathoms	Reef fish	Spring	288 stations
SEAMAP groundfish	Gulf of Mexico from Texas to Alabama, 5-60 fathoms	Shrimp/groundfish	Summer Fall	240 stations 240 stations
Shark survey	Gulf of Mexico/Caribbean from Florida to Texas, 10-40 fathoms; Cuba, 10-300 fathoms	Coastal sharks	Summer	250 to 300 stations
Gear comparison	Gulf of Mexico, Texas to Florida, 5-60 fathoms	Shrimp/groundfish	Fall	240 stations
Caribbean marine mammals	Caribbean-Puerto Rico to Venezuela—nearshore to 5,000 fathoms	Humpback whales	Winter	80 to 100 sightings
SEAMAP plankton/marine mammals	Gulf of Mexico, 100 fathoms to EEZ for spring survey 5-600 fathoms for fall survey	Bluefin tuna/cetaceans Mackerel/red drum/cetaceans	Spring Fall	196 plankton stations 240 mammal stations 118 plankton stations 200 mammal stations
Snapper longline	Gulf of Mexico, Texas, 35-80 fathoms	Red snapper	Summer	88 stations
Sperm whale	Gulf of Mexico from Louisiana to Alabama, 100-2,000 fathoms	Sperm whales	Summer	52 sightings

Pacific Surveys

Bottom trawl	Bering Sea shelf	Groundfish, king crab, Tanner crab, snow crab	Annual	380 stations
Bottom trawl	Bering Sea continental slope	Groundfish, crab	Biennial	70-100 stations
Acoustic/trawl	Bering Sea–Bogoslof Island	Pollock	Annual	1,500 nautical miles of trackline
Acoustic/trawl	Bering Sea shelf	Pollock	Biennial	6,000 nautical miles of trackline
Bottom trawl	Aleutian Islands	Groundfish	Biennial	425 stations
Bottom trawl	Gulf of Alaska shelf and continental slope	Groundfish	Biennial	776 stations
Acoustic/trawl	Gulf of Alaska–Shelikof Straits	Pollock	Annual	900 nautical miles of trackline
Surface trawl	Bering Sea–Bristol Bay	Juvenile salmon	Semi-Annual	55 stations
Surface trawl	Gulf of Alaska	Juvenile salmon	Annual	Variable
Surface trawl	Southeast Alaska coastal monitoring	Young salmon, sablefish, other epipelagic species	Annual (five 7-day cruises)	24 stations (250 km spread)
Longline with limited surface gillnets	Gulf of Alaska (GOA), Bering Sea (BS) and Aleutian Islands (AI) outer shelf and slope	Sablefish and rockfish	GOA Annual BS and AI Biennial	74 to 76 stations
Egg and larval survey	Gulf of Alaska	Pollock	Annual	100 to 120 stations
Bottom trawl	West Coast shelf	Groundfish	Triennial	620 stations
Bottom trawl	West Coast continental slope	Groundfish	Annual	200 stations
Acoustic/trawl	West Coast	Pacific whiting	Triennial	6,700 nautical miles of trackline

(continued)

TABLE 3-3 Continued

Type of Survey	Area	Species or Species Complex	Frequency or Seasonality of Survey	Number of Stations
Bottom trawl (4 charter vessels)	West continental slope	Groundfish	Annual	400 stations
CalCOFI	Monterey, California to Mexico 200-360 nautical miles offshore	220 (20 in groundfish FMP; 3 in pelagic FMP)	Quarterly	66 stations
Shark	Point Conception to Mexican border, offshore 120 nautical miles	4 species of pelagic sharks	1 to 3 years	40 Stations
Sardine	San Francisco to Mexican border, offshore 300 nautical miles	Sardine, jack mackerel, anchovy	2 to 6 years	1,000 Stations
Rockfish	Central California to 50 nautical miles offshore	Groundfish (rockfish, whiting, lingcod)	Annual	99 Stations
Antarctic	Antarctic Peninsula	Krill, groundfish	Annual	100 Stations
Lobster	Northwestern Hawaiian Islands	Spiny and slipper lobster	Annual	N/A

SOURCE: National Marine Fisheries Service.

NOTE: CalCOFI = California Cooperative Oceanic Fisheries Investigations; CTD = conductivity-temperature-depth recorder; FMP = fishery management plan; N/A = not applicable; SABRE = South Atlantic Bight Recruitment Experiment; SEAMAP = Southeast Assessment and Monitoring Program.

the current gear is working properly and whether the selectivity of the gear is well known, rather than on whether that type of gear still is being used by the fishery (see Box 2-2 for details on gear selectivity). A further consideration is that when there is an important change in gear, both old and new gear should be used in parallel for a long-enough period of time to establish the conversion factor needed to use historical data.

A question needs to be answered before a survey is changed by updating the gear: Does the existing survey produce data of quality adequate for use in an assessment? If the answer is "no," the existing survey should be phased out and replaced with a better survey. If the answer is "yes," and the data are still needed, the associated cost of changing gear is the cost of an adequate intercalibration experiment. This consists of many parallel samples using the new and existing gear. Many intercalibration studies have been conducted as part of regular surveys and the sample size for the number of parallel tows usually corresponds to the number of tows in the actual survey. Pelletier (1998), however, has presented results from a directed intercalibration study where 30 parallel tows appeared to be quite adequate to estimate conversion factors for most species observed. Managers and fishermen must decide fishery by fishery whether the gain in precision and compatibility with existing gear types is worth the cost of the transition.

Advanced Sampling Technologies for Fisheries

A complement to standard surveys based on fish capture are newer remote-sensing methods in which fish are counted or the biomass of a school estimated using sound (hydroacoustics), lasers (lidars [light detection and ranging]) and laser line scanners), and other optical techniques. Optical and acoustic techniques also can be used to estimate the number of larval fish and the phytoplankton and zooplankton biomass available as food. However, like any remote sensing technique, it is important to include a "ground truthing" component in sampling, in which samples of fish are collected from the remotely sensed areas for species identification to calibrate the sensors and validate the values obtained remotely.

Advantages of hydroacoustic techniques are that they cover more area in less time and can offer better three-dimensional coverage of the water column and can give a better indication of the total population, not just those fish that can be captured in nets. Hydroacoustic surveys transit an area with either single acoustic beams faced directly below the vessel or multiple beams arrayed from straight down to extending horizontally away from the vessel. Multibeam systems can sample a swath many kilometers wide. The vessels used for hydroacoustic surveys must be specially designed to be quiet. Limitations of hydroacoustic techniques include

1. difficulties in species identification in mixed populations,
2. inaccurate determination of size distributions of schools with individuals of many sizes,
3. inability to estimate benthic species,
4. lack of sensitivity to estimate species with no swim bladder, and
5. requirement for special fishery research vessels.

Hydroacoustic surveys are used extensively by other nations and are used for a few U.S. surveys (see Table 3-3). Hydroacoustic techniques are most useful for estimating the abundance and biomass of single-species schools of mid-water fish, such as pollock and whiting in the North Pacific Ocean and herring, mackerel, butterfish, and squid in the North Atlantic Ocean. A new generation of fishery research vessels is being designed to meet standards developed by the International Council for the Exploration of the Sea (ICES, 1995) for acoustic quieting (see following section on survey vessels).

Lidar, which is more experimental than acoustic techniques at this time, is implemented using a laser projected from a small aircraft, so it has the advantage of surveying an area 20 times faster than a trawl or hydroacoustic survey carried out by a fishery research vessel. It is most useful for species that school and spend a significant amount of time in the relatively transparent upper layer of the ocean (e.g., sardines, herring, squid). It suffers from the same target identification problems as acoustic methods.

Survey Vessels

NOAA has a fleet of 16 ships, staffed by officers of the uniformed NOAA Corps and by merchant marines. Nine of the vessels are designed for fishery surveys and fisheries oceanography. NMFS' estimates of ship time needed to fulfill its mission exceed the amount of time available on its limited and aged fleet. This situation has been highlighted in a number of reports (discussed below).

NMFS presently charters 40 percent of its days at sea on commercial and university vessels to supplement work done by its own vessels, but it must maintain its own survey capabilities, particularly for trawl surveys. Reasons to maintain strong survey capabilities within the agency include the need for consistency of survey vessels to reduce variability, the present lack of acoustically quiet ships in the commercial fishery and university fleets, the availability of proper equipment for surveys and fisheries oceanography on NOAA vessels, and flexible scheduling for NOAA vessels. New NOAA fishery research vessels are designed to be quieter than commercial vessels (to minimize the influences of vessel noise on the distribution of fish during sampling) and are equipped to take measurements that are not normally taken from commercial vessels (e.g., environmental variables, depth, acoustic measurements). When using private vessels for trawl surveys, NMFS prefers to enter long-term charters so that the time and expense related to intercalibrations are minimized.

NOAA has developed a series of fleet replacement and modernization (FRAM) plans that have been reviewed by the General Accounting Office (1986), National Research Council (1994b), and others. Many of these reports have considered whether NOAA should use chartering more extensively. The three interstate marine fishery commissions[2] convened a workshop in 1996 to "discuss the fishery survey capabilities and the critical need to ensure compatibility and continuity with historic data sets" (ASMFC, 1996). The commissions issued a joint call for replacement of the aging NOAA fisheries fleet with quiet, modern, multipurpose fishery research vessels, either by purchase or long-term charter (ASMFC, 1996). They recommended that new

[2] Three interstate marine fishery commissions—the Atlantic States Marine Fisheries Commission, Gulf States Marine Fisheries Commission, and Pacific States Marine Fisheries Commission—were created to coordinate the management of stocks within their coastal waters, including 3 miles from shore on open coasts, bays, and estuaries. Congress passed the Interjurisdictional Fisheries Act in 1986, directing the Secretary of Commerce to "apportion funds for interjurisdictional fisheries research projects among the States." The act was extended by the Atlantic Coastal Fisheries Cooperative Management Act in 1993: "the responsibility for managing Atlantic coastal fisheries rests with the States, which carry out a cooperative program of fishery oversight and management through the Atlantic States Marine Fisheries Commission. It is the responsibility of the Federal Government to support such cooperative interstate management of coastal fishery resources." (Sec. 802[a][4] of the Coast Guard Authorization Act of 1993). All the interstate commissions collect data and provide some level of coordination of fishery management activities. The Atlantic States Marine Fisheries Commission is the most active in coordination of management. The Pacific and Gulf commissions are less involved in management, focusing on data collection and data management. Data are sent from individual states to the commissions and NMFS. The commissions generally do not conduct their own commercial and recreational sampling although Congress established a budget line item in fiscal year 1999 for a Gulf of Mexico Fisheries Information Network (GulfFIN) to establish a data collection and analysis program for both commercial and recreational fisheries. Presently, the Gulf commission continues to conduct the MRFSS intercept surveys and newly adopted charter vessel data collections under a fiscal year 2000 cooperative agreement with NMFS. The same data are used for fisheries management within individual states and for federal management through NMFS stock assessments and regional fishery management council decisions.

vessels, gear, and methods be calibrated adequately to be comparable with existing methods and gear. Calibration is important, particularly for trawl fisheries, because each combination of vessel, gear, and skipper fishes differently and thus has a different catchability coefficient. This difference in catchability occurs because each vessel has a distinct acoustic signature (which may affect fish behavior) and each vessel pulls its nets through the water or over the bottom differently. An unpublished memo from Dorman (1998) to the Office of Management and Budget also supports the construction and deployment of new fishery research vessels. Dorman endorsed the proposed design, although he also recommended that new vessels be planned in the context of a national plan for their use with other national assets for fisheries research, fisheries monitoring, and oceanography. A great deal of debate has ensued about whether NMFS should increase its charter of academic and industry vessels.

Congress has appropriated funding for four fishery research vessels to meet ship time needs for fisheries surveys and research. About 86 percent of the funds for the first fishery research vessel were appropriated in fiscal year 2000 and the remaining funds to complete the ship have been requested by NMFS in its fiscal year 2001 budget request. NMFS plans to award the contract for the first vessel in the 4th quarter of fiscal year 2000 and expects to take delivery of it in fiscal year 2003. This ship will be stationed in Alaska. If funding is appropriated, the second, third, and fourth ships will be delivered in fiscal years 2005, 2006, and 2007. They will be stationed in Woods Hole, Massachusetts; the Pacific Northwest; and Pascagoula, Mississippi, respectively.

Sampling Designs

The sampling design of a fishery-independent survey determines where the fishing stations are to be located in the area or domain of the survey. Ideally, the survey domain should cover the stock area(s) for the target species at the same time each year. In practice, surveys gather information for a number of species, many of which may have overlapping but not identical spatial and temporal distributions. This requires the survey to cover a very large area to include the stock areas of all species. For example, the biannual East Coast bottom trawl survey covers the entire continental shelf from Cape Hatteras to Nova Scotia (4 to 200 fathoms depth). Traditionally, survey stations are located either randomly within stock areas each year (e.g., groundfish surveys off the east coast of Canada and the United States, acoustic surveys for pelagic fish off South Africa) or at sites fixed over time in the stock areas (e.g., longline survey for halibut off western North America, groundfish surveys off Iceland and in the North Sea). No matter how the sampling sites are selected, a standard fishing procedure is used at each station in terms of the length of time that gear is deployed or distance through the water that trawls are towed. The number and weight of each species caught at each station are recorded, along with size, sex, maturity, and age of fish caught. Measurements of depth, water temperature, salinity, and other environmental factors are often obtained at each fishing station.

The type of survey design chosen determines how the mean and total estimates (e.g., catch rates, abundance) are calculated. The most commonly used design is stratified random, for which the survey area is divided (stratified) into subareas based on such factors as depth ranges, stock areas, habitat, and management areas. Survey stations are located randomly within each subarea, which is called a stratum. A mean or total estimate is calculated within each stratum and an overall estimate is calculated by weighting each stratum estimate by the relative area covered by it. The validity of estimates of means, totals, and variances does not require any assumptions about the properties of the statistical or spatial distribution of organisms within strata. These kinds of estimates and their associated properties are called design based. The main advantage of a design-based strategy is that a property such as unbiasedness depends only on how the survey is designed, not on assumptions about the frequency distributions of fish populations (e.g., normal or log-normal), spatial independence, or the existence of covariance of any particular form. Design-based methods do require selection of the sample according to the design, involving random or probability selection at some stage.

The second common survey method uses the same stations year after year. There are no definitive methods for calculating broad-area estimates from such fixed-station survey data. The methods that have been applied range from using design-based methods to complex spatial models. The latter methods generally model spatial patterns as a function of distances among stations and are not tied to specific locations. Indeed, fixed-station designs may be the most appropriate if a spatial model is the objective of the eventual analysis because these kinds of designs ensure that stations selected constitute an adequate range and distribution to construct an efficient estimate. The properties of estimates from survey data that are derived from a model are referred to as model based. Model-based estimation methods make statistical assumptions about the distribution of fish, such as a normal or log-normal spatial model, with an assumed spatial covariance structure. Under the assumed model, best linear unbiased or maximum likelihood estimation methods can be used. Such methods can provide good estimates even when the location of sample sites in the study region is uneven. These two methods have competing advantages and disadvantages, but generally, a model-based method should be chosen if, and only if, the assumptions in the model are approximately correct. Since this can never by fully assured, it is important to use both types of surveys in a balanced, self-correcting manner, but with an expectation that the balance will tilt more toward model-based surveys as increasing knowledge builds confidence in the model assumptions.

Limitations—Most fishery-independent surveys collect data in a short time period, on the order of weeks. It is important to consider timing in the interpretation of survey data because many species change their distribution over the seasons and seasonal movements may differ as a function of fish age. Surveys should be short enough in time and large enough in spatial extent to minimize possible biases due to large-scale movement or migration by some of the fish over the survey period. Conducting a survey at the same time each year may ensure that light regimes (the number of daylight versus night hours) are comparable from year to year, but fish migrations and biology are highly dependent on water temperatures, which can vary greatly on a specific date from year to year.

Many species move in the water column each day in response to light levels and vary in their availability to the fishing gear, particularly bottom trawls, over the course of each day. Thus, many large-area surveys estimate abundance from catches over the entire 24-hour period. Evaluating the possible effects of daily movements on estimates from survey data is difficult, because limits on sampling capability and the need to move vessels to new locations require that differences among catches made at different times of the day are also made at different locations. Studies of the same area over 24-hour periods are needed to determine whether and how daily migration affects estimates for various species (Jones and Pope, 1973; Atkinson, 1989; Engås and Soldal, 1989; Walsh, 1991; Jones et al., 1995; Korsbrekke and Nakken, 1999; Somerton et al., 1999).

Another limitation of fishery-independent surveys is that their high costs, demands for intensive labor, and major requirements for capital investment in vessels restrict the sampling intensity of a survey. In terms of days fished, surveys may represent a small fraction of the effort expended on the grounds by the commercial fleet. For example, for summer flounder, survey days equal about 0.1 percent of the days fished by commercial fishermen. Although the commercial information represents directed effort, it may be fruitful for scientists to develop ways to make the best use of both sources of information, rather than argue over the benefits of one versus the other or reject commercial data outright.

Evaluation of Survey Design—The design of each survey should be evaluated, just as the performance of the sampling gear and the coverage of the stock area of a survey are evaluated. Within the context of a random survey, this evaluation should determine whether the design is the most efficient attainable with available resources. For stratified random designs, precision is a function of how well the strata correspond to the distribution of the target species and whether the higher sampling intensities were assigned to the more variable strata, as they should be. Smith and Gavaris (1993) provide an evaluation of a stratified random design for a groundfish survey. Appendix C presents an evaluation specifically for the NMFS winter bottom trawl survey, including summer flounder. This evaluation also indicates how multispecies surveys will not necessarily be optimal for any single species.

Evaluation of fixed-site designs should be focused on ensuring that the number of fishing stations and their locations are adequate to estimate all the parameters of the model with adequate precision. For statistically optimal (best linear unbiased estimates) methods such as kriging,[3] this evaluation may involve determining whether the range of distances between stations has been sampled adequately so that the variogram[4] usefully describes both the small- and large-scale variations.

[3] Kriging is a minimum-mean-square-error method of spatial prediction (Cressie, 1993), a method of interpolation using data observed at known locations to predict unknown intermediate values (http://www.tc.cornell.edu/visualization/contrib/cs490-95to95/clang/kriging.html). Kriging has some problematic properties for estimating distributions and abundances of living organisms, particularly mobile species. Kriging is the best linear unbiased estimate for the mean of abundance, but can greatly underestimate the variance of estimated abundance.

[4] The variogram is defined as the variance of the difference of observations at different sample sites, and thus summarizes the spatial variation. This measure is used to predict observations at the unsampled sites when using geostatistical spatial models (Cressie, 1993).

Options for Increasing Survey Precision—Careful survey design can improve the precision of design-based estimates. For example, stratified random designs can be improved by modifying the existing stratification scheme, the station-to-strata allocation scheme, or both. Evaluation of completed surveys (see Appendix C) may provide insights into how these modifications can be made for future surveys, especially if the spatial distribution of the target species is reasonably constant from year to year. For multi-species surveys, optimal allocation or stratification for one species may not be optimal for others and a compromise allocation or stratification scheme may be needed (see Appendix C for an example).

Optimal survey design can also improve the precision of model-based estimates. Such designs typically have stations distributed more or less uniformly over the survey grounds (MacLennan and Simmonds, 1992). Additional stations may be warranted to distinguish small-scale from large-scale variation, or to characterize the influence of auxiliary variables on sampling density. The structure implied by the additional assumptions made in a model-based approach does not eliminate the careful attention that should be paid to traditional concerns about good survey design, and continuing validation of the assumptions under possibly changing conditions may require additional data.

Adaptive Sampling—The spatial distribution of fish can change in ways that are not entirely predictable and changes in the distribution of species that school or aggregate can have major effects on estimates. As a result, survey biologists may need to adapt their surveys to encountered patterns of abundance. In theory, adaptive sampling could reduce the standard error and coefficient of variation compared with a stratified random sampling design. In an adaptive sampling design, the selection of sites at which samples are taken depends on what has already been caught or observed during a survey. Adaptive survey design allows the optimization of survey data collection in light of current knowledge rather than past knowledge, but such designs can be complex to implement, suffer from possibly poor data at earlier steps in the current survey, may not provide data needed for time series, and appear to be unusually subject to human error in implementation. Thus, they must be introduced and evaluated with some caution before they are adopted widely.

Types of adaptive sampling designs include adaptive allocation and stratification, adaptive cluster sampling, and optimal Bayesian designs. In adaptive allocation, the final sample sizes in each strata are determined during the survey based on observed abundances. Examples include the adaptive allocation designs for anchovies described by Jolly and Hampton (1990), and for mackerel and orange roughy described by Francis (1984), the design-unbiased adaptive-allocation shrimp survey design described by Thompson et al. (1992), and the adaptive allocation of effort for scallop surveys (Smith et al., 1999). In adaptive stratification, the stratum boundaries are determined adaptively during a survey, for example, by drawing a new stratum for intensive sampling where high concentrations of fish have been observed during the survey. In adaptive cluster sampling, neighboring units or sites are added to the sample whenever high abundances are observed (Thompson, 1990; Quinn et al., 1999). Optimal model-based designs rely on a spatial model of distribution of fish abundance and take into account abundances encountered at initial sites in the survey in selecting subsequent sampling sites, for the purpose of maximizing overall estimation precision for a given amount of effort (Chao and Thompson, 1997). Hanselman et al. (in press) demonstrated that adaptive sampling in their experiment with rockfish reduced variance by about the same extent as stratification of sampling by habitat. The combination of adaptive and stratified sampling yielded an even lower variance. Hanselman et al. suggested that adaptive sampling might be especially appropriate when not enough information about a species' habitat preferences and extent of specific habitats in an area is available to stratify by habitat. Adaptive sampling may decrease the travel time in a survey because similar sampling effort is clustered into fewer locations (Quinn et al., 1999). Adaptive designs of all types are described in Seber and Thompson (1993) and Thompson and Seber (1996).

An important application of adaptive sampling could feature commercial fishermen as adaptive samplers in joint NMFS and industry sampling activities. Generally, commercial fishing vessel skippers do not fish at pre-determined locations but respond to encountered fish abundances. When few fish are caught at one location, the vessel may travel some distance to try a new area, and when many fish are caught the vessel may spend more time at nearby locations. Because of aggregation tendencies in fish abundance, considerably more fish may be caught this way than could be caught through any non-responsive (random or fixed station) strategy. Thus, the commercial fishing pattern is adaptive, and conventional estimators that ignore the non-random spatial distribution of the fishing effort, such as a sample mean of catch or CPUE, can have a significant bias. This would be the case, for example, if the skill of vessel skippers together with the aggregation tendencies of the fish allow CPUE to remain high even while total stock abundance declines substantially. These features have rightly been recognized as invalidating standard procedures for estimating abundance that assume a proportional relationship between CPUE and population size. Recognizing the adaptive nature of commercial fishing practices, better estimates of fish abundance might be obtained by working with industry to conduct cooperative adaptive sampling using likelihood model-based estimates, taking into account the spatial locations of catches. Fishing patterns can be taken into account in several different ways. The simplest would be to partition the catch into much smaller geographic regions than is usually done, so that averaging of CPUE would be in areas of more uniform effort than is the case when a larger geographic unit is used. A more sophisticated approach would use a spatial statistical model together with accurate geographic location information on where each catch was made, and ideally would take into account any auxiliary information available, such as acoustic measures of fish biomass or numbers.

Adaptive commercial surveys seem to be catching on in some places for a number of practical reasons. In the surveys described by Quinn et al. (1999), in which the commercial fleet was used to survey marine fish in Alaska, the adaptive design was well received because fishermen were reimbursed for their costs with fish caught in the survey and most of the catch came from adaptively added sites. Even though more fish were being caught, unbiased or nearly unbiased estimates of fish abundance were available, since the adaptive survey had been well designed by the biologists involved—whose primary concern was increasing precision—by focusing more of the sampling effort in areas with greater fish abundance. To implement such adaptive sampling programs widely, more research is needed regarding the most effective designs to use in cooperation with commercial fisheries, data needs (including accurate geographic information and use of acoustic and other auxiliary data), data analysis methods, and the social and economic aspects of cooperation of commercial fishermen in adaptive surveys. The methodology for adaptive sampling will have to be developed on a fishery-by-fishery basis, so the results can be incorporated into assessments appropriately.

Interpretation of Survey Results

Uncertainties concerning selectivity, catchability, and availability of fish to survey gear are usually cited as reasons for interpreting survey estimates of abundance as relative indices rather than estimates of actual population size or biomass (Somerton et al., 1999). Users generally assume that the relationship between the survey index and actual population abundance is constant, although both fluctuate in absolute terms over time.

Survey indices are rarely used by themselves to evaluate the impact of fishing on a population of fish, but are instead incorporated into integrated fishery models along with other information about the fishery. Although design- and model-based strategies can provide variance estimates for the survey indices of abundance, these variance estimates are rarely incorporated into the fishery models. This increases the difficulty of assessing gains in precision versus improvements in design on the estimates and predictions from the fishery models (e.g., Nandram et al., 1997). With today's modern integrated fishery models, estimates of uncertainty can and should be incorporated directly into the assessment. Appropriately weighting likelihood components is one way to incorporate uncertainty, although some research will be needed to figure out the best way to do this. Determining appropriate weights is not a trivial task and if an analysis weights the components incorrectly, the uncertainty of the results will increase. Knowledge of the biology of a species and characteristics of the data available (e.g., its precision) can help in the choice of weights to apply to individual data sources. Once incorporated, precision in the estimates of interest should be carried through the methods of approximations to statistical distributions (e.g., Gaussian), bootstrap, or Monte Carlo techniques. Carrying precision through analyses will provide an incentive to increase or optimize the precision of the various input data so that they yield the required precision of the estimates (e.g., the TAC) in the most cost-effective fashion. Such an approach is desirable to avoid the expense of increases in sampling precision that result in insignificant increase in the precision of stock assessment advice. There is little reason to make one measurement with a micrometer if it is to be added to another measurement made with a meterstick, since the precision of the sum will reflect the precision of the meterstick.

Ecosystem Data

Fisheries impact the wider ecosystem in various ways (Hall, 1999; Kaiser and deGroots, 1999) because they

- Generate mortality on target and non-target species by direct capture and by less obvious interactions with the fishing gear (e.g., escape or bycatch mortality).
- Provide food to scavenging species.
- Generate litter and cause "ghost fishing" from broken, lost, or discarded fishing gear.
- Disturb the seafloor or other habitat.
- Change the life history characteristics of target and possibly non-target species (e.g., mean weight at age, growth rates, size composition) by selectively removing larger (and faster growing) elements of the population.
- Change behavior of target and non-target species.

These effects can result at any intensity level of fishing. In addition, severe overfishing can alter the structure of marine food webs (Hall, 1999) and one species may replace another in the same ecological niche (e.g., spiny dogfish and other sharks and rays for cod and haddock on Georges Bank; Fogarty and Murawski, 1998). The prevalence of niche replacements is a crucial issue in regards to the feasibility and desirability of stock rebuilding programs as mandated by the Magnuson-Stevens Act, because such replacements could make stock rebuilding ineffective. Alterations in food webs increase in significance as gear selectivity increases (Wainright et al., 1993). Because the abundance of species depends in part on the effects of fishing but also on the effects of climate, pollution, habitat loss, and natural fluctuations (e.g., Hofmann and Powell, 1998), understanding the relation between a species' abundance and fishing activity requires more than merely monitoring these latter two factors. For some species that are not harvested, such as marine mammals and seabirds, dedicated surveys may be needed to determine their abundance and distribution. Additional studies, such as predator-prey and gear bycatch studies, may be needed to verify correlations seen in these observations. In other cases, routine fisheries monitoring may provide adequate data.

Five main routine sources exist for ecosystem-level fisheries data:

1. Commercial catches and catch rates
2. Fishery-independent surveys
3. Onboard observer programs
4. Recreational catches
5. Plankton surveys
6. Surveys of marine mammals and seabirds

In each of these types of monitoring, sampling is focused on target species or species assemblages. However, monitoring the status of the ecosystems more inclusively requires maintaining or enhancing the collection of abundance data for non-target species (e.g., the California Cooperative Oceanic Fisheries Investigations) and conducting process studies such as the Global Ocean Ecosystems Dynamics program. Likewise, essential fish habitat must be identified and its status monitored (Benaka, 1999).

Another way that fisheries monitoring could be of value is in providing overviews of ecosystem health. Ecosystem health is a somewhat ambiguous concept, but includes such things as an ecosystem's productivity, the diversity of its organisms (in terms of species and age classes), its stability and resilience in the face of disturbances, the general health and fecundity of component species, and the maintenance of unimpaired ecosystems processes. Size spectra of all species, resulting from fishing surveys, could be a robust measure of the integrated effect of fisheries on a marine ecosystem. Other integrative measures, such as fish fecundity, liver condition indices (e.g., Marshall et al., 1999), and growth and condition factors may also be possible indicators of ecosystem health if they are measured for representative sets of species. Taking samples to detect fish diseases and measure immunity levels have been proposed as approaches for monitoring the effects of contamination of the marine environment. All of these indicators are, or could be, byproducts of the ongoing monitoring of fish stocks.

Fisheries sampling could, therefore, serve

wider requirements than those they were set up to achieve. It is important that, while rendering the basic monitoring effective and efficient, these other uses are properly recognized and maintained.

Environmental Data

The strength of a given year-class of fish depends on both the size of the spawning stock biomass and environmental conditions from the time of spawning through recruitment. Ocean temperature, salinity, dissolved oxygen concentration, prey type and abundance, availability of essential fish habitat, and predation can have significant effects on recruitment and subsequent year-class strength (Murawski, 1993; Hixon and Carr, 1997). Important examples of fish and shellfish species that may be particularly vulnerable to environmental conditions are those that exhibit a small number of dominant year classes (e.g., surf clams, ocean quahogs, West Coast rockfish, Pacific whiting) or short life spans (e.g., shrimp).

Most stock assessment models and most management control rules still assume that recruitment is related to spawning stock biomass and that the major dynamics of populations are driven by fishing-induced mortality, not by environmental factors. Unfortunately, methods that focus on spawning stock biomass to the exclusion of environmental factors may ignore regime shifts (Steele, 1998) occurring in these systems. Therefore, collection of environmental data and use of such data in future stock assessments is an important adjunct to estimation of spawning stock biomass.

Regime shifts can change the productivity of major ocean basins and there is evidence that this has occurred in the North Pacific Ocean (Francis and Hare, 1994; Francis et al., 1998; Hare et al., 1999) and other areas. If ecosystems have shifted to conditions less favorable for the growth of one or more fish species, the Magnuson-Stevens Act's requirement for rebuilding fish populations to historic high levels may not be attainable because such levels may have occurred under more favorable environmental conditions.

In many cases, environmental data are being collected by other government agencies (e.g., other parts of NOAA, the National Aeronautics and Space Administration [NASA], the Environmental Protection Agency [EPA]) and are available from these sources. It will be necessary to provide environmental data on a scale and reference grid that is compatible with fisheries data. High-quality, long-term, spatially referenced data sets will continue to be required to assess the influence of environmental effects on fish populations. There is also a need for the development of scientific methods to facilitate recognition of regime shifts and to incorporate such recognition where necessary (e.g., incorporating realistic expected recruitment in short-term projections that consider existing environmental regimes). Continued research on the effects of the environment on fish population size, individual growth, survivorship, and spawning potential in conjunction with density-dependent effects is crucial for determining the kinds of data collection and stock assessments that should be used for any given species.

Fishery-Dependent Data

Data gathered from commercial and recreational fisheries are essential for assessing the mortality and other stresses that result from fishing and these data provide a direct measure of the effectiveness of management regulations. Such data can also provide information on many other aspects of a fishery, including fish population structure, gear selectivity over time, and the behavior of fish and fishermen. Fishery-dependent data can sometimes be used to provide a measure of relative abundance that can be compared with that determined from fishery-independent sources. Data costs for fishery-dependent data tend to be lower than for fishery-independent data because the former are a byproduct of commercial fishing activity, but fishery-dependent data are substantially more subject to bias and do not

TABLE 3-4 Information Available from Commercial and Recreational Fisheries

	Source	
Information	Commercial	Recreational
Total catch by species (biomass, number)	Logbooks, observers	Intercept surveys, angler diaries[a]
Total catch and discards	Logbooks, observers	Intercept surveys, angler diaries
Proportion caught—by age, size class, market category, sex, and life history stage	Port sampling of landings, observers	Limited sampling from intercept surveys
Harvest location (GPS, latitude-longitude, loran)	Logbooks, landing surveys, VMS, observers	Intercept surveys, angler diaries
Landing location (port)	Landing sales receipts	Intercept surveys, angler diaries
Effort (number of hooks, tow length, angler hours, boat days)	Logbooks, landing surveys, observers	Telephone and intercept surveys
Cost of fishing	Logbooks, landing surveys	Surveys, angler diaries
Catch per unit effort	Logbooks, landing surveys	Intercept surveys, angler diaries
Gear type	Logbooks, landing surveys, observers	Intercept surveys, angler diaries
Vessel size and power	Logbooks, landing surveys	N/A
Crew size	Logbooks, landing surveys, observers	N/A
Fisherman or fishing vessel identifier	Logbooks, landing surveys	N/A

NOTE: GPS = global positioning system; N/A = not applicable; VMS = vessel monitoring system.

[a] Angler diaries are a method of obtaining these data, but are more frequently used in freshwater fishing.

include all important data elements. Thus, their collection, use, and integration with fishery-independent data require some care. Table 3-4 lists the kinds of information that can be gathered from commercial and recreational fisheries. The nature, scope, and reliability of this information are influenced by the time and effort required to gather, process, and store them. As a consequence, clear objectives are required if the data-gathering processes are to run effectively. To assist in specifying objectives it is necessary to understand the costs of precision and the likelihood of avoiding bias. In that context, the following issues should be considered.

Many commercial fisheries in the United States are required by law to report fish landings. For these fisheries, total landings are estimated from the landing reports, and thus are really a census of the harvest rather than a statistical estimate. Recreational fisheries generally are characterized by smaller landings, often of one or two fish by each angler, and usually are not required to be reported; thus, statistical sampling approaches must be used to estimate total harvest and discards for recreational fisheries. Statistical sampling approaches are also used to collect and process the auxiliary data that come from landings (e.g., age and size composition). Therefore, it becomes important to use appropriate random sampling protocols (Cochran, 1977) and recognize that (1) certain steps can improve statistical efficiency (e.g., stratification), (2) variation decreases with increasing sample size (Figure 3-1), (3) bias may exist in the data, and (4) the error associated with this process, whether random or systematic, will propagate itself into assessment estimates.

Commercial Fisheries

Stock assessment scientists have long been concerned about biases and other inadequacies in commercial fishery-dependent data of all types (Fox and Starr, 1996). These concerns include uncertainty about reporting accuracy, the component of the population encountered by the fishery, and the scale on which the observations are taken. Although it may be easy to react to such concerns by ignoring fishery-dependent data, such an approach is not prudent because information contained in such data cannot be obtained from other sources. Moreover, fishery-dependent data frequently cover a broader geographic area and a greater portion of the year than do survey data. Finally, fishery-dependent data are necessary for both scientist and fishermen to understand how data from fishery-independent sources relate to situations in managed fisheries (see Chapter 4 for recommendations about increased user group participation in data collection).

The following sections describe the variety of commercial data sources listed in Table 3-4. Many of these sources are in transition from manual methods to methods that are automated or otherwise take advantage of new technologies. The application of appropriate technology to collection of commercial fisheries data requires an

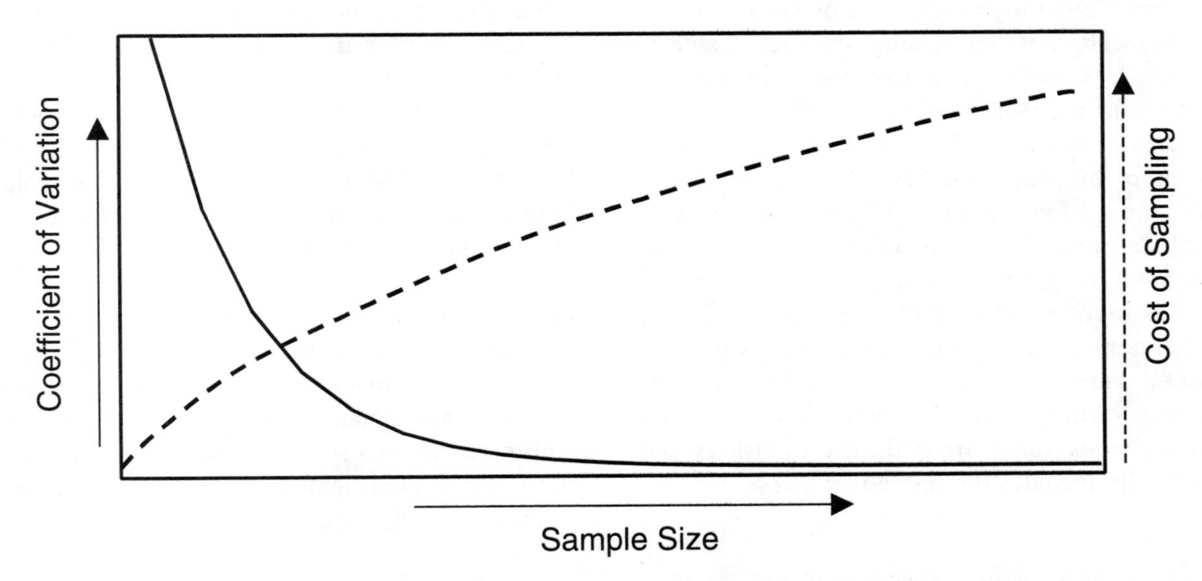

FIGURE 3-1 Coefficient of variation and cost as a function of sample size. Note that certain levels of precision are unattainable for limited cost levels. Above certain sample sizes not much additional precision is obtained while costs continue to rise.

intimate knowledge of fisheries management and of the equipment available to fishermen and other data providers. Equipment that might be sensible in an office may be totally inappropriate on a fishing vessel that may encounter high winds, waves, and corrosive seawater. Nonetheless, fishing vessels today often carry and use personal computers, global positioning systems (GPS), cellular phones, vessel monitoring systems that work with GPS, electronic chart systems, sonar, and fishing net sensors. The value added by any new technology must be balanced against the incremental cost to fishermen, regulatory agencies, and managers. Any proposal for use of a new technology must take into account the impacts on existing systems, the potential for data fouling,[5] constraints imposed by the need to use historic data, resistance to change by data managers and users, and costs. Managers wanting to encourage implementation of new technologies should consider incentives, particularly cost sharing, provision of free equipment, or provision of value-added services. Tax credits are another possibility.

Logbooks—In many fisheries, commercial vessel operators are required to submit detailed records, called logbooks, logs, or vessel trip reports (VTRs), of their fishing activities. Table 3-5 lists U.S. federal fisheries that have logbook requirements. Some additional federal and state fisheries are managed by the interstate commissions or by states and have their own logbook programs. For example, Washington, Oregon, and California have a mandatory tri-state trawl logbook program coordinated by the Pacific States Marine Fisheries Commission. This and other non-federal logbook programs may provide useful data for regional and national fishery data management systems. The information required in logbooks varies from fishery to fishery but generally includes the following:

- The vessel's identity
- Date, time of day, and position (longitude and latitude) of fishing activity
- Information about weather
- Details of fishing gear used
- Amount of fishing activity (e.g., tow length, number of hooks, or trips)
- Catch of target species
- Catch of other species, including protected species

Logbooks for the commercial summer flounder fishery contain the same general information. VTRs are to be completed at sea and submitted to NMFS within 15 days after the end of the month in which the fishing occurred. In some regions, logbooks are collected at dockside. This practice ensures that data are submitted in a timely fashion, adds a level of verification to the data, and establishes a point of contact between NMFS personnel and fishermen. Logbooks can also be validated—at least in terms of catch, discards, and misreporting—by comparing data from unobserved trips with observed trips employing the same fishing strategy[6] for the same species, and with dealer records. Data from paper logbooks often are entered into a computer database without verification or double entry, and are considered confidential.

The Electronic Fish Catch Logbook Project represents an innovative effort to automate and standardize logbook data (Box 3-1). Electronic logbooks are also being tested or used operationally in other countries (e.g., South Africa).

Dealer Reports—In some states, fish dealers are required to report the amounts of fish bought and sold. This information can be used to adjust retained catch estimates from logbooks, but is not considered to be a substitute for the detailed information contained in logbooks. For summer flounder and other species taken with federal per-

[5] Data fouling is a phenomenon in which changes in regulations diminish the ability of managers to compare new and old data or that make the newly available data represent the fishery less accurately or precisely.

[6] A fishing strategy involves the type of gear used and the location at which it is fished.

TABLE 3-5 U.S. Fisheries[7] With Mandatory Federal Logbook Requirements

Region	Fishery
Alaska	Groundfish Halibut
Northeast	Sea scallops, summer flounder, squid, scup, mackerel, butterfish, black sea bass, and northeast groundfish Atlantic purse seine Bluefish
Northwest	None
Southeast and Gulf of Mexico	Shrimp Pelagic longline Snapper/grouper and Gulf reef fish King mackerel Golden crab Wreckfish Shark Headboat Charterboat
Southwest and Western Pacific	Pacific pelagic longline Lobster Precious corals Eastern Pacific purse seine Western Pacific purse seine (South Pacific Tuna Treaty) Antarctic (Convention for the Conservation of Antarctic Living Marine Resources)

SOURCE: National Marine Fisheries Service.

mits, dealers in the mid-Atlantic and Northeast regions submit data of two types by two different methods. First, dealers submit weekly reports (weighout slips) of fishery landings by vessel to either NMFS statistical port agents or to states first (for Delaware and Connecticut) and then to port agents. The port agents enter dealer data in electronic form and transmit them to their regional office and the Northeast Fishery Science Center. Dealers are also required to report by phone (via the Interactive Voice Response [IVR] system) weekly landings by species, either directly to federal scientists or via states (for Massachusetts). NMFS received funds in fiscal year 2000 to establish a coastwide IVR system for all state and federal voice reporting requirements. On the U.S. West Coast, dealer reports are collected by the states, rather than by NMFS.

The weekly and IVR reports are used for seasonal monitoring of total allowable catch and other quotas. These reports provide timely verification of reports obtained by other methods. They can be compared with logbook data, biological data, days-at-sea estimates made from call-in systems, and observer data. The weighout

[7] Fisheries can be defined in different ways, as illustrated in this table. Fisheries can be defined in terms of species or species complex targeted, gear/area combination, area alone, and type of vessel. In each case, the fishery encompasses a fleet and its associated activities. A specific vessel may be involved in more than one fishery.

BOX 3-1
Electronic Fish Catch Log

The Northwest Fishery Science Center designed a prototype electronic logbook for commercial fisheries operating out of California, Oregon, and Washington. West Coast fishermen requested such a system to reduce the effort required for data entry with paper logbooks and to make logbook data more useful. The Innovation Fund Committee of the National Performance Review provided funds for this project. The project uses personal computers that are already onboard many fishing vessels, combined with ship-to-shore communications, and a secure onshore database. The project will standardize methods of logbook data collection and will seek opportunities to improve data quality, quantity, and timeliness. Both logbook and fish ticket data will be reported electronically, thus allowing for independent verification of the data and combination of biological and economic data. System design is such that the Electronic Fish Catch Log will be transferable to other regions and can be integrated with vessel monitoring systems. The technology exists for electronic logbook data to be transmitted to shore-side data servers while vessels are still at sea. The timeliness of data analysis can certainly be increased where near-to-real time tracking of a fishery's performance is required. In addition, reducing the number of steps and handling of data from the recording step by fishermen to the availability of data to analysts in electronic form can improve the timeliness of resulting analyses and provide quicker feedback to respondents on the quality and content of their submissions. This compares to the current situation, in which the majority of logbook data typically are unavailable for analysis until weeks or months after landings are made because of time needed for mailing/faxing, handling/sorting, data entry, and quality control of the data entry. Currently, other technologies using form-based scanning/imaging of key logbook data and Interactive Voice Response systems are in widespread use on the Atlantic coast for tracking fisheries managed by total allowable catch. NMFS is engaged in a wide variety of initiatives related to planning, research, and implementation of data collection that seek to capitalize on the speed and cost-efficiency of new technologies for capturing, analyzing, and disseminating data.

The Electronic Fish Catch Log development process has included three stages:

Stage 1: Review existing data collection processes and identify potential uses of the new system. It was determined that an Electronic Fish Catch Log should allow electronic reporting of logbook and observer data and electronic reconciliation of catch data with fish ticket information. It was hoped that such a system would make more and better data available to fishermen and processors to manage their businesses more effectively and to improve the quality, quantity, and timeliness of fisheries data available to them. Stage 1 identified a number of other goals, such as improving confidence in the data; reducing the labor and cost involved in collecting, analyzing, and reporting data; and making logbook data more accessible to a wider range of users.

Stage 2: Analyze available technical alternatives and develop a design.

Stage 3: Develop a field-ready prototype partnership with commercial entities.

The program is presently in stage 3 and has identified potential commercial partners.

slips and IVR data also can be analyzed and used to monitor vessel compliance with trip limits. Most importantly, they provide the principal source of price data of landings by market category and species.

Port Sampling—Dealers and auction houses in ports can add to the reports of onboard observers (see below) about biological information such as size and sex of captured fish, and they can be a source of samples of tissues (scales, fins, otoliths) for age determination. NMFS port samplers or their representatives are authorized to gather biological information from state- and federally permitted dealers (CFR 648.7, Section G, Additional Data and Sampling). Access to the fish, however, often is restricted because the dealers are concerned that the sampling process will diminish product quality by disfiguring the fish and/or delaying shipping. The authority of the port samplers varies in different regions and with different management systems. For example, most U.S. individual fishing quota systems require advance notice of landing so a port agent can be present when the catch is landed. Other management systems use a more ad hoc approach to port sampling. On the U.S. West Coast, port sampling is conducted by the states, with partial support from NMFS for such activities. For the states of Washington, Oregon, and California, port sampling is conducted by the states and the Pacific States Marine Fisheries Commission, with partial support from NMFS for such activities. The laws of all three states provide the state with the right to sample the fish being landed, even if this means disfiguring the fish to remove otoliths. In some cases, the states purchase the damaged fish.

In the U.S. Northeast region, suggested sampling schedules (by species, quarter, gear, area, market category) are provided to port agents by NMFS stock assessment scientists (Burns et al., 1983). For West Coast groundfish, the suggestions for sampling schedules come primarily from the Pacific Fishery Management Council's Groundfish Management Team. NMFS stock assessment scientists are involved, but so too are scientists from the state agencies that do almost all of the port sampling. On-site biological sampling procedures are governed by port agent sampling manuals. Port agents can supplement the sampling suggested by stock assessment scientists with opportunistic sampling. The sampling unit is a box, which may not be representative of the entire catch of the vessel from which the box is obtained. Fish are measured and hard parts are collected from the sampled fish. Some concern exists regarding possible biases in the samples because of non-random sampling and because samplers may not be permitted access to the highest quality fish.

Onboard Scientific Observers—Commercial fisheries can also provide data from independent observers onboard commercial vessels. Scientific observers are placed on fishing vessels for a variety of purposes, including monitoring of (1) interactions with protected species under various fishery management plans, (2) catch, and (3) bycatch and discards (Table 3-6) (Karp and McElderry, 2000). Some observer programs are mandated (e.g., all foreign vessels in the U.S. EEZ [Sec. 201(h)] and some North Pacific vessels [Sec. 313]) under the Magnuson-Stevens Act and for marine mammals under the Marine Mammal Protection Act (e.g., Secs. 114[b][3][B], 114[e]). Even when the primary purpose of observers is to monitor fishery interactions with protected species, they can often keep records of bycatch and discards, and occasionally take biological samples of the catch.

The extent to which vessel operators misreport bycatch and discards is unknown. Observer data can be used to verify log sheet information and to provide correction factors for bycatch on unobserved trips. Observer data provide information that makes it possible to manage by what is caught, not merely what is landed and reported. This is important because the difference—bycatch and unreported catch—affects the assumptions about the mortality of non-target species or age classes. Observers are also important for fisheries in which fish are processed at sea and cannot be sampled by port agents.

TABLE 3-6 U.S. Fisheries in Federal Waters That Use Observers (excludes state programs and salmon observations)

Region	Agency	Fishery	Percentage of Catch Observed	How Observers are Placed	Purpose	Funding	Mandatory vs. Voluntary
Alaska	NMFS	Groundfish (at sea and shoreside)	30 percent for 60-125 ft. vessels 100 percent for vessels > 125 ft	Contractors	Landings	Industry (80 percent) NMFS (20 percent)	Mandatory
	NMFS	CDQ Fisheries	100 percent	Contractors	Total catch monitoring	Industry	Mandatory
	ADF&G	Aleutian Island King Crab Fishery	100 percent	Contractors and ADF&G	Landings/bycatch biological data	Industry ADF&G	Mandatory
	ADF&G	Bering Sea King/Tanner Crab Fishery	14 percent	Contractors and ADF&G	Landings/bycatch biological data	Industry ADF&G	Mandatory
	ADF&G	Scallop Dredge Fishery	100 percent	Contractors and ADF&G	Landings/bycatch biological data	Industry ADF&G/AKFIN	Mandatory
West Coast & Western Pacific	NMFS	Offshore Pacific Whiting Fishery	100 percent	NMFS-certified contractors	Landings/bycatch	Industry/NMFS	Voluntary
	NMFS	Shoreside Landings of Pacific Whiting	13 percent	ODFW Admin.; contractors placed by PSMFC	Landings/bycatch	Industry/NMFS/ ODFW/WDFW/ CADFG	Voluntary
	NMFS	California/Oregon Drift Gillnet Fishery	20 percent	NMFS and contractors	Landings/bycatch, marine mammal interactions	NMFS	Mandatory
	NMFS	Monterey Bay Halibut Set Gillnet Fishery	30 percent	NMFS	Landings/bycatch, marine mammal interactions	NMFS	Mandatory
	NMFS	Hawaii Pelagic Longline Fishery	4 percent	NMFS	Landings/bycatch biological data	NMFS	Mandatory
	NMFS	Northwestern Hawaiian Islands Lobster Fishery	100 percent	NMFS	Landings/bycatch biological data	NMFS	Mandatory
Northeast	NMFS	Atlantic Sea Scallop Dredge Fishery	25 percent (in 1999)	NMFS/National Fish and Wildlife Foundation	Landings/bycatch biological data	Industry/NMFS	Mandatory

	NMFS	Giant Bluefin Tuna Purse Seine Fishery	96 percent	Contractors	Fishery interactions	NMFS	Mandatory
	NMFS	Large Pelagic Drift Gillnet Fishery	81 percent (program ended after 1998 season)	Contractors	Catch, management, and economic data	NMFS	Mandatory
	NMFS	Lobster Pot Fishery	<1 percent	Contractors	Scientific, economic, compliance, and management data	NMFS	Mandatory
	NMFS	New England and Mid-Atlantic Gillnet Fishery	2-4 percent	Contractors	Scientific, economic, compliance, and management data	NMFS	Mandatory
	NMFS	Northwest Atlantic Trawl Fisheries	<1 percent	Contractors	Scientific, economic, compliance, and management data	NMFS	Mandatory
Southeast	NMFS	Southeastern Shrimp Otter Trawl	<0.1 percent	Contractors	Bycatch and turtle interactions	NMFS	Voluntary
	NMFS	Swordfish and Pelagic Longline Fishery	2.5-5 percent	Contractors	Bycatch, effort, species interactions, biological data	NMFS	Mandatory
	NMFS	Southeast Atlantic Shark Drift Gillnet/Strike Net Fishery	<100 percent	NMFS/Contractors	Landings/bycatch	NMFS	Mandatory
	NMFS	Directed Large Coastal Shark Fishery	4 percent	NMFS/University of Florida	Landings/bycatch biological data	NMFS	Mandatory

SOURCE: National Marine Fisheries Service.

NOTE: ADF&G = Alaska Department of Fish and Game; AKFIN = Alaska Fisheries Information Network; CADFG = California Department of Fish and Game; CDQ = community development quota; NMFS = National Marine Fisheries Service; ODFW = Oregon Department of Fish and Wildlife; PSMFC = Pacific States Marine Fisheries Commission; WDFW = Washington Department of Fish and Wildlife.

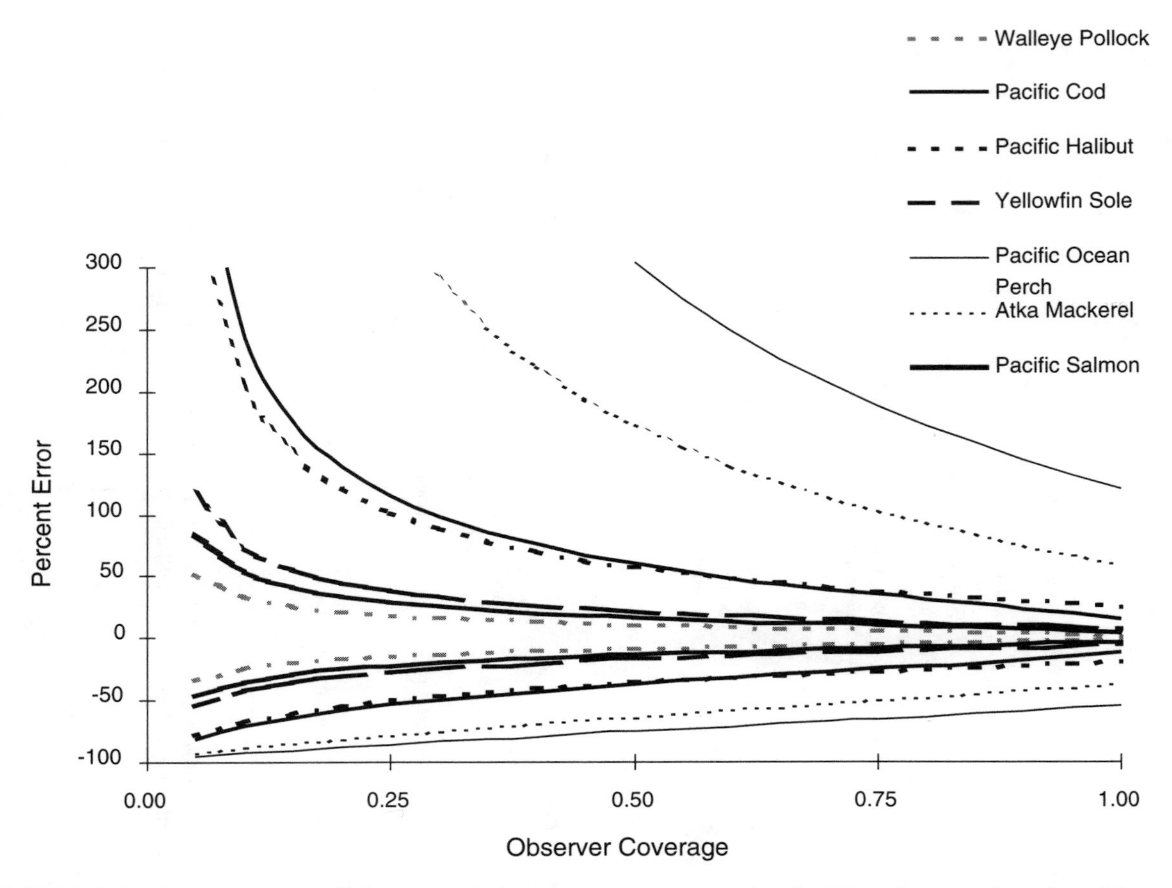

FIGURE 3-2 Percent error as a function of observer coverage associated with estimates of catch and bycatch species in the autumn 1996 trawl fishery for walleye pollock in the Bering Sea/Aleutian Islands region. Reprinted from Figure 3-1 in *Precautionary Approach to Fisheries* with permission from the authors and the Food and Agriculture Organization of the United Nations.

SOURCE: Dorn et al. (1997).

Sampling designs allocate observer effort at three levels: among vessels, among hauls taken by a given vessel, and among samples in a given haul. Random selection of vessels can be difficult to achieve (Karp and McElderry, 2000). Random sampling among hauls is easier to achieve, but sampling within a haul can be difficult, depending on the size of the haul, the commonness of the species of interest, and operating constraints of the shipboard working environment.

Low levels of observer coverage yield high levels of uncertainty in the observed variables for most species (Dorn et al., 1997). For target species and some non-target species, most reductions in uncertainty are achieved by the time 30 percent of the effort is observed (Dorn et al., 1997; Vølstad et al., 1997) (Figure 3-2). However, Turnock and Karp (1997) showed that variability of estimates remain high until 50-70 percent of hauls are sampled and for rare bycatch species, uncertainty can remain high even when 100 percent of vessels are covered.

Observer programs are expensive and tend to focus on fisheries for which the probability of interactions of fishing operations with protected species (e.g., marine mammals, marine turtles, and seabirds) is high. Observer programs are more common in large-scale "industrial" fisheries because vessels in these industries are most able to fund such programs and most likely to have bunk space for observers. Observer programs have been set up for many specific fisheries (Table 3-6), yet many of the largest U.S. fisheries have inadequate observer coverage. In 1998 and 1999, onboard observers monitored fishing operations that resulted in 0.6 percent and 0.8 percent respectively, of landings of summer flounder.

Most observer programs are funded solely by the government, although the Alaskan groundfish fishery observer program—the nation's largest—is funded 80 percent by industry. Some other West Coast observer programs have also received industry and state support. This differs from the system generally used in Canadian fisheries, in which the government pays program administration costs (about 30 percent of the total) and the industry pays for the costs of the observers. The total U.S. observer coverage is approximately 60,000 to 70,000 days per year (Karp and McElderry, 2000), equal to about 300 yearly full-time equivalents. The direct costs of these observers, plus overhead, thus account for a substantial fraction of all data-generation costs of monitoring fisheries.

Data from observers are considered proprietary company information and such data are usually collected and maintained separately from vessel logbook data by NMFS. Observer data are kept confidential to preserve proprietary information reflecting the location of high-quality fishing grounds. Although such an effort is appropriate to protect the rights and needs of fishermen, in some instances confidentiality has proved to be a barrier to the fishing industry in its attempts to reduce bycatch. Dave Fraser, a commercial fisherman, noted that one innovative approach taken by fishermen in the Gulf of Alaska and Bering Sea pollock fisheries is for the captain to ask observers for their bycatch records as they are collected during the trip. The captain can then share that information with fishermen on other vessels (this is done through a commercial firm that produces and distributes bycatch maps [see Figure 3-4]), so they can jointly identify and avoid areas of high bycatch—an effort that benefits all, because the entire fishery is closed when the bycatch limits for a protected species are reached. More widespread voluntary sharing of bycatch data could help reduce bycatch and keep bycatch-limited fisheries open for a greater part of their allotted seasons.

Observer programs are being used increasingly in innovative management plans in exchange for program features desired by industry. For example, the community development quota (CDQ) pollock fisheries in the Bering Sea require two observers on each vessel so the catch of these non-stop fisheries with at-sea processing can be fully observed. Another example is the 25 percent observer coverage of the scallop harvesters in the 1999 experimental opening of Georges Bank closed areas designed by the New England Fishery Management Council. This provision was included in the management package to obtain better information related to the bycatch and discard of yellowtail flounder and other species that the closed areas were designed to protect, and to help gauge the impact of scalloping on the seafloor habitat. Observers were randomly assigned to 25 percent of the harvesting trips. Vessels with an observer were allowed to land an additional 200 pounds of scallop meats for each day at sea in order to pay the cost of carrying the observer. The New England Fishery Management Council has extended this program to other closed areas in 2000.

The use of prohibited species bycatch limits as a management tool requires observers, which is why many of the Alaskan fisheries are observed. This region features these limits for halibut, king crab, and salmon, and zero quotas for endangered species such as Stellar sea lions and short-tail albatrosses. The observer program in the northeastern United States has focused on marine mammal-fishery interactions, but also produces useful fishery data.

Vessel Monitoring Systems—Vessel monitoring systems have been used since 1988 for foreign

TABLE 3-7 Fisheries Using Vessel Monitoring Systems

Fishery	Year Implemented	Number of Vessels
Foreign high-seas drift net vessels	1988	1,000
Hawaii/Western Pacific pelagic and lobster	1994	120
Western Pacific foreign fleet	1995	24
Atlantic sea scallops	1999	250
Georges Bank scallop fishery	1999	185
Atlantic swordfish	2000	20
	Planned	400
Alaska pollock	2000	28
	Planned	250
Alaska Atka mackerel	Planned	12
Southeast Calico scallop	Planned	30
Southeast Gulf of Mexico shrimp	Planned	50
Southeast gag grouper	Planned	100
Highly migratory species/Atlantic pelagic longline	Planned	800
Northeast groundfish	Planned	400
Northeast herring	Planned	10
Northeast monkfish	Planned	50

SOURCE: National Marine Fisheries Service.

high-seas drift-net vessels and since 1994 on a variety of U.S. vessels (Table 3-7). Vessels are tracked by either the ARGOS or INMARSAT satellites. The on-board VMS unit periodically reports its position to a central tracking facility and can be controlled by the tracking facility to change the frequency of reporting. VMS programs are used by treaty organizations—for example, the International Commission for the Conservation of Atlantic Tunas, the North Atlantic Fisheries Organization, and the Commission for the Conservation of Antarctic Marine Living Resources—to help regulate several international fisheries.

Congress asked NOAA in the NOAA Authorization Act of 1992 (P.L. 102-567) to conduct a study that would consider "active, transponder-based systems and passive, vessel signature-based technologies capable of localizing or identifying individual vessels without the use of vessel-carried transmitters." NOAA reported back to Congress on satellite capabilities for fisheries enforcement (NOAA, 1993), concluding that no single satellite possesses all the necessary attributes for passive monitoring. NOAA also tested the ARGOS and INMARSAT systems in 1991 for transponder-based monitoring and found that both systems provided the necessary characteristics. Of the two available satellite systems, the ARGOS system is somewhat cheaper, but less capable, than the INMARSAT system. The INMARSAT system has a 5-10 minute delay versus 2 hours for ARGOS, and ARGOS is a one-way (ship-to-satellite-to-shore) system versus INMARSAT's two-way capability. INMARSAT units collect information from GPS satellites and transmit position and other information to INMARSAT satellites. INMARSAT also provides more precise position information (because it uses the GPS positioning) and enables vessel operators to keep in close touch with fishing company staff to increase coordination of shore-side and at-sea operations. The two-way communication feature allows near-real-time reporting of daily catches for quota monitoring purposes.

VMSs are used most often for monitoring vessel position, but in their most advanced form they also can be used to enhance a fisherman's business practice and to automate catch and effort reporting. VMSs have the potential of producing reliable and useful information for both management and research and at a minimum can be used to verify times and dates reported on both fishing and observer logbooks and can provide efficient monitoring of vessel position (e.g., to enforce closed areas and seasons) and days at sea. VMS programs can reduce the costs of certain types of surveillance and allow targeting of enforcement resources.

VMS programs have some limitations. They are unable to monitor the fishing gear being used, or whether vessels stay within their catch quota; neither can they detect unlicensed fishermen or arrest violators. Therefore, VMS units must be augmented by vessel-, aerial-, port-, and observer-based monitoring programs. The costs associated with purchasing and installing VMS units may be prohibitively high for some fisheries or fishing vessels (estimated at $8,000 per vessel plus $100,000 for the shore station).

The NMFS Office for Law Enforcement is sponsoring a National Vessel Monitoring Project. To date, approximately 375 U.S. commercial vessels have been equipped with a VMS unit as part of this project. For the most part, NMFS paid for the necessary equipment, although there have been a few cases in which fishermen paid for the on-board equipment. Acceptance of and reactions to the VMS implementation have varied considerably, with fishermen in the Hawaii region being most accepting and those on the East Coast least accepting (S. Yin, NMFS, personal communication, 1999). VMS programs are more likely to be accepted if costs are offset by incentives or significant benefits to fishing businesses. NMFS believes that sophisticated VMS programs can provide value-added services in the form of weather and environmental data, market conditions, currency exchange rates, fleet management information, port services information, automated licensing, and automated transfer of quota shares.

Confidentiality restrictions designed to limit the public availability of proprietary business information prevent sharing of VMS data among different branches of NMFS, thereby hindering research. Although these systems have been perceived by some as an unwanted intrusion into confidential business information, the benefits to the fishery as well as to the personal safety of fishermen may well outweigh other concerns. NMFS should work with fishermen to jointly develop and implement this critical fisheries science and management information technology.

Misreporting and Data Fouling—Gallagher (1987) reviewed several factors that may contribute to commercial fishermen's lack of cooperation with mandated reporting and other management requirements, including allowable gear. He noted that efforts to eliminate "impasses between fishermen and regulatory authorities with regard to reporting catch data" were needed to create a cooperative atmosphere in which fisheries data could be generated, shared, and used.

Restrictions on gear may not be observed by fishermen for several reasons. Kaplan (1998) observed that conch fishermen did not abide by gear limits and marking requirements because of the problems associated with frequently replacing lost gear and being "unable" to keep up with the marking requirements for that new gear.

Gallagher (1987) suggested that management systems based on total allowable catch reduce the reliability of commercial catch data because fishermen may believe it is in their best interest to misreport or underreport catches or to discard sub-optimal catches. Gallagher characterized many problems associated with a logbook system that make it difficult to match catch data with individual vessels accurately, such as: (1) concerns related to financial privacy; (2) the extent to which knowledge about a fisherman's preferred area would become public knowledge; and (3) misreporting of amounts and locations of fish catches to evade regulations or restrictions that might be based on such data. Gallagher also noted that certain regulatory restrictions, such as species-based vessel quotas, may reduce the commercial fisherman's motivation to report catch data accurately.

Extent of bycatch discarding and underreporting are important parameters in many stock assessments. Discards can be increased by vessel quotas (Palmer and Sinclair, 1996), individual fishing quotas (Squires et al., 1998), and trip limits (Chambers, 1998); Kennelly and Broadhurst (1996) asserted that it is usually in the fishermen's best interest not to report discards. At-sea discards are estimated by comparing commercial fishermen's logbooks and data from fishery observers. Measures of bycatch are notoriously difficult to collect, and both processors and fishermen have financial incentives to underreport the true discard rate and catch (Palmer and Sinclair, 1996; Chambers, 1998). Hanna and Smith (1993) observed, however, that fishermen are concerned with the discard waste resulting from minimum fish sizes and trip limits.

Kennelly and Broadhurst (1996) suggested that a partnership between fishermen and scientists, beginning with observer programs on commercial vessels, is the best means to reduce bycatch and the data fouling that results because the bycatch amount and/or species composition is either unknown or misreported. The Alaskan pollock fishery has incentives for timely and accurate reporting of bycatch and other data, and vessel-specific bycatch rates of prohibited species in the North Pacific are available on the World Wide Web at www.fakr.noaa.gov/1999/pscinfo.html. (Information provided elsewhere in the report describes how fishermen, observers, and commercial data management firms have teamed up to help pollock and other fishermen avoid high bycatch areas.)

Although the work of Kennelly and Broadhurst was focused on identifying the best bycatch-reduction gear and encouraging its adoption by industry, their success in establishing partnerships between fishermen and scientists bodes well for such partnering efforts to address stock assessment needs. Kennelly and Broadhurst (1996) reported that the initial phase, involving scientists as observers on commercial vessels, helped establish productive working relationships that served as the basis for improved credibility, respect, communication, and collaboration in other efforts. Ongoing conferences and workshops to discuss the data and their implications continue to support this notion of shared responsibility for the fishery.

Kaplan (1998) reported that fishermen's perceptions that government is not interested in their suggestions about managing a fishery contribute to the fishermen's lack of compliance with and attitudes about enforcement of regulations. Incorporating fishermen more actively into aspects of the management and data-gathering process could help alleviate some of these concerns and increase incentives to comply with regulations to report catch and discard data accurately. Kaplan (1998) emphasized that fishermen were concerned that their experiences and knowledge were not recognized and incorporated during decision-making processes and that they were asked only to react to policies that had already been formulated.

Smith (1995), however, suggested that problems of noncompliance by commercial fishermen are rooted in fundamental differences in worldview between fishermen and scientists. These differences, she argued, center on differences in their beliefs regarding (1) what data are important or relevant; (2) how data should be gathered; (3) how data should be analyzed; (4) how those analyses should be interpreted; and (5) what responses should be crafted.

Recreational Fisheries

Survey Series—Historically, mortality from recreational fishing was thought to be low compared to that of commercial fisheries. This perspective changed for some marine fisheries as an increasing number of persons moved to coastal counties and as growth in disposable income resulted in increased marine recreational fishing. The importance of recreational fishing is illustrated in Table 3-8; in 1998 and 1999 the recreational catch of summer flounder was more than half the total landings for this species (see also Figure 2-1). The Marine Recreational Fisheries Statistics Survey (MRFSS) was established in 1979 to provide information on marine recreational fishing to complement data collected by NMFS from U.S.

TABLE 3-8 Comparison of U.S. Recreational and Commercial Catches for Popular Recreational Species in 1998

Fish Species	Recreational Catch (thousand metric tons)	Commercial Catch (thousand metric tons)
Bluefish (*Pomatomus saltatrix*)	5.80	3.77
Red snapper (*Lutjanus campechanus*)	1.98	1.85
Spotted seatrout (*Cynoscion nebulosus*)	4.33	0.27
Summer flounder (*Paralichthys dentatus*)	5.68	4.99

SOURCE: data from www.st.nmfs.gov/st1/commercial/landings/annual_landings.html and www.st.nmfs.gov/st1/recreational/database/index.html, accessed 02/03/00.

marine commercial fisheries. Recreational data are obtained from a broad-scale survey of coastal residents. According to NMFS the volume of MRFSS records handled and processed is more than an order of magnitude greater than for commercial fisheries. Three significant problems with the survey data are (1) the imprecision of estimates for some fishing modes,[8] (2) the lack of funding for collection of data for some states, and (3) the lag of data availability and perhaps lack of interest from the regional councils that hinders in-season management of recreational fisheries.

MRFSS was conceived and designed to provide accurate and reasonably precise estimates of fishing for specific regions (e.g., mid-Atlantic states). The survey has to cover large geographic areas with numerous access sites throughout the year for all major and most minor target species. No single survey can achieve all of these goals without some compromises. The precision of estimates is best at the regional scale; fisheries that are heavily targeted over a broad season provide the best estimates of catch and effort. Precision is lower at finer geographic scales and for species that have shorter seasons of exploitation (typically migratory species) or are not targeted as heavily. For many states, the survey contracts with samplers to collect recreational data. Other states, such as Texas and Alaska, do their own sampling. Texas has collected recreational data since 1976 and funds collection of such data at a higher level than MRFSS could afford to do. Recreational fisheries data from Texas are provided to NMFS for stock assessments, but are not included in the MRFSS database. The survey does not conduct data collection for Hawaii, where a large portion of the population is involved in non-commercial fishing; the recreational catch is relatively unknown there. This situation persists even though the Western Pacific Fishery Management Council has requested survey coverage. Other states share responsibilities with the survey (Figure 3-3).

To accomplish a broad regional estimate of marine recreational fishing, MRFSS uses two complementary surveys to estimate total catch and effort annually: (1) a telephone survey and

[8] MRFSS uses the term "fishing modes" to indicate different forms of access to the fishery by recreational anglers. MRFSS modes include fishing from (1) shore, (2) party and charter boat, and (3) private and rented boat.

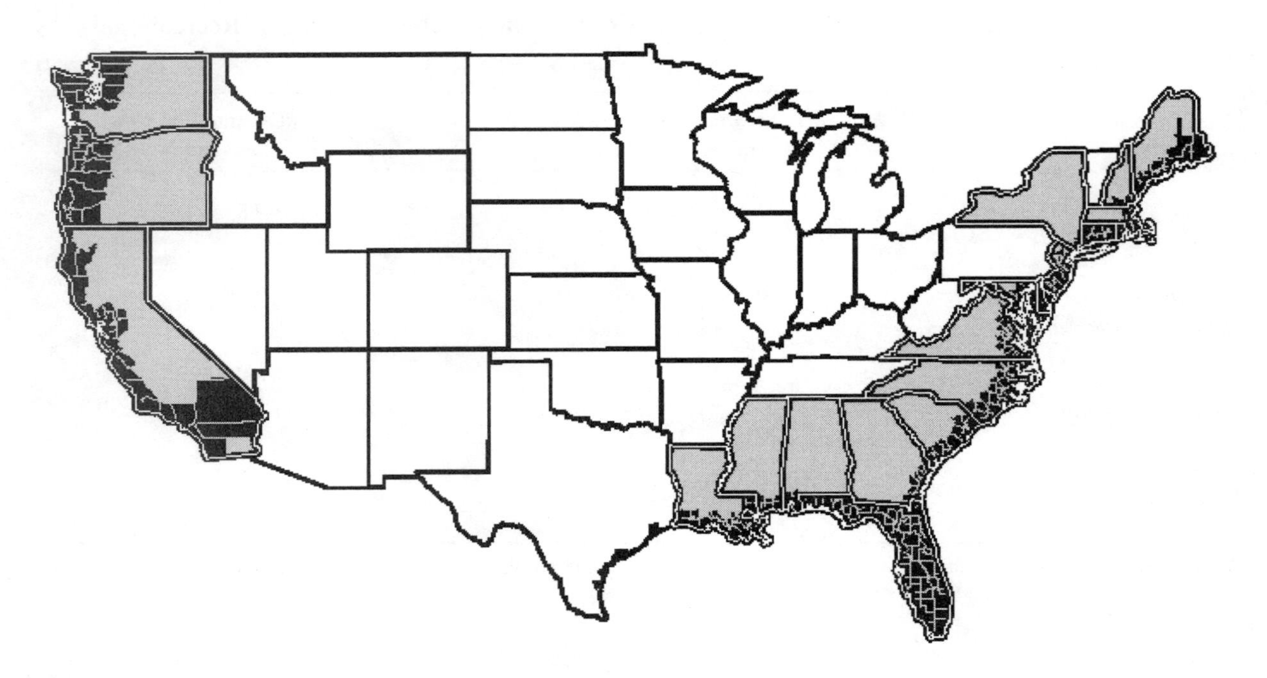

FIGURE 3-3 States with light shading are included in MRFSS. Darker shading indicates counties sampled with telephone surveys.

SOURCE: http:/www.st.nmfs.gov/st1/recreational/survey/coverage.html, accessed 06/12/00.

(2) an intercept survey. The telephone survey was designed to assess effort by estimating numbers of fishing trips per year, categorized by mode, where mode is defined as shore, party and charter vessel, or private and rental boat access. Each survey is conducted during a two-month period, called a *wave*, to reduce errors in anglers' memories. The intercept survey was designed to estimate catch and effort by species, length, and weight of fish caught by sending interviewers into the field to speak with anglers and observe their catch. The two data sources are combined to develop an estimate of total catch for each species in each wave for each access mode, as follows: catch = $\text{effort}_{\text{telephone survey}}$* $(\text{catch/effort})_{\text{intercept survey}}$. Thus, the intercept survey is intended to supply the ratio needed to expand effort estimates from telephone surveys to total catch. Intercept surveys also provide the ratio estimate of directed effort that is expanded from the telephone survey to estimate total directed effort.

Telephone Survey—The telephone survey is used to sample coastal counties to determine fishing patterns, for example, how many people actually go fishing. Coastal counties are those in which some part of the county is within 25 miles of the coast, except in North Carolina, where the limit is 50 miles (Figure 3-3). The survey interviewer calls households using a system of random-digit dialing and computer-assisted telephone interviewing, with telephone prefixes that include all the coastal counties. Only a small percentage of households contacted actually participate in fishing activities, even though this proportion is higher in the coastal counties than farther inland. The telephone survey also samples anglers who use both public and private access to the fishery to provide a more accurate estimate of total fishing effort than can be gained from on-site surveys alone (Pollock et al., 1994). Given the great difficulty in obtaining unbiased estimates of access mode using on-site surveys, the telephone survey provides a more efficient and less biased estimate of this angling characteristic.

For the telephone survey, sample size is allocated on the basis of the square root of population size in a coastal county. In some cases, less than 100 phone calls are made in a county. This means that a few successful calls (i.e., those households with anglers) greatly influence county totals. It is unclear why each county is analyzed separately, thereby acting as de facto strata. It would be more efficient to group counties and to stratify on region within the state, based on similar angling patterns. Overstratification in sampling generally decreases statistical efficiency (Cochran, 1977). MRFSS has a wealth of historic data that would allow for a successful reduction in the number of strata.

Intercept Survey—The proportion of effort (i.e., number of trips) made by anglers from non-coastal counties is estimated from the intercept survey; the survey does not account for possible correlations between distance of residence from the coast and frequency of fishing. The intercept surveys are also the only source of information on the species and size composition of marine recreational harvests. This is used to correct effort-to-state totals. For this survey to be reasonable an assumption needs to be made that the intercept survey equally measures coastal and non-coastal fishermen within each mode, wave, and state. Another assumption is that fishing patterns with respect to types of fish caught are the same within these categories for coastal and non-coastal counties. It is not economically feasible to sample all non-coastal counties; therefore, a study designed to compare coastal and non-coastal fishing patterns may be warranted. Electronic data entry in the field is not cost effective at this time because of the power requirements of available computers and their susceptibility to contamination with salt water, sand, and biological material. However, the MRFSS team and the prime contractor are focusing on other uses of technology to improve the MRFSS process, including supplying all regional representatives with laptop computers, creating interactive Web applications for scheduling interview assignments, data review and clean-up, and providing on-line weekly tallies of quota status.

The intercept survey is conducted on-site, largely at public access points such as boat ramps, piers, and marinas, and yields catch rates for targeted species. Lists of access sites were compiled initially for MRFSS and are periodically updated as new sites are developed and old sites fall into disuse. The magnitude of use is evaluated by observation, size of the site (e.g., number of ramps), and historic use patterns. The amount of on-site sampling at access points is allocated proportionally to their use. Thus, heavily used sites are sampled more than lightly used sites. Interviews are obtained as anglers complete their fishing trips, at which time total catch and trip length are known.

In general, little sampling is directed toward private access fishing, which is difficult to survey because of private-property concerns and because the fishing trips in this mode are diffuse and difficult to observe. Unless otherwise indicated, it is assumed that target species and catch rates are similar for public and private access fishing. However, catch rates can differ between public and private access when fishing patterns are influenced by factors that correlate with access type. For example, if larger boats predominate at private access points and if these boats fish farther offshore where fish species and abundance change, estimated catch rates could be biased. This is a special concern in Maryland, Virginia, and New Jersey, which have appreciable private-access use.

Interviews are more difficult to obtain when anglers fish from shore. To increase the number of anglers interviewed, the survey agent walks along the shore or pier and interviews anglers encountered. Shore and pier surveys have a large percentage of "incomplete trips" because anglers are not finished for the day. Typically, the angler is in the midst of fishing and this type of sampling (roving survey) assumes that the catch rates are the same before and after the interview. Another assumption is that the total catch can be estimated from the catch at that time and the proportion of the trip finished at the time of the interview (e.g., if a trip is half over, one can double the number of fish caught in the first half to estimate the angler's total catch for the day). The committee was not presented with evidence to support this assumption, which may not be valid if weather changes during the day. In addition, it was unclear how time of day impacts survey outcome. For example, do early morning catch composition and numbers differ from afternoon catch? Is sampling across time of day conducted in an appropriate fashion? How interviewers attempt to sample a cross-section of fishermen within each mode is not clear from MRFSS documentation. This may be a larger issue with some fish species than others because some species or certain size classes within a species are not as likely to be caught from shore. Because the probabilities of encountering an angler differ between the access and roving survey types, different calculations for estimating CPUE must be employed (Jones et al., 1995; Hoenig et al., 1997; Pollock et al., 1997).

Interviewers occasionally ride on party boats to interview anglers and to examine their catch. Private and rental boat anglers are interviewed while recovering or cleaning their boats at ramps or docks. It is unclear how interviewers choose one interviewing method over another. A small study comparing dock interviews with onboard observations would be useful to assess bias. Clearly, surveying on a party boat is apt to be more accurate but considerably more expensive than dock interviews. Either method must make the assumption that fishing practices are not influenced by the presence of an observer.

MRFSS analyses treat data from each fisherman on a single boat as being independent (ASFMC, 1994), implying that the skill of each fisherman is more important than the fact that a group of fishermen is fishing in the same location. But, the data cannot be assumed to be completely independent. If fishermen coming off the same boat have similar catches, the estimated standard error will be too low. The MRFSS document does indicate a cluster effect for charter vessel fisheries for anglers who have similar demographic characteristics (ASMFC, 1994, pp. 1-10). Adjustments are made case by case. Adjustments for clustering effects should be better documented with objective criteria. Clustering has a significant impact on uncertainty in the estimates because the formulas are based on a random sample of fishermen within strata.

MRFSS recognizes that survey intensity impacts estimates of numbers of fishing trips. MRFSS is designed to assess national landings and coastwide harvest limits, and may give reasonable estimates for each coast, but sampling effort is not sufficient to provide precise totals of effort for each coastal state for some fisheries. Several states have chosen to augment the MRFSS recreational survey effort, primarily through additional questions, rather than additional interviews. More interviews can be added if needed, but at additional cost. The average cost for intercept interviews nationwide ranges from approximately $22 to $51 per interview, and the average cost of a telephone interview is $2.87 per interview, with a three-fold range among regions of the United States.

MRFSS deals with missing and unusual values in a way that is not explained and justified in its documents. For example, when a fisherman is not home, the telephone survey method assumes that the absent person has the same patterns of fishing as those already surveyed at that home. Also, "[a]ny household [that] reports more fishing trips than the 95th percentile for the five-year distribution is reduced to the value of the 95th percentile." (ASMFC, 1994, pp. 1-9). It is not clear why these observations are altered.

Marine Recreational Licenses—Some coastal states require licensing of marine anglers, although many states exempt various categories of anglers—primarily the elderly, young, disabled, and individuals fishing from shore (Table 3-9). Selecting individuals to contact in telephone surveys using sampling frames based on licenses or

TABLE 3-9 Salt Water Recreational Fishing License Requirements and Exemptions

State	License Required (Yes/No)	License Exemptions
Maine	N	
New Hampshire	N	
Massachusetts	N	
Rhode Island	N	
Connecticut	N	
New York	N	
New Jersey	N	
Delaware	N	
Maryland	Y[a]	Residents under age 16 or older than 65 years, disabled veterans and former prisoners of war, the blind, fishermen on charter boats, and persons fishing outside Chesapeake Bay.
Virginia	Y[a]	Children under 16 years of age; persons 65 years or older, individual purchasers of a saltwater recreational boat license; persons fishing on a recreational boat, charter boat, headboat, partyboat, commercial fishing pier, or rental boat which possesses a valid Virginia recreational fishing license covering all persons using such craft or structure; persons fishing with commercial gear licensed by the Virginia Marine Resources Commission; persons fishing in coastal salt waters and ocean waters outside of the easternmost boundary of the Chesapeake Bay; and organized groups of individuals with physical or mental limitations, veterans in veterans hospitals, and school groups (K-12) with permission from the Virginia Marine Fisheries Commission.
North Carolina	N	
South Carolina	Y[b]	Residents under the age of 16 and over the age of 65; residents fishing from a public fishing pier; and disabled residents.
Georgia	Y	Residents under the age of 16 and over the age of 65, blind residents, and disabled residents.
Florida	Y	Florida residents fishing from land or a structure fixed to the land—a pier, bridge, dock, floating dock, jetty or similar structure—but not from a boat; residents older than 65; residents who are a member of the U.S. Armed Forces stationed outside of Florida and home on leave for 30 days or less; residents under age 16; anyone fishing from a boat that has a valid commercial saltwater products license or a valid recreational vessel saltwater fishing license; and disabled persons.

(continued)

TABLE 3-9 Continued

State	License Required (Yes/No)	License Exemptions
Alabama	Y	Residents under the age of 16 and over the age of 65; residents fishing from a public fishing pier; and disabled residents.
Mississippi	Y	Residents who are blind, paraplegic, a multiple-amputee, adjudged totally disabled by the Social Security Administration or totally service-connected disabled by the Veterans Administration.
Louisiana	Y	Residents under the age of 16 and over the age of 65; citizens of the state who are on active military duty; veterans having a permanent service-connected disability; a resident who is totally and permanently disabled; and residents who have artificial limbs or permanent braces or are mobility impaired.
Texas	Y	Residents who are disabled veterans, under the age of 17 or over the age of 65, or who are mentally disabled.
Alaska	Y	Residents who are blind, under the age of 16, who are 60 years old or older and have a permanent ID card, and residents who qualify for a disabled veterans license.
Washington	Y	Residents under the age of 14 and over the age of 70, and disabled veterans.
Oregon	Y	Residents under the age of 14; disabled persons qualify for special permits.
California	Y	Residents who are under the age of 16, blind; low-income American Indians; wards of the State residing in a State hospital; developmentally disabled persons receiving services from a State regional center; residents who are so severely physically disabled that they are permanently developmentally disabled; persons receiving services from a State regional center; and residents who are so severely physically disabled that they are permanently unable to move from place to place without the use of a wheelchair, walker, forearm crutches, or a comparable mobility-related device.
Hawaii	N	

SOURCE: National Marine Fisheries Service.

NOTES: Only basic saltwater licensing information is indicated. Some states also require additional permits or tags (e.g., snook, rockfish, salmon) and may have differing exemptions for these special permits. Some states, such as Oregon, have an "angling" license, which is for both fresh and saltwater fishing. Other states sometimes issue special recreational gear licenses or single-species licenses. In some states "exempted" persons don't have to buy a license, but must carry a license they receive at no cost.

[a] In Virginia and Maryland a license is only needed to fish in Chesapeake Bay waters, not for ocean waters.

[b] South Carolina officially doesn't have a saltwater license, but they do require that fishermen purchase a "stamp" that functionally is equivalent to a license.

re-contacting known recreational anglers could be more cost-effective and efficient, although many factors need to be considered in making such an assessment. Random-digit dialing is most effective (in terms of cost and precision of estimates) when the target group forms a substantial portion of the general population. This is not true for marine anglers because they make up only a small percentage of the contacted households.

Requiring licenses for marine angling (even free licenses) could improve data collection efforts by making a comprehensive sampling frame available and eliminating the inefficient random-digit dialing approach for the telephone survey (but not the expensive intercept surveys). In theory, recreational effort assessments could be less costly (in terms of time, money, and staff) than current methods if based on license sample frames because anglers would be identifiable and sampled more easily. MRFSS compared the use of a license sampling frame from the State of Oregon with random-digit dialing to explore increasing precision and efficiency of estimates of recreational effort (Gray, 1997). The comparison yielded mixed results. Precision was improved in the shore and private boat modes at no additional cost by using the license frame, but the precision of data from the party/charter mode was not improved. This pilot study indicates that a more detailed test of license frames is in order.

States have jurisdiction over implementing marine recreational fishing licenses, over who must be licensed, and over the administration and structure of the resulting databases. Inconsistencies in the data that are included in state sampling programs, insufficiency of the frame, their tendency to become outdated quickly, and issues of confidentiality temper the value of license sampling frames. Further testing should include the standard measures of precision and accuracy, along with a detailed study of the costs involved in maintaining the license frames, access to the frames if they are maintained by the states, adequacy of the frames, and the data management issues that will ensue from combining separate frames. The use of the license frame should also be compared with the use of partial frames (random-digit dialing used with a partial license frame) and with longitudinal retention of anglers previously contacted in the random-digit dialing component of MRFSS. The issue of saltwater recreational licenses is controversial because many states do not presently require licenses, and anglers in those states do not want to face additional bureaucratic hurdles and a perceived intrusion of government into their private lives. Also, for a licensing system to achieve statistical and cost efficiencies, it would probably need to feature more uniform criteria for licenses among states. Different exemptions on licenses in each state would require NMFS to determine the fishing characteristics of unlicensed fisherman, requiring reversion to a system like MRFSS.

Social and Economic Data

Collection of social and economic data is important for several reasons. As described in Chapter 1, using social and economic information to better understand fishery-dependent data can improve the process of estimating the current status of the stock. This may be the most cost-effective method of improving stock assessments.

The Magnuson-Stevens Act requires the use of social and economic information in fisheries management. Section 301 of the Magnuson-Stevens Act contains the national standards for fisheries management. National Standard 1 requires that a fishery be managed for optimum yield. The definition of optimum yield is " . . . the maximum sustainable yield from the fishery as reduced by any relevant economic, social, or ecological factor." Conservation and rebuilding of fish stocks are required by National Standard 1. National Standard 4 requires that allocations among fishermen be "fair and equitable." National Standard 5 requires that management ". . . consider efficiency in the utilization of fishery resources. . . . " National Standard 7 requires that management measures ". . . where practicable, minimize costs. . . ." National Standard 8 requires that management ". . . minimize adverse economic impacts on such communities." Most of the other standards are not absolute, but instead recognize a trade-off between competing goals. Social and economic data are required to analyze the trade-offs required by the Magnuson-Stevens Act.

Changes in harvest levels designed to conserve and sustain fishery resources can usually be accomplished by several alternative combinations of regulations. Different regulations have different economic impacts, and it is incumbent on fishery managers to minimize the costs of conservation and the adverse economic impact on fishing communities (National Standards 7 and 8). GAO (2000) notes that the regional councils and NMFS seldom consider the economic impact of alternative conservation measures. Typically, conservation measures are enacted and the economic impacts are calculated, but there is no consideration of the economic impacts of a range of alternatives so that the least costly alternative may be chosen.

Studies of social and economic factors that affect summer flounder and other fisheries could benefit fisheries management by

- making is easier to create more cost-effective regulations that encourage accurate data reporting,
- improving knowledge of which fishermen (commercial and recreational) and processors would be affected by new regulations and how seriously,
- increasing the equity of regulations among different parts of a council region and distributing data collection efforts to best meet data needs, and
- improving communication among fishery stakeholders.

Social and economic assessments should evaluate a number of factors, including

1. distribution of costs and benefits of current and proposed regulations;
2. incentives driving data-reporting accuracy, with the goal of increasing accuracy of reported data;
3. numbers of vessels, fishermen, crew, and processors and their economic dependence on the fishery;
4. regional differences in socioeconomic status of fishermen throughout the fish stock range;
5. information and communication needs perceived by affected fishermen, other industry members, environmental groups, recreational fishermen, and other stakeholders;
6. numbers and distribution of recreational anglers and associated industries; and
7. need for communication among decision-makers at all levels, including councils, NMFS, states, and legislatures.

In addition, it would be useful to conduct systematic social research to understand the factors that motivate compliance with both the spirit and the letter of conservation regulations. The results of such research could be the basis for improved management.

Cooperation, Communication, and Review

The committee attempted to determine whether data are collected cooperatively, how communication related to data collection is handled, and how the methods and results of data collection are reviewed.

Cooperation

Cooperation with Industry—Cooperation between commercial fishermen and fishery managers to improve fishery management, broadly called co-management, is not new. Hersoug and Ranes (1997) provided an overview of the co-management concept and why it is receiving increasing attention: "By engaging fishermen in management, they would act more responsibly towards the long-term goal of sustainability. In other words, by being partly responsible for the management of 'their own resources,' the need

BOX 3-2
Examples of Assistance Fishermen Have Provided to Research Efforts

Gallagher (1987) mentioned several examples of assistance that fishermen have provided to research efforts:

- specimens for analysis and identification of new species (e.g., at the Smithsonian Institution)
- observations that aid in behavioral studies
- information about the habits, distribution, and abundance of commercially exploited fish species based on observations on the water
- manpower for research and experiments (e.g., effects of fish-handling methods and evaluation of harvesting gear types)
- vessel time and equipment for research efforts
- aiding in research efforts to develop fisheries for underutilized species (e.g., red crab on Georges Bank)
- tagging fish
- providing stomach contents for feeding studies or hard parts for age and growth studies (e.g., swordfish).

for costly control, monitoring, and surveillance could be substantially reduced."

Arguments for co-management also are based in democratic theory about the fairness and legitimacy of decision-making processes (McCay and Jentoft, 1996), including the need to involve those affected by the decisions. Jentoft et al. (1998) offered a compelling argument why fishermen should have an opportunity to become involved more actively in the entire management process, and the reasons included (1) improved management decisions, because those with hands-on knowledge of the fishery are involved; (2) improved communication among all parties; and (3) increased consideration of the socioeconomic aspects of the fishery in decisions.

In this report, discussion of co-management is limited to partnerships involving commercial fishermen in the data collection and research aspects of fisheries management (i.e., selection of methods, data gathering, and data interpretation), rather than in the policy and regulatory aspects of management (e.g., Jentoft, 1989; Pinkerton, 1989a,b). Gallagher (1987) described different types of assistance fishermen have provided to research efforts (Box 3-2). With co-management, fishermen have provided local knowledge of fish stock movements and distribution; funding for research; and stimulus for academic scientists to become more involved in generating scientific data (Pinkerton, 1989a,b). Johnston (1992) hypothesized several benefits that would accrue through improved partnerships among fishermen, scientists, and managers. These include "(1) a greater commitment to success by fishers who plan an active role in the design of management strategies; (2) reduced public cost through the use of data provided by fishers (through their organizations); and (3) a better understanding of the behavior of fishers." Other benefits include enhanced credibility of the management process, and sensitizing scientists and other decision-makers to the sociopolitical context in which the fishing industry is embedded, leading to acceptance of fishing regulations (Palmer and Sinclair, 1996). Hersoug and Ranes (1997) suggested that "by involving fishers more directly into the decision-making process and by bringing the management process closer to those fishers who are affected, their willingness to come to an agreement and comply with the rules and regulations is enhanced." They also noted that many authors have hypothesized or anticipated benefits of co-management, with little empirical demonstration of the benefits or costs of co-management in fisheries.

Fox and Starr (1996) suggested that information derived from commercial logbooks could be a valuable complement to survey data and could "improve estimates of the distribution and relative abundance of commercial fish species." They advocated the use of commercial data, particularly given the growing availability of GIS tools to help analyze the large spatial databases provided through logbooks. Thus, they suggested that logbooks could be used to redesign the stratification scheme used for the NMFS triennial trawl survey. Like any data used in stock assessments, however, logbook data are only useful if they are reported accurately and completely for all fishermen, are subjected to appropriate quality control, and are used in the stock assessments with appropriate adjustments and assumptions (Fox and Starr, 1996). However, the extensive data available from West Coast trawl logbooks from the groundfish fisheries is virtually unused in management of these species, so that potential cooperation is not realized. Because the logbooks are mandatory, fishermen maintain them, despite the fact that only a handful of scientists have used the data contained in them.

Furthermore, information gathered and recorded by fishermen in logbooks is sometimes questioned even when it is quantitative in nature. The reason for this is that logbook information can be compromised in at least two ways. First, inaccuracies in reporting may take place with respect to species caught, area fished, and amount caught if such statistics are viewed as intrusive, irrelevant, invasive of proprietary information, or of low priority in the daily operation of the vessel. Even when the information is accurate, it may not be useable as changes in technology, experience, targeting practices, or regulations can all modify catch rates in ways that makes it difficult to track changes in species abundance. For these reasons, stock assessment scientists often feel they must rely solely upon fishery-independent measures of stock abundance (Fox and Starr, 1996; Rose and Kulka, 1999).

Partnerships between the commercial fishing industry and the agencies responsible for scientific research and management can take several forms. Cooperative sampling efforts can involve researchers actually employing commercial vessels and crews to gather data (e.g., Kennelly and Broadhurst, 1996). Use of commercial vessels is most feasible with long-term charter situations or non-trawl fisheries.[9]

Benefits from improved cooperation accrue not only to managers and scientists but also to commercial fishermen, who may benefit through improved stock assessments, leading to more comprehensive and appropriate management plans and ultimately ensuring the continuation of the fishery that provides their livelihoods (Gallagher, 1987). Hanna and Smith (1993) demonstrated that fishermen supported increased adoption of cooperative fishery management approaches, involving fishermen, processors, and managers, particularly with regard to ensuring "regulations that are compatible with the economic aspects of the fishery." They also characterized fishermen as concerned with long-term sustainability of the fishery rather than a sole concern with short-term profits (Hanna and Smith, 1993). Such perspectives help justify the effort needed to develop long-term partnerships and cooperation.

Using commercial vessels and crew (in addition to research vessels) can improve data-gathering efforts by providing a crew with important

[9] Trawl fisheries drag a trawl net through the water in such a way that a net, trawl lines, and vessel are an integrated system with a unique acoustic signature and physical action that affect the catch of fish. Fixed gear systems are not integrated with a vessel. Of course, standardized gear is important for both trawl fisheries and fixed-gear fisheries.

BOX 3-3
New England Council Fisheries Research Program

In the past decade, the U.S. Congress has provided financial support to New England fishermen in several forms, including vessel buyouts, as well as a $5 million appropriation in 1998 and a $21 million appropriation in 1999 for cooperative research between NMFS and industry. The funding is intended to engage commercial fishermen in cooperative research and management as a disaster relief measure. The New England Fishery Management Council, through its new Experimental Fisheries Research Steering Committee, is responsible for reviewing proposals and selecting research that will most appropriately fit the needs of managers. The NMFS regional administrator will make the final decision on what research is funded. Fishermen who receive disaster assistance in payments based on equivalent days at sea will be required (upon request) to contribute an equal amount of time assisting in research. In addition to providing disaster relief and building cooperation, this program resulted from a desire by fishermen and the New England Council to be able to direct some research support to issues they believe are a priority.

local knowledge specific to the sites or species of concern, providing data based on gear similar to what is or will be used in the commercial fishery, and improving confidence in the data among those commercial operators not directly involved in the research because they can observe the commercial research vessel in known fishing areas, using familiar gear (Kennelly and Broadhurst, 1996). Commercial fishermen believe strongly that data provided by commercial fishermen can improve the information on which stock assessments are based, by providing a sample size much larger than can be obtained with standard surveys conducted by dedicated research vessels (J. Easley, Oregon Trawl Commission, personal communication, 1999).

Examples of partnerships among commercial fishermen and research scientists can be found in the United States and other nations. Canada has made considerable strides in forging partnerships between scientists and managers.

Researchers at Oregon State University (funded by NMFS) are currently analyzing the potential for industry-scientist cooperative fisheries research programs (G. Sylvia, Oregon State University, personal communication, 1999). The initial stage of this work included a summary description of fisheries incorporating industry-scientist partnerships for fisheries research, including examples from Australian inland and trawl fisheries, Canadian marine fisheries on the east and west coasts, and a new initiative by the Oregon Department of Fish and Wildlife to begin a cooperative research effort (Sylvia and Harms, undated).

In another example, Northeast Fishery Science Center (NEFSC) scientists completed an at-sea survey in December 1998 as part of the Advanced Fisheries Management Information System (AFMIS), using commercial trawl vessels, paid in part with the scallops they caught. In 1999 the New England Fishery Management Council received new funding for council-identified research carried out with industry (Box 3-3). The Magnuson-Stevens Act allows councils the discretion to develop fishery management plans that "reserve a portion of the allowable biological catch of the fishery for use in scientific research" (Sec. 303[b][1]). Some councils have taken advantage of this provision to engage commercial fishermen in research activities, allowing them to harvest additional fish. The Northwest Fishery Science Center is in its second year of conducting surveys using commercial vessels and paying cooperating commercial fishermen with fish caught.

In Norway, partnerships show potential to lead to improved knowledge about local fish stocks, environmental influences on fish distribution and abundance, and potential implications of alternative management regimes on the fish stocks as well as on local communities (Maurstad and Sundet, 1994).

In Canada, the Department of Fisheries and Oceans (DFO) and the Pacific Blackcod[10] Fishermen's Association (PBFA) have established a co-management relationship that, in their own assessment, "has benefited both DFO and the PBFA, and has improved the overall research, assessment, monitoring, enforcement, and management of the fishery" (DFO and PBFA, 1998). PBFA members assist by providing vessels and crew to conduct tagging operations for mark-recapture studies, returning tagged fish through a tag-return incentive program, collecting and synthesizing the data from the tagged fish program, collecting biological samples based on sampling procedures provided to each vessel, and contracting with scientists to conduct both the annual assessments and longer-term research. PBFA has provided "all of the costs directly associated with the research and assessment of the sablefish resource," including DFO salaries, benefits, and operating expenses. PBFA also coordinates the dockside monitoring program, including reporting compliance problems to enforcement authorities. DFO and PBFA (1998) concluded that "the additions of dockside monitoring, dedicated enforcement resources, industry funding, and a more responsible attitude by sablefish fishery participants have resulted in improved compliance with the sablefish fishery rules and regulations. This is supported by a study of the sablefish fishery released by DFO's Internal Audit and Evaluation Branch in 1993." Key to this co-management relationship is clear and frequent communication among all participants.

Also in Canada, the Fishermen and Scientists Research Society (FSRS) operates in the Atlantic region. Since 1994, the society has supported participation of fishermen in the stock assessment process, enhancing the process by making available information that only the fishermen can obtain on a daily basis, and encouraging the participation of fishermen in developing a sound information base to lead to more effective resource management (see http://www.fsrs.ns.ca). Funding originally came through DFO and other government sources, although some funding since that time has derived from contracting out society expertise. Communication among fishermen, scientists, and the general public is an essential part of the mission of the society (King et al., 1994). The society produces a quarterly newsletter—*Hook, Line, and Thinker*—that is distributed internationally to fishermen, scientists, academics, and others interested in sustainability of fishery resources.

Although the regional fishery management councils established under the Magnuson-Stevens Act were designed to provide a vehicle for cooperation and outreach between NMFS and stakeholders, council activities have sometimes not been successful in this goal.[11] The problem may be inherent in the present management structure, the state of regulated fisheries, and regulator-regulated dynamics of conflict, and may be solved only by making "local incentives compatible with global goals" (Ecosystem Principles Advisory Panel, 1999). Greater attention to co-management ideals of involving stakeholders more closely in data collection and decisionmaking may help. The Canadian development of a joint industry-science Fisheries Resource Conservation Council committed to wide consultation and the fullest use of fishery-dependent data could be a pattern for U.S. regional fishery management councils to follow (Box 3-4).

[10] Blackcod and sablefish are interchangeable names for *Anoplopoma fimbria.*

[11] Center for Marine Conservation, *Missing the Boat*, www.cmc-ocean.org/missingboat.

BOX 3-4
Fisheries Resource Conservation Council[a]

The Minister of Fisheries and Oceans has established the Fisheries Resource Conservation Council as a partnership between government, the scientific community, and the direct stakeholders in the fishery. Its mission is to contribute to the management of the Atlantic fisheries on a "sustainable" basis by ensuring that stock assessments are conducted in a multidisciplinary and integrated fashion and that appropriate methodologies and approaches are employed; by reviewing these assessments together with other relevant information and recommending to the minister total allowable catches (TACs) and other conservation measures, including some idea of the level of risk and uncertainty associated with these recommendations; and by advising on the appropriate priorities for science.

Council Objectives

- To help the government achieve its conservation, economic, and social objectives for the fishery. The conservation objectives include, but are not restricted to (1) rebuilding stocks to their "optimum" levels and thereafter maintaining them at or near these levels, subject to natural fluctuations, and with "sufficient" spawning biomass to allow a continuing strong production of young fish; and (2) managing the pattern of fishing over the sizes and ages present in fish stocks and catching fish of optimal size.
- To develop a more profound understanding of fish-producing ecosystems, including the interrelationships between species and the effects of changes in the marine environment on stocks.
- To review scientific research, resource assessments, and conservation proposals, including, where appropriate, through a process of public hearings.
- To ensure that the operational and economic realities of the fishery, in addition to scientific stock assessments, are taken into account in recommending measures to achieve the conservation objectives.
- To better integrate scientific expertise with the knowledge and experience of all sectors of the industry and thus develop a strong working partnership.
- To provide a mechanism for public and industry advice and review of stock assessment information.
- To make public recommendations to the minister.

Activities

- Review appropriate Department of Fisheries and Oceans science research programs and recommend priorities, objectives, and resource requirements.
- Consider scientific information—including biology, and physical and chemical oceanography, taking into account fisheries management, fishing practices, economics, and enforcement information.
- Conduct public hearings wherein scientific information is presented or proposed and conservation measures and options are reviewed and discussed.
- Recommend TACs and other conservation measures.
- Prepare a comprehensive, long-term plan and a work plan for the council that are reviewed annually at a workshop with international scientists and appropriate industry representatives.
- Ensure an open and effective exchange of information with the fishing industry and contribute to a better public understanding of the conservation and management of Canada's fisheries resources.

---

[a] Excerpt from http://www.ncr.dfo.ca/frcc/Baseinfo/terms_of_reference.htm, accessed 10/14/99.

O'Boyle et al. (1995) describe joint government-industry surveys in Atlantic Canadian waters, which outnumbered the government surveys there in 1995. New Zealand has an extensive system of data collection and uses an open stock assessment process in which all stakeholders participate; fishermen are increasingly involved in data collection and research (NRC, 1999b).

Palmer and Sinclair (1996) cautioned that cooperative management and research efforts may be difficult to implement because local knowledge and perceptions are rarely uniform. Others have also warned that lack of unity among commercial fishermen and differences in motivation and perceptions can reduce the potential usefulness of partnerships (e.g., Felt, 1994; Jentoft and McCay, 1995). Many of these concerns relate to broad co-management proposals or arrangements, rather than more limited partnerships involving the employment of commercial fishermen and vessels to complement or augment fishery-independent data collected by research vessels. Additionally, differences in perceptions between fishermen and scientists must be overcome. Harms and Sylvia (1999), studying the West Coast groundfish fishery, found that scientists and fishermen share very few opinions about the extent to which NMFS currently encourages industry-scientist cooperation, the strength of a conservation ethic among industry, and the degree of interest among scientists in sustaining a long-term, profitable fishing industry. In that study, majorities of both fishermen and scientists agree that "there is currently very little or no trust between industry and scientists" (Harms and Sylvia, 1999). On the positive side, most scientists and industry respondents agreed that "industry-scientist cooperative research has 'significant' or 'moderate' potential for improving the science used in management of groundfish" (Harms and Sylvia, 1999).

Effects of Existing and Potential Fishery Regulations on the Quality of Fishery-Dependent Data—For many fisheries, harvest capacity grows to the point where it substantially exceeds the maximum sustainable yield of the fish stock. In this case, increasingly stringent regulations are required to keep total catches within acceptable bounds. As regulations decrease fishing opportunities, fishermen's short-term interests may become very different from the long-term interests of fishery managers. This can have profound effects on how fishermen react to existing and future regulations, and may reduce the possibilities for cooperation and co-management discussed earlier. Decreases in fishing opportunities can also have substantial effects on fishery-dependent data, including:

- strategic bias—regulations can create a strategic bias in unverifiable fisheries data. For example, catch and bycatch may be underreported or misrepresented in terms of the species, weight caught, or locations fished.
- increased costs—regulations that require expensive data collection will discourage such collection and create sparse data sets. This is true whether the government or industry is paying.
- refusal to cooperate.
- data fouling—different regulatory systems can have unintended effects on the quality of data collected from the fishery. Data fouling can be manifested as increased bias in the size distribution of the catch and in the nominal CPUE due to shortened trips and seasons and changes in fishing practices.
- regulatory discards—implementation of new management programs, such as trip limits and closed seasons, can increase discards of specific species in a multispecies fishery.

Managers should anticipate such effects when designing new management schemes. Regardless of the way a fishery is regulated, efficient management should achieve the desired effect at the least cost to the individuals being regulated, to the management system itself, and to the remainder of society. When new regulations are developed and implemented, their costs and effectiveness in terms of levels of compliance and whether they are having their intended effect(s) can only be determined through concomitant monitoring of the biological, economic, and social results of the regulations. Monitoring systems should also be able to detect unexpected and unintended effects, which would require supplemental research to determine the cause(s) of such effects.

Cooperation with Recreational Fishermen—Fishery managers and scientists have rarely developed cooperative programs that include marine recreational anglers, largely because this sector is composed of many more individuals than the commercial sector and is more dispersed. These factors make it harder to identify contact points and to coordinate scientifically designed projects. One level of cooperation that does occur is anglers participating in MRFSS and state-sponsored surveys. Another occurs when anglers return tagged fish.

Government agencies, private-sector organizations, and scientists have attempted to include anglers in fish-tagging programs. However, the data provided from these programs can be problematic when anglers tag and release fish in an ad hoc fashion, because the value of the tag returns are limited to general patterns of movement (at least for life stages and in areas where anglers fish), and inspiring conservation interests. For example, the non-profit American Littoral Society has a 35-year-old program that now includes approximately 1,200 marine recreational anglers and 95 clubs who buy tagging kits and tag over 30,000 fish each year. Most of the tagged fish are striped bass, but summer flounder and bluefish are also commonly tagged. About 4% of the tags are returned and data are supplied to the Northeast Fishery Science Center. This tag return rate is on the lower end of the range listed in the literature for scientifically designed tagging studies. For example, tag return rates for tropical tunas range between 2 and 13 percent (Hampton and Gunn, 1998; Kaltongga, 1998; Itano and Holland, 2000). Most summer flounder are recaptured within months of being tagged. The number of tags returned does provide a very rough indication of the fishing mortality, but accurate measures depend on random introductions of tags, measures of tag loss and tagging mortality, and knowledge of reporting rates that are not available from such ad hoc surveys. Many in the academic community doubt the usefulness of tagging programs such as this that are conducted non-systematically.

Tagging studies must be implemented from scientifically-designed procedures to obtain information on the behavior of individual fish, estimate mortality rates, and determine the mixing between two populations of fish. Recent advances in tagging models now allow the estimation of natural and fishing mortality in mixed-use fisheries (see for example Brooks et al., 1998 and Pollock et al., 1991 for creel surveys combined with tagging studies). An important recent example of cooperation of scientists and anglers is the "Tag a Giant" program, in which recreational anglers and charterboat operators are assisting scientists in placing archival, acoustic, and pop-up (Block et al., 1998) tags on medium and giant Atlantic bluefin tuna. The New England Aquarium is conducting a similar program.

Communication

Data, and the information associated with them, must flow from fishermen to scientists and managers and vice versa for use in stock assessments and for implementation of fishery management actions. What makes communication difficult is the disparity in how data are perceived, the difference in the language used to communicate the issues, and the preconceptions that exist among scientists and fishermen as to the other's motivations for gathering or presenting certain types of data. Even when fishermen, scientists, and managers agree on the data available, they may not understand them in the same way, and they may disagree about interpretations, hypotheses, and implications. This was obvious during the committee's public sessions, as issues were raised by fishermen concerning, for example, exactly how and why NMFS conducted their surveys. Surveys are a natural focus of concern for fishermen when they perceive that TACs are lower than justified by their observations of fish populations because a survey is the information component they can most closely relate to their own fishing activities.

NMFS personnel informed the committee that opportunities had been provided for fishermen to participate in surveys, but that these opportunities were seldom used. As discussed previously, co-management arrangements may facilitate better communication, directly and indirectly, through closer ties between agencies and industry. It is difficult to determine what steps would improve communication and change the adversarial relationships that exist between scientists and managers and fishermen and that hinder communication in many U.S. fisheries. Meetings mandated by an agency obviously will not work, but neither will missed industry opportunities to participate in NMFS cruises. Some communication about data collection occurs at council meetings, but the council setting may be inappropriate for the extended presentation and discussion that may be necessary. The most effective communication could involve a coordinated program of meetings, lectures, and demonstrations at sea.

Communicating data gathered from commercial vessels engaged in research can face challenges similar to those of sharing data from survey research vessels. Kennelly and Broadhurst (1996) suggested that photographs, videos, and other visual presentations at meetings or other interactive arenas may be most successful at convincing commercial vessel operators who were not involved directly in the data-gathering efforts. For example, a scallop fisherman participated in a joint U.S.-Canada research trip in the Bay of Fundy during the summer of 1998. The scallop fisherman was taken down in a submersible to look at the bottom he had dragged for scallops and was completely surprised by the deserted appearance of the bottom. The view from underwater did not match his perception while scallop dragging with his own boat (S. Smith, Bedford Institute of Oceanography, personal communication, 1999). In addition, public presentations to concerned stakeholder groups and involvement of both scientists and fishermen in mass media outreach (print, radio, television) are critical to fostering greater public understanding of fishery issues (Kennelly and Broadhurst, 1996). Ticheler et al. (1998) emphasized the essential role of proper communication of data gathered from commercial fishermen back to the larger commercial fishing communities, including associated interpretations by scientists. They suggested that improved information exchange could enhance community awareness about harvest patterns, management implications, and the reasons for management decisions. Kaplan (1998) noted that fishermen have emphasized the importance of timing such information exchanges to coincide with non-fishing hours or seasons to enable fishermen to participate.

Review

NMFS conducts annual and semi-annual reviews of many of its assessments and associated data collection practices (e.g., Stock Assessment Review Committees panels on the East Coast and STock Assessment Review panels on the West Coast). Because of constraints arising from scheduling and financing such reviews, and the scarcity of stock assessment scientists outside NMFS, most participants in these review panels are employed by NMFS. Such reviews provide a useful quality assurance function, can be completed quickly, and can produce high-quality results that are comparable from year to year. Because of the limited number of stock assessment scientists outside NMFS and the large number of stock assessments that must be reviewed, the reality is that NMFS employees will *have* to form a major part of most regular reviews.

The National Research Council's only previous review of stock assessments in the U.S. northeast region found that the assessments for cod, haddock, and yellowtail flounder used methods accepted worldwide and yielded results that rightly guided the New England Fishery Management Council to make conservative management decisions (NRC, 1998b). The committee has no reason to doubt that NMFS scientists are just as capable and professional as academic scientists and that the SARC and STAR panels provide an appropriate means of routine reviews. The committee found no evidence of bias in the summer flounder assessment.

One consequence of reviews that are not conducted by groups of external scientists is that they may inhibit the influx of new ideas into the system and another is that the industry and environmental groups may perceive the assessments as being biased. NMFS itself has recognized this problem in setting up its Center for Independent Experts. Academic institutions and departments recognize the same issue when they occasionally appoint committees of visitors. The key is to make sure that review panels work independently, and are perceived to be giving independent advice. There are ways to do that, such as making the review meetings more open and transparent, with industry participation (under strict rules of engagement) and non-NMFS chairspersons.

The committee could find few examples of truly independent reviews of data collection and stock assessment procedures for U.S. fisheries. Congress asked the NRC to review the assessments of Atlantic bluefin tuna (NRC, 1994a), northeast groundfish (NRC, 1998b), and summer flounder (this study), including various aspects of data collection. The Pacific Council commissioned an independent review of the groundfish assessments (West Coast Groundfish Assessment Review Panel, 1995). Such extensive reviews are typically brought on by concerns that the review system is somehow not operating properly. What is needed are thorough, periodic, independent reviews of assessment and data collection procedures as part of the ongoing data collection and assessment process for each fishery. These reviews do not need to be conducted at each assessment, but periodically, especially when there is a change in methodology or a significant change in the fish population or fishery. Periodic, rather than assessment-by-assessment, reviews could overcome logistical and other constraints. Independence of such reviews might be enhanced if they were conducted under the direction of the regional fishery management councils, perhaps through ad hoc review panels overseen by the scientific and statistical advisory committees to these councils (assuming these committees themselves are made up primarily of scientists outside NMFS).

An alternative could be the Center for Independent Experts, in which NMFS is funding a pilot project at one of its joint institutes (the University of Miami) to put together teams of independent reviewers. The center's steering committee, formed by the University of Miami, selects reviewers, oversees the reviews, and transmits the review results to NMFS. The center has been active since 1998 and has conducted reviews of a number of stock assessments (by individual reviewers) and of the science conducted pursuant to the International Dolphin Protection Act (by a team of three reviewers). It remains to be seen whether the reviews of this type will be acceptable to stakeholders.

Finally, stakeholders should play a more substantial role in the review process than the 30 minutes or so of public testimony currently allowed at assessment and management meetings. Commercial fishermen, recreational fishermen, and other stakeholders need to take the incentive in developing that role. Many successful examples exist of stakeholder involvement in the review process (e.g., the International Pacific Halibut Commission, and New Zealand fisheries management). Stakeholder involvement may require that independent scientists be brought into the process to facilitate communication, although this has not been uniformly successful.

For recreational data, MRFSS provides somewhat more nationwide structure, standardization, and control. MRFSS has had extensive review by the scientific and statistical communities (Hayne, 1977; Jones et al., 1990; Bolstein, 1992, 1993; Osborn and Lazauski, 1995), although many of the recommendations of earlier reviews have not been implemented.

DATA MANAGEMENT

Confidentiality

Fishery management agencies, both state and federal, usually have procedures to ensure data confidentiality. NOAA Administrative Order 216-100, dated July 16, 1994, defines the conditions under which information collected by NMFS may be shared with others. Confidential data are defined as data identifiable with any individual, corporation, or other entity. In other words, if the identity of the fisherman or processor can be determined from the data, the data are confidential and cannot be shared without permission from the fisherman or processor. If the identity of the fisherman or processor can be obscured by aggregation into strata by time or space or both, the data are not confidential and can be shared. There is no sunset clause on data confidentiality; data are confidential in perpetuity. These constraints make it illegal to share data under some circumstances even with a scientific and statistical committee (SSC) of a regional fishery management council. In the case of summer flounder, the number of fishermen or processors in some states is so small (e.g., Delaware and Maryland) that landings by state cannot be made public in some years. Confidentiality provisions can hinder development of models of how management measures work, except on a very large scale.

Institutional Arrangements for Data Management

Fisheries data management systems are maintained at state, regional, federal, and international levels. Additional systems are under development, including at least one research system. The following section describes the range of existing and incipient systems.

Federal Data Sources

NMFS collects a variety of data needed for stock assessments and other fishery management purposes, as described in the first part of this chapter. NMFS has the primary responsibility for collecting biological, social, and economic data to monitor activities in the U.S. exclusive economic zone. The U.S. Coast Guard also provides data related to enforcement activities; such information can be useful in evaluating the effectiveness of different management approaches. The federal government is responsible for gathering data from states, interstate commissions, and international commissions, as needed, for federal fisheries management activities.

The federal government pays for most of the data collection and management needed for its uses, whether done directly or through the commissions or states. Decentralized data collection activities create extra requirements for coordination and quality control. Ensuring that the data needed are available and accurate can be difficult when NMFS can exert only limited control over collection, standardization, and quality control.

States

Coastal states generally have fisheries management responsibilities within state waters (up to 3 nautical miles from shore in most states; up to 9 miles in Texas, the west coast of Florida, and Puerto Rico; and in bays and estuaries). States issue commercial fishing licenses and may require marine recreational fishing licenses. States are also responsible for regulating ports and for maintaining environmental quality in state waters. Some states maintain hatchery programs for fish and shellfish to enhance natural stocks. States must collect their data and have access to data from adjacent states and NMFS. However, states may conduct their fishery surveys in different ways and for different purposes. Many types of state-federal cooperation in data collection and fisheries management can be found around the United States, including sole state involvement in nearshore species (e.g., sea trout and mullet), state management of federal fisheries in adjacent federal waters (e.g., spiny lobster in Florida), and federal management of fisheries that cross state-federal boundaries (e.g., shrimp in the Gulf of Mexico).

Interstate Commissions

Since fish stocks don't respect political boundaries, states need a mechanism to coordinate the collection of data and to manage stocks shared among states and among regional councils. This need became particularly apparent at the time of the decline of the striped bass stocks in the mid-Atlantic region in the 1980s, which led to the Interjurisdictional Fisheries Act of 1986. The three interstate commissions play a major role in coordinating management and use of data from individual states.

International Commissions

The United States participates in several binational and multinational commissions that manage fish stocks that span the national waters of two or more nations or extend into international waters. Examples of international commissions include the International Pacific Halibut Commission and Inter-American Tropical Tuna Commission in the Pacific region and the International Commission for the Conservation of Atlantic Tunas. Most commissions combine data collected from national fishery agencies with data they collect themselves. NMFS and the councils must obtain data from the commissions for fisheries under U.S. jurisdiction. Management may be joint with commissions, councils, and NMFS (e.g., for North Pacific halibut).

Fisheries Data Management Systems

As fisheries data have become more voluminous and complex, fisheries data management systems have been developed. State-federal collaborative data management efforts were started as early as the 1970s and 1980s, for example, the state-federal cooperative statistics program in the southeast and Gulf of Mexico and the Northeast Marine Fisheries Information System. Systems today include the Alaska Fisheries Information Network (AKFIN) and the Pacific Fisheries Information Network (PacFIN) for the Pacific Coast States, FIN in the Gulf of Mexico, and most recently, the Atlantic Cooperative Coastal Statistics Program (ACCSP). A nationwide umbrella Fisheries Information System (FIS) has been proposed.

PacFIN (Daspit et al., 1997) *and AKFIN*—Commercial data from fisheries in the ocean areas off Washington, Oregon, and California submitted by state fishery agencies are central to the PacFIN database. The states are primarily responsible for data quality control. PacFIN also includes limited Alaska groundfish information, which is supplied in aggregate form. The PSMFC sponsors AKFIN to provide the "framework needed to consolidate collection, processing, analysis and reporting of a variety of information essential to management of Alaska fisheries" (NMFS, undated). AKFIN system partners are coordinating the development of data element standards and coding systems in concert with other Pacific area fisheries data. Thus, AKFIN is adopting the PacFIN code sets with some modifications to accommodate unique Alaskan requirements. Currently AKFIN data reports are available through PacFIN.

PacFIN includes a publicly accessible Web site with summary reports and a restricted access site accessible by dial-up or telnet. The PacFIN restricted area is run on a system named ack1 (formerly ORCA). The initial setup requires new users to enter Unix commands to complete setup of the environment. This procedure might be difficult for new users with no Unix experience. Using the ack1 database requires some knowledge of Structured Query Language (SQL).[12]

[12] Structured Query Language (SQL) provides basic language constructs for querying and processing data in a relational database. SQL is a standard of both the International Organization for Standardization and the American National Standards Institute. Over time, SQL has evolved to provide additional facilities such as those for schema manipulation and data administration and language for the definition and management of persistent, complex objects.

However, over 4,000 scripts are available to users on the system that can be used to generate a number of different reports. On-line help is available for both Unix and SQL. Additionally, staff members offer support to those who need help mastering ack1's command line interface.

Ack1 includes tables that contain summary data; therefore, users writing their own queries rather than using scripts must take care to ensure that they are not capturing both the detail and the summarized data in custom reports. Summary data were initially included to avoid dealing with the complex issue of confidentiality. As the database evolved, finer detail levels were included, but some data providers, such as the State of Alaska, provide only summary data for inclusion in PacFIN. The SGI server that supports PacFIN was replaced by a new Sun Microsystems server, ofis450a, on December 6, 1999. The primary applications on the new server will be Oracle 8i, SPSS, SPLUS, IMSL, and FORTRAN software libraries. Also included are linkages to applications such as Oracle/ArcInfo/Arcview and Oracle/SAS.

FIN—The Fisheries Information Network (FIN) is a state-federal cooperative program to collect, manage, and disseminate statistical data and information on the commercial and recreational fisheries of the southeast region, including the Gulf of Mexico. FIN encompasses two separate programs: the Commercial Fisheries Information Network (ComFIN) and the Recreational Fisheries Information Network (RecFIN-SE). FIN came into being as a result of a memorandum of understanding signed in 1996 that resulted from efforts by state and federal agencies to develop a cooperative program for the collection and management of commercial and recreational fisheries data in the southeast region. GSMFC (1996) indicates that the ComFIN and RecFIN-SE are comprehensive programs comprised of coordinated data collection activities, an integrated data management and retrieval system, and procedures for information dissemination, as outlined in the mission, goals, and objectives of its Framework Plan. Databases to be included are the MRFSS files, the NMFS trip interview program (TIP) files (which primarily hold biological data on catch from commercial fishing trips but also include length-frequency data from recreational trips), the NMFS headboat survey, and a variety of state and federal databases.

In his June 1999 presentation to the committee, David Donaldson of the Gulf States Marine Fisheries Commission told the committee that only Louisiana and Florida have put in place the trip ticket programs needed to implement ComFIN, and that trip ticket programs are still needed in Texas, Mississippi, and Alabama. The FIN Committee emphasizes in its framework document how communication with the Pacific and Atlantic coasts will also be established and maintained to coordinate with and benefit from their data management efforts and to ensure compatibility with a planned national commercial and recreational fisheries database system (GSMFC, 1996). The Framework Plan mentions the need to develop standard protocols and documentation, including quality assurance and quality control standards for data formats, data element definitions, input, editing storage, access, transfer, dissemination, and applications.

ACCSP—The Atlantic Coastal Cooperative Statistics Program is a "cooperative state-federal marine and coastal fisheries data collection program. It is intended to coordinate present and future marine and coastal data collection and data management activities through cooperative planning, innovative uses of statistical theory and design, and consolidation of appropriate data into a useful database system."[13] Planners believe that the program must serve not only the needs of fishery scientists and managers but also fishermen, fishing companies, and the public. The program is based on the assumptions that (1) the migratory nature and transboundary distribution of many Atlantic coastal stocks require coastwide

[13] http://www.safmc.nmfs.gov/ACCSPHM/accsp.html, accessed 10/20/99.

cooperation in management, and (2) good data and statistics are needed for effective management. Toward this end, ASSCP plans include elements related to (1) collection, management, and dissemination of statistical data and information for the conservation and management of fishery resources of the Atlantic coast in a cooperative manner, and (2) support of the continued development and operation of a national data collection and data management program.

Mike Cahall, Information Systems Program Manager for the ACCSP, indicates that the ACCSP is built "according to the best commercial practice" rather than emphasizing federal or other standards. However, many vendors incorporate Federal Information Processing Standards (FIPS) and other federal data standards in their products. If the ACCSP does not adhere to federal data standards, it should not serve as a model for a federally funded national data collection and data management program for marine fisheries. OMB Circular A-130 Section 8b(4) requires that "Federal agency management and technical frameworks for information resources should address agency strategies to move toward an open systems environment." Standardizing data formats is one aspect of creating an open-systems environment, rather than a proprietary, non-accessible, and non-standardized environment. The commercial approach has not precluded the use of some FIPS, such as the codes for state and county. The program has standardized on the emerging Integrated Taxonomic Information System (ITIS) to provide scientifically credible taxonomic information.[14]

The ACCSP is testing its system concept through its Prototype Data System, which began with multiple years of data from the NMFS Northeast Region and 1 year of data from the Florida Department of Environmental Protection. The program initially is focusing on fishery-dependent data, both commercial and recreational, but will eventually include fishery-independent data. Recreational catch and effort data were available beginning in December 1999. A proposed cornerstone of the program will be an ability to track individual vessels through time and space. This vessel information could be linked with biological, social, and economic information, and joined with catch and landings data to provide trip summary information for each fishing vessel. ACCSP will limit access to unaggregated data to specific "named users" who are members of the ACCSP partner organizations. "Unnamed users" from the general public and non-partner organizations will be limited to viewing highly aggregated, non-confidential data. Although no definite date is available, the ACCSP system is likely to be available to users in a preproduction mode between October 2000 and January 2001 (M. Cahall, NMFS, personal communication, 1999).

The ACCSP prototype architecture consists of three layers: the operational layer, the reconciled data layer, and the information layer. A data reconciliation component and a data derivation component are the bridges between the layers (ACCSP, 1997). The operational layer consists of the disparate systems maintained by the partners that provide data for the system. The data reconciliation component includes those processes that are necessary to transform the data from the source operational system into a consistent base in the reconciled data layer. The reconciled data layer is the standardized or common data repository. This layer is derived from the minimum critical data elements and associated standards required by the system. The data derivation component consists of processes required to derive the data from the reconciled data layer to the information layer. The information layer consists of the information repository and data

[14] ITIS is a taxonomic database for terrestrial and aquatic plants and animals developed through a partnership of federal agencies and systematists from government and the private sector. This database is a partnership of U.S., Canadian, and Mexican agencies, other organizations, and taxonomic specialists to develop an online, scientifically credible, list of biological names for the biota of North America. ITIS is also a member of Species 2000, an international project to index the world's known species.

manipulation tools that support end-user data access and analysis.

The conceptual architecture provides an effective representation of the planned system. According to the documentation, the components described above can be addressed in terms of views of the architecture, such as the data view, the tools view, the hardware and communications view, and the administrative and operational view. As presented in the otherwise comprehensive planning documentation, none of these views specify the federal, national, or international standards with which the system will or should comply. Although the program is a regional and not a federal system, providing information on standards would promote a better understanding of the system and aid those hoping to use it as a model, or to interface with it.

Fisheries Information System—In the Sustainable Fisheries Act of 1996, Congress requested that the Secretary of Commerce "develop recommendations for implementation of a standardized fishing vessel registration and information management system on a regional basis." (SFA, Sec. 401). The act further specified that the recommendations should "integrate information collection programs under existing fishery management plans into a non-duplicate information collection and management system" (Sec. 401[a][2]); that the recommendations should be implementable through cooperative agreements with state, regional, and tribal entities; and that it should "establish standardized units of measurement, nomenclature, and formats for the collection and submission of information" (Sec. 401[a][6]).

The NMFS response to this congressional mandate (NMFS, undated[15]) recommends that the Fisheries Information System (FIS) create a nationwide umbrella for ongoing regional activities such as ACCSP, FIN, and PacFIN. Furthermore, the Vessel Registration System (VRS) would be implemented as the Vessel Information System being developed by the U.S. Coast Guard that provides a system for national registration of commercial and charter vessels. Access to the data would be controlled to balance ease of access and confidentiality. Use of common codes or bridges among coding systems was recommended. The combined system would associate individual vessels with a record of their fishing activity and would cost an estimated $51.9 million, 80 percent of which would recur annually. This funding would be applied in three major areas: data collection ($43.9 million), information technology and architecture ($7.2 million), and institutional arrangements ($1.65 million). The recurring cost related to information technology and architecture includes networking and infrastructure costs (routine upgrades of hardware and software), data quality assurance and quality control, technology and electronic reporting, and database integration and harmonization (which includes archiving and integration of "a regional detail/central summary" VRS-FIS) in excess of existing activities.

NMFS emphasizes the "umbrella concept of building links among existing regional statistics systems," rather than creating an entirely new national system. In building the links, the FIS would reconcile data across the regional repositories into interregional and nationwide summary information. Instead of presenting a detailed architecture at this very early conceptual period, NMFS provided a very high-level summary of plans to indicate that "a nationwide view of summary-level data implemented in regional data 'warehouses' is the likely model for a nationwide FIS" (NMFS, undated). Rather than describe interfaces and specific standards that are not yet available in this conceptual level document, NMFS emphasizes that the FIS will provide a single, complete view of the data, providing consistency by eliminating regional data differences, and providing data to users in a consistent, understandable way. The FIS conceptual model presents the information flows, along with sources, systems, and repositories at the state, regional, and national levels.

[15] This report will be referred to as the VRS-FIS document.

AFMIS/LOOPS/AOSN—The Advanced Fishery Management Information System, a consortium of investigators from the University of Massachusetts, Harvard University, and a private corporation, has obtained funding from NASA for a research project to develop a prototype fisheries management information system that incorporates physical, biological, and fisheries data.[16] Fishery managers are responsible for setting regulations governing where and when fishing can occur and the types of gear that can be used to fish for different species. The goal of the AFMIS is to construct, validate, and demonstrate a prototype system that will apply state-of-the-art technology and expertise from multiple disciplines to help fishery managers improve the allocation of fishing effort in time and space. This system also will be used in attempts to understand how environmental factors affect the distribution of fish and influence recruitment of young fish. Major system features are the involvement of the fishing industry in data collection and research and the inclusion of a fishing vessel subsystem. The system uses satellite-borne sensors (sea-surface temperature and ocean color) and will synthesize and analyze previously unanalyzed, novel, very large data sets with modern statistical techniques. The system will cover the Atlantic shelf off the New England states.

The Littoral Ocean Observing and Prediction System (LOOPS) and Autonomous Oceanographic Sampling Network (AOSN) are funded by the Office of Naval Research and are coordinating with AFMIS. The LOOPS concept links models, observational networks, and data assimilation algorithms. LOOPS activities include (1) development of a modular structural concept for linking models and measurements through data assimilation,[17] using adaptive sampling; (2) observational system simulation experiments for the design of quantitative sampling strategies; and (3) sea trials to demonstrate the concepts of system integration and real-time implementation. The AOSN couples autonomous underwater vehicles with moorings and existing remote sensors monitor the ocean, using acoustic and other means.

[16] http://afmis.cmast.umassd.edu, accessed 3/3/00.

[17] Data assimilation is the use of actual measurements of model parameters to constrain the range of input parameters in a model.

Commercial and Cooperative Data Management

New sources of data management services are companies that specialize in collecting voluntary data from fishing fleets and the rapid processing and distribution of NMFS data for specific purposes. In some cases, individual vessels waive their confidentiality rights to provide data to and use data from other vessels in their fishery cooperative or their fleet. These systems emphasize direct fisherman contact with the commercial data manager and illustrate the kind of services NMFS could provide if it had the flexibility and resources to do so.

The committee was provided several examples of commercial data management from the North Pacific region (Dave Fraser, commercial fisherman, personal communication, 1999). The first example is SeaState,[18] which provides two primary services to the North Pacific catchers and processors: (1) real-time monitoring of catch and bycatch quotas and (2) provision of bycatch hotspot maps based on GIS (Figure 3-4). SeaState obtains its data directly from observers or indirectly through NMFS. These data are used by the general fleet, members of fishery cooperatives, and vessels engaged in the community development quota fisheries. For the latter fisheries, access to data is available through password-protected Web sites.

The Alaska Groundfish Data Bank (AGDB) provides a variety of services to fishing fleets operating in the Gulf of Alaska, ranging from analyzing and disseminating NMFS data to coor-

[18] The committee and the NRC do not necessarily endorse the companies mentioned in this section.

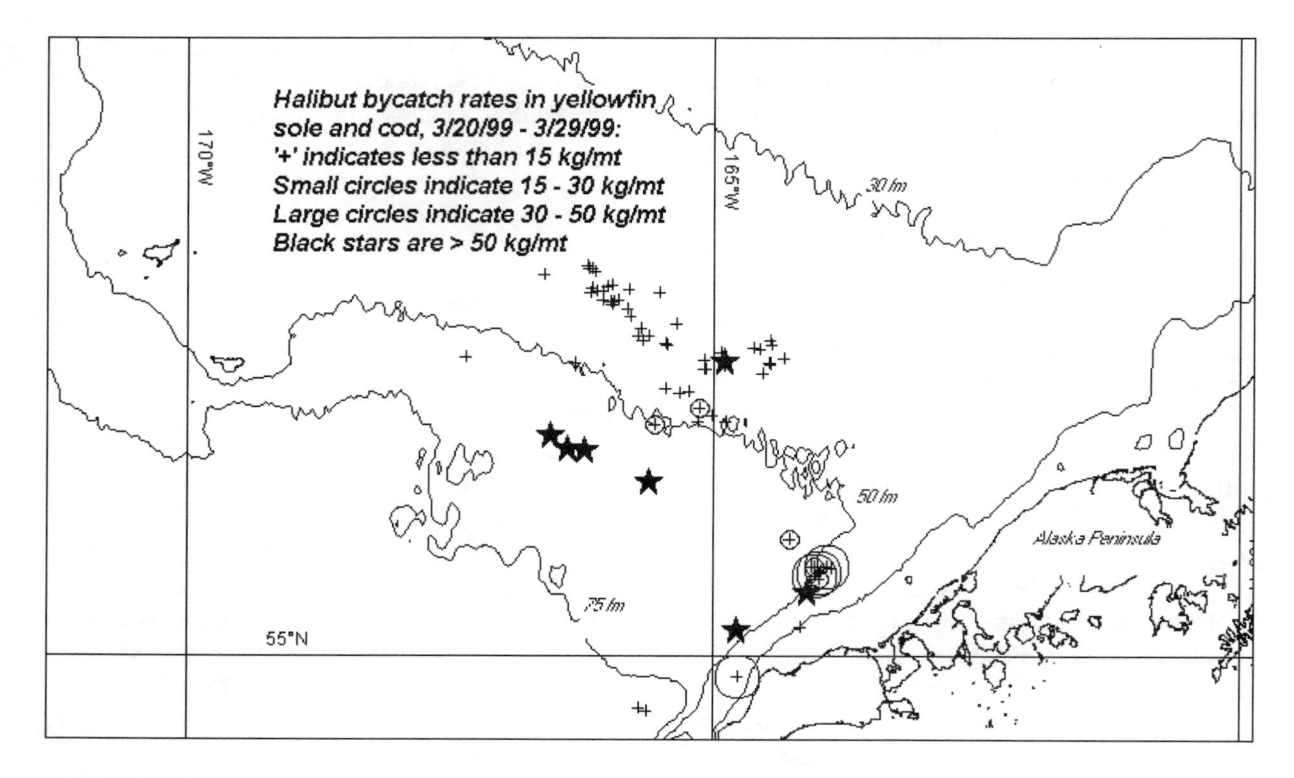

FIGURE 3-4 Example of SeaState product for halibut bycatch in southern Bering Sea yellowfin sole and cod fisheries. Used with permission from SeaState, Inc.

dinating the flow of information from harvesters and processors to NMFS. The AGDB works mostly with smaller catcher vessels (less than 125 feet in length) and monitors catches delivered to processors in Kodiak, Alaska. It also works with NMFS to set closure dates and with the fleet to advise fishery managers about trip limits. The AGDB serves as a conduit of information between fishery biologists and the fleet.

A final example from the North Pacific region is Fisheries Information Services, which provides bycatch hotspot reports, based on observer data, to the freezer longliner fleet. Fisheries Information Services is also involved in a logbook program for the Alaskan sablefish individual fishing quota fishery. Other companies provide software to help fishermen manage their own data and remote sensing products to target their fishing activities.

Data Quality Control Procedures

The slogan appearing on the cover of the ACCSP outreach brochure is "Good Data, Good Decisions For Fisheries Management." This slogan epitomizes the need for good data quality for fisheries management and emphasizes that uncertainties in fisheries data—and the assessments for which they are used—make it difficult to balance the needs of fishermen with the conservation requirements of the resource (ACCSP, undated a).

The "Glossary of Quality Assurance Terms"[19] defines data quality indicators as "quantitative statistics and qualitative descriptors that are used to

[19] Developed by the Quality Assurance Management Staff (QAMS) in the Office of Modeling, Monitoring Systems, and Quality Assurance in the EPA's Office of Research and Development.

BOX 3-5
Data Exchange Standards

CORBA - Common Object Request Broker Architecture is a standard from the Object Management Group (OMG) for communicating between distributed objects. CORBA provides a way to execute programs written in any language without regard to network location or platform.

HTML - HyperText Markup Language is a document format used on the World Wide Web. HTML is a document type of Standard Generalized Markup Language (SGML) that uses fixed tags to describe pages. SGML is a standard of the International Organization for Standardization (ISO) for defining the format in a document. More information on SGML can be found at: http://www.oasis-open.org/cover/sgml-xml.html.

XML - Extensible Markup Language is a subset of SGML used on the World Wide Web that provides flexibility by defining the codes used in each document. More information on XML can be found at: http://www.oasis-open.org/cover/sgml-xml.html.

HTTP - HyperText Transport Protocol is the communications protocol used to connect to servers on the World Wide Web.

TCP/IP - The Transmission Control Protocol/Internet Protocol is the protocol suite used by the Internet. TCP/IP handles network communications between network nodes.

Z39.50 is ANSI Z39.50 (also ISO 10161/62) which is a standard for information retrieval that provides a standard means for a search application to submit a query to databases without regard to the kind of hardware or software the database uses.

interpret the degree of acceptability or utility of data to the user." The principal data quality indicators are bias, precision, accuracy, comparability, completeness, and representativeness (EPA, undated). A number of data management systems at regional and national levels have grappled with the issues surrounding maintenance of sufficient data quality to allow effective management.

Standards for Exchanging Data

Although some consider the World Wide Web to be the answer to all data exchange concerns, an examination of existing fisheries data management systems highlights the continuing need to address standards for exchanging data and information. Internet computing is based on universal standards, but these standards still fall short in certain areas. Data held in disparate distributed databases must be integrated to facilitate single-search access and retrieval. Concerns linger regarding security, privacy, recall, and precision of retrieval methods. Groups such as the Open GIS Consortium are just beginning to address access to Web-based geospatial information. Important data exchange standards are defined in Box 3-5.

In its VRS-FIS document, NMFS (undated) recommends establishment of a temporary liaison office in OMB to facilitate reviews and approvals of collection of the additional federally sponsored information needed to implement the VRS and FIS. The VRS-FIS is planned as a partnership of states, interstate marine fisheries commissions, and federal agencies, including NMFS. The latter intends the FIS to complement existing data collection and management planning efforts by "providing a common thread among programs to take advantage of opportunities in technology, economies of scale, and efficiencies in re-use of survey and information management experiences, and to develop a context for assessing how to pay for these activities." (NMFS, undated).

Legal Requirements—The Paperwork Reduction Act and other federal laws, regulations, policies, and guidance are important considerations in choosing data standards. For example, Federal Information Processing Standards (FIPS),[20] Records Act provisions, the Clinger-Cohen Act, General Services Administration (GSA) regulations, and OMB Circular A-130 and other OMB guidance and memorandums may prove helpful to developers of fisheries data management systems. The regional information systems discussed earlier that are expected to participate in the FIS addressed the Privacy Act provisions and the Magnuson-Stevens Act confidentiality regulations codified at 50 CFR part 600. It is not clear how the FIS plans to provide information technology access for persons with disabilities as required by 41 CFR 201-20,103-7. In addition, the U.S. Access Board, the federal agency set up to implement the Americans with Disabilities Act, has recently completed a set of standards for access to government Web sites.

OMB Circular A-130 Section 8b(4) instructs agencies to use strategies that consist of "one or more profiles (an internally consistent set of standards), based on the current version of the NIST's (National Institute of Standards and Technology's) Application Portability Profile. These profiles should satisfy user requirements, accommodate officially recognized or de facto standards, and promote interoperability, application portability, and scalability by choosing interfaces that are broadly accepted in the marketplace to allow for as many suppliers as possible over the long term."

Software Compatibility—Daspit et al. (1997) stated that in May 1992, NMFS announced that all of its computing resources would be replaced by the UNIX operating system and the Oracle relational database management system. An example of an implementation of this standard is PacFIN's use of Silicon Graphics workstations running IRIX, a version of the UNIX operating system, along with Oracle 7 Server Release 7.3.3.5.0. The ACCSP currently under development complies with the above organizational standard by including Oracle as the backend with Businessobjects Webintelligence software providing the front-end user interface and query capabilities (M. Cahall, NMFS, personal communication, 1999).

Development of organizational standards (e.g., the core set of data elements, standardized data quality assurance and quality control procedures, coding standards, and metadata) and earlier standardization on the proprietary Oracle relational database management system facilitate data interchange within NMFS. However, caution is in order when standardizing on proprietary systems rather than on open standards that facilitate interchangeability among products. For example, Oracle 7.x SQL conforms to the first or entry level of SQL-92 rather than the intermediate or full level. Entry level SQL-92 is similar to SQL-89 to which Oracle has added enhancements or extensions (Harrison, 1997). These noncompliant enhancements make migration to a competing vendor's SQL-compliant database management system more difficult. When a proprietary product is used, the organization often becomes locked into the single vendor's products. The vendor is then under little pressure to reduce cost or improve and differentiate the product. The single vendor approach also limits systems design flexibility because the user of the proprietary product must

[20] The FIPS Program was established in the 1960s to standardize federal computer usage for federal agencies and organizations.

wait until the vendor is willing to provide the new products that the customer needs. The latest version of Oracle (8i) includes both proprietary SQL features and somewhat greater open systems support than in previous versions. The open systems approach achieves interoperability by defining interfaces, services, protocols, and data formats favoring the use of non-proprietary specifications (OMB Circular A-130). Whether Oracle represents the best commercial practice of today, users who may someday wish to change vendors must deal with migration to a new system that will not support Oracle's proprietary SQL enhancements. Such migration issues are best handled by advance planning.

Compliance with Interagency Data Standards—NMFS summarized for the committee the status of its standards of compliance by saying that "due to historical regional and programmatic autonomy, NMFS and its partners do not have a single, national, integrated system or standard for data collection." At the same time, NMFS acknowledges in the FVR-FIS document the need for coordination in the design of data collection forms, quality assurance and quality control, coding standards, and metadata. NMFS is participating in the NOAA Biodata Working Group to ensure that all NOAA staff members document data sets, make them available, and ensure that the data are available for use in the future according to the National Archives and Records Administration's requirements for archiving data (G. Barton, NOAA, personal communication, 1999).

To implement a national FIS, NMFS will provide the extensive coordination necessary to define and implement organizational standards. Since the FIS is at the conceptual stage of development, NMFS has an opportunity now to expand its current plans for developing national standards for certain data elements or coding systems to include development of an architectural framework for interoperability among fisheries data management systems. In fact, the Clinger-Cohen Act of 1996 instructs federal agency chief information officers to take responsibility for developing, maintaining, and facilitating the implementation of a sound and integrated information technology architecture (ITA). An ITA is an integrated framework for evolving and maintaining existing and new information technology to achieve an agency's strategic goals and information management goals.

A great deal of effort must still be expended to complete the critical task of translating incompatible data formats now used by the regional systems to allow the level of interoperability necessary for the proposed FIS umbrella system. The ACCSP's use of standardized nomenclature provided by the Integrated Taxonomic Information System (ITIS) is an example of using an emerging standard to help ensure successful biological data discovery and retrieval.

Documentation of how data were collected and analyzed (*metadata*) provides a way to understand data sets. Formal metadata standards employ a controlled or common set of terms to use when describing data. Metadata standards of potential importance to fisheries include the Federal Geophysical Data Committee (FGDC) standard implemented by the National Geospatial Data Clearinghouse and the National Biological Information Infrastructure (NBII) biological profile of the FGDC's content standards for digital geospatial metadata.

The NBII biological profile includes fields for analytical tools and methodologies. It also includes a supplemental information data element in the FGDC format for input of additional relevant information. Inclusion of information supporting the robustness of the data, such as the following, could be made available through supplying metadata along with fisheries data sets to promote understanding of the data and how it should be used: methodologies used; how much information has been directly measured as opposed to inferred, extrapolated, or produced by models; and level of uncertainty (do all included parameters hold true?), assumptions made, and uncontrolled variables. If estimates of uncertainty are incorporated directly into the assessment, this should be explained in the metadata.

The biological profile also includes a taxonomy data element that would be of particular use to biological data sets, such as fisheries data. More complete metadata would also facilitate peer review of the data and results and foster better quality control.

The U.S. Fish and Wildlife Service has developed a process for identifying, defining, monitoring, and promoting data standards to ensure the compatibility of data management and usage throughout the agency. The standards and the process for their development are described at http://www.fws.gov/stand/.

Information Management Architecture—Although the documents that describe existing and planned regional fisheries data management systems address many issues of great importance to information sharing, they need to expand on their description and justification of information management architecture. For the most part, the planning documents for fisheries data management provide comprehensive discussions of information system organization and policy and emphasize data collection and data element standards, but could benefit greatly from more detailed computational information, particularly on the topic of compliance with open-system standards. For example, the Reference Model of Open Distributed Processing (RM-ODP) ISO/IEC 10746 models an architectural hierarchy of viewpoints that is being used by the U.S. Geological Survey, NASA, NOAA, and other participants in the interagency Digital Earth Project to serve as a guide for project participants. The technology viewpoint is defined as the specific collection of technology products implemented in the system. The Digital Earth Reference Model provides a listing of interoperability standards by infrastructure category. Emphasis is placed on identifying relevant federal, national, and international standards rather than on defining proprietary products or organizational standard products that must be used by participants. As a final step, the organization identifies the specific technologies that can be used to meet the standards. For example, the FIS designers could decide at this time that they require only SQL-92 level-1 compliance, so that Oracle would be a viable technology choice to fit the fisheries information system architecture. Vendors other than Oracle, however, may meet the required standard. In fact, in its VRS-FIS document, NMFS (undated) indicates that a process will be designed to identify and evaluate candidate technologies according to specific criteria.

The Raines Rules (OMB Memorandum 97-02) indicate that agencies must report how well they developed information architectures and evaluated prototypes. Executive Order 12906, issued in 1994, established in the Executive Branch of the federal government a National Spatial Data Infrastructure (NSDI) and a National Geospatial Data Clearinghouse. This executive order directed the FGDC to develop standards for implementing the NSDI and directed individual agencies to use such standards or require their use by entities from which they obtain data. The executive order also directed the FGDC to submit a plan for implementing a national digital geospatial data framework.

Earlier in this section, the statistical aspects of data quality were discussed. The following section describes the approaches to quality control used by planned and existing fishery data management systems.

PacFIN

The content of each PacFIN data file is the responsibility of the agency that provides the data; thus, the current PacFIN system does not include comprehensive validation routines, though some data validation routines are in place. An example of a current PacFIN validation routine is the standard duplicate check. If a transaction is a duplicate or includes out-of-range values, it is flagged as an error and rejected (B. Stenberg, PSMFC, personal communication, 1999).

Sampson and Crone (1997) documented data collection procedures for U.S. Pacific Coast groundfish. Some data are subjected to rigorous quality control before submission to PacFIN. The groundfish trawl logbooks used in Washington, Oregon, and California provide examples of the data quality procedures followed before transmission to and entry into PacFIN (Tagart, 1997). The trawl logbooks in Washington State typically are collected by the port sampler each time a fisherman completes a trip. The port sampler records the logbook data to disk with the aid of custom software that conducts cursory error checking. Coded logbook data are then sent to the state Marine Resources Division, where the data are stored and transmitted to a data specialist, who ensures that individual files are filtered through a comprehensive error-checking program, after which they are considered to be processed data. An error-screening program checks raw logbook data for out-of-range errors in Loran or GPS coordinates, depth, fishing block, species, port, and trip type. It also screens for such errors as tow information entered in the wrong column and missing data. Records with errors are flagged in the database and a separate file is generated that describes the type of error and records the data line in the raw trawl data file in which the error occurred. The data specialist then rectifies the error by reviewing the raw data or returns the coded data to the port sampler for clarification. This procedure results in more than 95 percent of the logbooks being free of coding errors after two passes through the error-screening program (Sampson and Crone, 1997). Processed data are then aggregated into a single file and further processed into tow-expanded logbook data to account for tows that were not keypunched.[21] Tow-expanded data are next processed with fish ticket data to generate expanded trawl logbook data.

Different error-checking protocols are implemented in Oregon and California, with no single standard. Oregon accepts incomplete logbooks, but codes them with a number indicating their incomplete status. According to Sampson and Crone (1997), the Oregon local port biologist "evaluates every logbook for completeness and consistency. The process includes checking the logbook for incorrect temporal sequencing of the tows or inappropriate dates or times, and filling in the following items: (1) the ticket number(s) corresponding to each trip, (2) missing depths based on the tow location and the depths indicated on the nautical charts, and (3) missing target species based on the most prevalent species hailed. The port biologist assigns each logbook a code of 1, 2, or 3, depending on its degree of completeness." Sampson and Crone report that the port biologist will attempt to obtain missing information by interviewing the captain (the logbook has only partial information on the tow location or hail weights).[22] If the captain does not provide the missing information, the port biologist will assign the logbook a code 2 if only hail weights are missing, or a code 3 if tow locations are missing. Logbooks assigned a code 3 are excluded from further processing (Sampson and Crone, 1997).

ACCSP

The ACCSP database planning document notes that data should be checked for accuracy and consistency before being submitted to the coastwide database. The data form review re-

---

[21] According to Sampson and Crone (1997), Washington State did not record tow-by-tow logbook data until 1985. Originally, they keypunched every fourth tow from each logged trip, in effect subsampling tow-by-tow data. Thus, their system accommodates subsampling and provides for data expansion for subsampled trips, for example, the tows that were not keypunched.

[22] The hail weight is an estimate by fishermen of either tow-by-tow landings (entered into the logbook), or of the total weight the fishermen tells the fish buyer he has on the boat as he returns from a fishing trip. When the fish are subsequently weighed, this "landed weight" is usually used to adjust the estimated weight for each tow in the logbook. This is why it is so critical to link fish tickets (weight at the dock) with logbooks (estimated catch by depth and location).

quires that records should be checked for at least the following items:

- legibility
- completion of all necessary fields
- reasonableness of dates and times
- accuracy of species and gear combinations

The ACCSP recommends that:

• incorrect or inconsistent entries follow data protocols that include contacting the data provider. (Problem data providers who consistently make errors may be given additional training or legal action may be considered [e.g., fines, license revocation].)

• reviewed data forms be entered into the database by adequately trained data entry clerks, with an error rate of less than 0.5 percent for the set of all data entered, through use of a double entry system (each data point is entered twice and not accepted unless both entries are identical).

• someone other than a data entry clerk should perform a spot check for errors on 5-10 percent of a year's entries.

• the following standard computer data edit checks, at a minimum, are run:
—species ranges, lengths, and weights
—dates of catch
—fisherman and dealer licenses
—fishing gear used
—invalid codes
—outliers
—blank fields
—comparisons with tracking database

The edit checks flag errors and probable errors, alerting the data entry clerk and permitting changes before the data reach the database (ACCSP, undated b). The program design document also mentions that unannounced audits of dealers' and fishermen's records may be used as a data verification tool (ACCSP, undated b).

Version 1.5 of the application also advocates that summary reports, similar to monthly bank statements, be sent to fishermen on a periodic basis for data verification. Benefits of such reports include (1) allowing fishermen and dealers to see the data after they have been entered, increasing confidence that their data are being used; (2) giving fishermen an opportunity to correct erroneous data, thus improving accuracy; and (3) providing fishermen with an official record of what they have caught and their revenues.

All the data quality routines discussed here supplement those implemented by the data sources, since the ACCSP agreement vests responsibility for the quality and completeness of the archived records with the agencies that originally collect the information (ACCSP, 1999). ACCSP requests that the states implement standard operating procedures (SOP) and develop SOP manuals, and that members sign agreements to use a specified standard of data collection elements and reporting formats. Data will be collected by individual fishing trips, including a standardized collection of elements such as species, area fished, gear type, quantity and value of catch, and vessel identification number. Members will also use standardized units of measure, coding systems, and nomenclature whenever possible.

The Georgia program that will provide data to ACCSP is illustrative. The draft SOP manual notes that the Georgia Coastal Resources Division's statistics project is under the Commercial Fisheries Program in the Marine Fisheries Section. Historically, the project has been funded by the NMFS Cooperative Statistics Program. In 1999 the project received additional funding through the Atlantic Coast Fisheries Cooperative Management Act (ACFCMA) to implement a commercial fisheries trip ticket program that would comply with the ACCSP (Anonymous, 1999).

Commercial landings data are collected with each fishery's self-coded trip tickets, which contain pre-labeled columns for the market grade and condition of the predominant species. In some fisheries, the gear quantity and area fished are also labeled. Seafood dealers receive the trip ticket forms, postage-paid envelopes, and a plastic card imprinter engraved with their business name and dealer code. Seafood harvesters are provided with a personalized plastic card embossed with their name and commercial fishing license number. Port agents code the remaining fields using four-digit species codes and three-digit gear codes provided by NMFS and 6-digit area codes supplied by the Georgia Department of Natural Resources.

After all data have been reviewed for completeness, they are entered by the port agents and a data clerk into a customized database implemented with the PROGRESS relational database management system. The data are edited at three different points. First, while coding and conversion take place port agents identify illegible fields and odd species and gear combinations, and track down missing data. Second, as data are entered, the data entry software checks for invalid codes, gear combinations, and price ranges. Final data editing is accomplished by running edit programs to check for outliers and by spot checking data forms against the data set. All data are entered within 10 days of receipt by the statistics program (Anonymous, 1999).

According to the Georgia SOP, the PROGRESS database does not contain the fields that were added to bring Georgia into conformance with ACCSP data standards. A new database is under development in Oracle. The Oracle application is currently being tested.

Sometime later, scanning of the trip ticket form will begin and images of the form itself will be created and archived on read-only CDs and stored in a safe deposit box. This will eliminate the need to store paper documents and still meet Georgia's requirements for archived records.

After completed landings data are run through a PROGRESS program that converts all fields to the SEFHost format in ASCII code, data are then transferred to NMFS monthly by email (Anonymous, 1999).

Fisheries Information System

The VRS-FIS report to Congress (NMFS, undated) presents the following design principles for standards of measurement and quality for a future FIS:

- Establish standardized units of measurement and nomenclature, where possible.
- Establish standard coding systems, where possible, or build logical bridges or translations between separate coding systems, where necessary.
- Establish reasonable minimum data quality standards.
- Establish standard (minimum critical) data elements.
- Minimize number of coding systems.
- Develop processes to ensure the timely release of information to the public.

The FIS design principles indicate that the shortcomings of the existing fisheries data management systems are well known. The need for standardized data elements to allow comparability among systems is apparent. The proposed FIS includes funding of $1.575 million to establish and implement criteria and processes for evaluation of data quality and data quality standards. These funds would be used to:

- research and adopt nationwide data quality standards, with help from individuals from universities, other federal agencies, and private research contractors familiar with large-scale data quality issues.
- establish nationwide data quality control groups to provide continuous oversight and peer review of both data collection and data quality processes.
- research, design, and implement validation methods for self-reported statistical systems (e.g., logbooks) to measure and document the biases and accuracy of such data.
- create online metadata files containing system statistical information to improve avail-

ability of documentation on quality of the information.

Technologies for Data Management

Fisheries data collection, analysis, use, and archival storage present a challenge to the organizations responsible for providing accurate, timely, and easily accessible fisheries data. It is evident not only from existing systems, but also systems in various stages of planning and development, that the existing fisheries data management systems are heterogeneous, are often incompatible and/or duplicative, are developed by different organizational entities, have a regional focus (so far), do not necessarily share data elements, and may or may not comply with federal data management standards.

NMFS acknowledges that "despite some regional successes, it is clear that the current overall approach to collecting and managing fisheries data needs to be re-thought, revised and reworked. The quality and completeness of fishery data are often inadequate. Data are often not accessible in an appropriate form or a timely manner. . . ." (NMFS, undated).

Fisheries Data Integration

Since existing fisheries data management systems are heterogeneous, integration is of primary importance to enable data synthesis on regional and national scales. The lack of standardized data elements necessitates implementation of "translators" to allow incompatible data elements residing in different databases to be accessed through a single interface. The ACCSP is an example of a regional fisheries data management system that initially will employ translators to reconcile incompatible data elements now collected by participating states (ACCSP, undated b), including both commercial and recreational data. The FIS will take the process one step further by integrating ongoing regional fisheries data management activities (e.g., FIN and PacFIN) under a nationwide umbrella (NMFS, undated). The proposed system recognizes the need for implementing national standards for a core set of data elements, data quality protocols, coding standards, and metadata. The importance of fully defining and implementing these standards cannot be overemphasized. NMFS' current FIS plans could be enhanced by development of a framework that provides more detailed computational information with particular attention to compliance with open-system standards.

Another aspect of data integration is use of a format that allows full exploitation of the data asset, for example, verification of different data sources used in management and research. This could be as simple as matching logbook data against trip tickets. The existence of a unique identifier for each fishing vessel as planned in the FVR system—together with the date and time of each trip—will facilitate data verification by providing positive identification of vessels and trips that can be used across databases. Such a feature should enhance the confidence of managers in data from different sources.

Historically, fisheries data have focused on individual species and their population dynamics. In recent years, more attention has been given to the predatory and competitive relationships among species, as well as finding commonalties across species (e.g., meta-analysis, Myers et al., 1995). Research on biological interactions and other ecosystems research would benefit if data were collected in compatible formats and were integrated in ways that facilitate study of the complex interrelationships in marine ecosystems. This could be a useful adjunct to traditional (and still necessary) stomach content analysis.

Geographic Information Systems—Geographic information systems (GISs) allow spatial data from many sources, such as sampling tows, to be referenced to a single grid of spatial coordinates. The use of GIS techniques offers promise for combining data from many sources into a single spatial grid, with new possibilities for understanding how various factors affect fish populations at various spatial scales.

GIS applications and visualizations show promise as outreach tools to improve communication of the results of fisheries science and management activities to the general public. They could also promote better understanding among fishery scientists. Because these applications already exist, making them more widely available and easily used could benefit all stakeholders in fisheries management and promote cooperation among data providers and other stakeholders. The Open GIS Consortium states that "GIS information is available on the Web—in stovepipes.[23] Users must possess considerable expertise and special GIS software to overlay or otherwise combine different map layers. . ." (Open GIS Consortium, 1999). This limits the ability of scientists, managers, fishermen, and others to use such data.

The Open GIS Consortium is currently conducting a Web Mapping Testbed (WMT) project. This project is an "accelerated, multi-phase effort to meet the market's demand for interoperable geo-enabled Web technology. The project is advancing the state of Web technology to support diverse applications that access distributed geospatial information sources across the Web. Applications include environmental analysis and management." (Open GIS Consortium, 1999). Sponsors of WMT include the FGDC, United States Geological Survey (USGS), NASA, the U.S. Department of Agriculture (USDA), a group of 24 Australian government and commercial organizations, Pennsylvania State University, 23 vendor companies (including Microsoft, ESRI, Sun, Oracle, Cubewerx [Canada] and others [Anonymous, 1999; Open GIS Consortium, 1999]). Participants in the WMT are combining their expertise to make it possible for overlays and combinations of complex and essentially different kinds of GIS information to happen automatically over the Internet, despite differences in the underlying software (Open GIS Consortium, 1999). NMFS and other fishery organizations should consider responding to the Open GIS Consortium's call for participants in subsequent project phases.

[23] Stovepipes are systems that stand alone and do not inter-operate.

Visualizations—Most fisheries data are presented in tabular format in reports, papers, and Web pages. In some cases, plots on two-dimensional maps represent sampling results. Many opportunities exist to provide better visualizations of fisheries data to promote better understanding by scientists, fishermen, and the general public. These could include 3-D virtual underwater worlds highlighting bottom topography, currents, concentrations of fish, and other features. Examples using Virtual Reality Modeling Language (VRML) technology are available at the Web site of NOAA's Pacific Marine Environmental Laboratory (http://www.pmel.noaa.gov/home/visualization/visual.html). The Fisheries Oceanography Coordinated Investigations Web site (http://pmel.noaa.gov/foci/visualizations/visual.html) includes a virtual reality world, as well as other visualizations. The Pacific Fisheries Environment Laboratory (PFEL) Web site at www.pfeg.noaa.gov features a live access server that allows visitors to visualize and download selected PFEL data products. Visualizations could be used as a tool in fisheries simulations for both teaching and consensus-building purposes.

DATA USE

Data use is discussed only briefly here because it was the focus of the NRC's *Improving Fish Stock Assessments* (NRC, 1998a).

Uncertainties of Data in Stock Assessments

A good example of the potential complexity of data sources used in any fishery is the summer flounder fishery. All the forms of data discussed earlier in this chapter are available for this fishery (Table 3-10). Stock assessments are subject to uncertainties of various types, ranging from uncertainties in observations to implementation of management.

TABLE 3-10 Summary of Data Sources and Source Attributes for Summer Flounder

	Types of Data					Attributes of Sources	
	Spatial Information	Species Composition	Size Composition	Age, Sex Composition	Price	Confidential	Accessible[a]
Federal							
Mandatory logbooks	2	2	0	0	0	2	1
Scientific observers	2	2	1	1	0	2	1
Scientific survey	2	2	2	2	0	1	2
Recreational survey	1	1	1	0	0	2	2
Port sampling	1	2	2	2	1	0	2
VMS	2	0	0	0	0	2	0
States (see Appendix C for listing of states)							
Landings	1	2	0	0	2	0	1
Surveys	2	1	1	1	0	0	2

Key: 0 = none; 1 = some; 2 = complete.

[a] Accessible refers to the extent that data are shared between those with the responsibility to collect the data and those who may have use for the data (e.g., scientists, managers).

One level is the degree of uncertainty reflected in an observation, typically referred to as the observation error. This is the variation that would be seen under repeated sampling. The observation might be survey CPUE for a given year, and so would be a statistic summarizing a number of points gathered under a prespecified sampling design. A second level of uncertainty in information comes in through the natural variation that occurs in the environment, so that even if every individual in the population is measured, variation in size and abundance is still expected from one year to the next due to the natural variation in the individual growth and population dynamics. This variation is referred to as *process uncertainty* and does not represent our ability to measure but rather represents expected natural variation in the process.

The third level of uncertainty is reflected in how the system is characterized, which is typically done through specification of a model. The choice of model interacts with the previous two levels of uncertainty and reflects to a degree a choice between bias and variance in the estimation process. Models may characterize the system simply or with more complexity, but they may also mischaracterize the system through selection of a model that does not represent actual processes well (model misspecification). The uncertainty in model specification can be estimated only by challenging the data with different models and/or additional kinds of data, as was done through the committee's analysis of summer flounder data. Added to this is the fact that only a single realization of the data is available, a single series through time, and thus the overall uncertainty cannot be assessed with repeated time series. It is sometimes possible to get around this problem by using meta-analysis across similar systems or through adaptive management that allows informative exploration of the system. Finally, implementation uncertainty is an expression of "the ability of management to achieve a particular harvest rate in any one year" (Rosenberg and Brault, 1993) and institutional uncertainty arises from "the interaction of the individuals and groups (scientists, economists, fishermen, etc.) that compose the management process" (O'Boyle, 1993). In the end, the output from our assessments poorly represents the levels of uncertainty that exist, impeding an adequate assessment of risk in decisionmaking and informed development of research priorities.

Access to Data

Users of fisheries data management systems include fishery managers, scientists, fishermen, industry groups, and other interested parties. Because fishery databases contain sensitive business information, most restrict access and usage to those with a need to know. The specific approach chosen for authorization and access controls varies among existing fishery databases. PacFIN, for example, uses logins and a list of authorized or allowed Internet Protocol addresses. There is currently no interactive Web access to PacFIN. PacFIN and several other fishery databases provide summary reports on the World Wide Web for public access. The separation of general public access and restricted access to the data may impact the number of visitors recorded at fishery data Web sites.

Aside from the traditional sources of fisheries data, opportunities exist for overlaying data from other studies. GISs could help in accomplishing such overlays. For example, the U.S. Global Change Research Program is supporting field studies of the dynamics of fish and plankton populations and of the causes of variations of marine biological populations in the Global Ocean Ecosystems Dynamics (GLOBEC) and other activities (*Our Changing Planet*, fiscal year 1999). The California Cooperative Oceanic Fisheries Investigation (CalCOFI) has compiled a long time series of data related to fisheries for the California Current System. In these cases, fisheries data management could either gain from or contribute to overlaying data from other studies.

Management Information Needed by Councils

The committee sent a list of questions regarding fishery data issues to the executive directors of each of the eight fishery management councils. Some councils responded by stating that they wanted annual assessments for each species rather than the staggered subset of assessments available each year. The staggered schedule is a result of inadequate funding for assessment personnel and ship time, but councils with fisheries in an overfished and rebuilding mode have said that they need annual assessment updates, so they can manage the rebuilding process more effectively. Information relevant to management priorities was requested. Several councils noted that NMFS is limited by the availability of assessment personnel and that NMFS cannot go beyond its minimal fulfillment of legal mandates.

The councils identified several specific information needs (beyond what they already receive), including the following:

- Baseline information to manage the new national standards established by the Sustainable Fisheries Act, particularly in relation to bycatch and effects on fishing communities. Although fishery management plans must contain social and economic impact analyses, NMFS and states are still not doing an adequate job of collecting data and providing them to the councils.
- Economic data, such as would be required to estimate consumer benefits, construct bioeconomic models, analyze vessel costs and returns, and describe fishing community structure.
- Enforcement data from NMFS, for example, how many landings are examined for rates of compliance and non-compliance with management measures such as trip limits.
- Data on the stock status of artisanal fisheries.
- Tagging data to study stock interactions and fish movements.

Fisheries Data Discovery

Users must discover fisheries data in order to use it. Most fishery databases have a public component with Web access to standardized reports. These sites have home pages indexed by the common search engines that allow discovery by the public, made possible by metatags.[24] However, not all fisheries Web sites are currently using metatags. As of June 1999, among the AKFIN, PacFIN, RecFIN, and ACCSP home pages, only RecFIN included metatags in the document source. The use of metatags should be promoted because they are used by many search engines when indexing pages and thus provide an aid to document discovery.

Fishery databases also have a private component, with access restricted to researchers, managers, and others with a need to know (see earlier section on confidentiality). This restriction is related to the inclusion of sensitive business information in the database and in some cases is mandated by state or federal law or both. In general, persons who desire access to restricted fisheries information must sign a non-disclosure agreement and fill out a database access request form that must be approved by a database official.

The NOAA server (http://www.noaa.gov) provides unified access to fishery data sets held by the NMFS Northwest Fisheries Science Center, with discovery through a query of FGDC metadata. The server system is now being redesigned so that access to planning documents will require a password. The existing system allows a user to select a query term, but the terms are very general. There is also a spatial query interface based on either latitude-longitude or a map. This appears to be a secondary access method with only minimal metadata included about NMFS fisheries data sets. It is often necessary to contact the individual listed in the metadata to gain access to the data. Landings information is available from the NOAA Web site for commercial (www.st.nmfs.gov/commercial/index.html) and recreational (www.st.nmfs.gov/recreational/index.html) fisheries.

Cooperation and Communication

The committee did not investigate all possible communication links, but it did query the regional fishery management councils about how they obtain the data they need for management decisions and whether they wanted information in a different form or wanted different information.

It is obvious that cooperation in data use is essential for effective management. Different regional councils accomplish cooperation differently. In some cases, councils receive data and information on a regular basis. Most councils rely greatly on members from NMFS, the Coast Guard, states, and other organizations to ensure that the necessary information is transferred to the councils and used in management.

One council expressed the desire to gain access to non-aggregated data and requested a more efficient means of data access than transfer of data disks. Another council noted the need for better ways to "analyze and reduce information so that it may be readily assimilated by council members and the public during the decision process." The councils believe that greater efforts and resources need to be devoted to improving communication of the reasons why data are collected in specific ways, for example, using outdated trawl gear and random, stratified sampling. It was also suggested that information communication specialists be enlisted to improve communication between councils and stakeholders. Most of the councils have extensive Web sites that include their fishery management plans, other reports, meeting schedules, committee rosters, and other information. These Web sites can be very efficient communication tools, if kept current.

---

[24] Metatags are information in World Wide Web documents that have a number of functions, including providing keywords and descriptions of the document that are accessed by search engines to categorize Web documents.

4

Findings and Recommendations

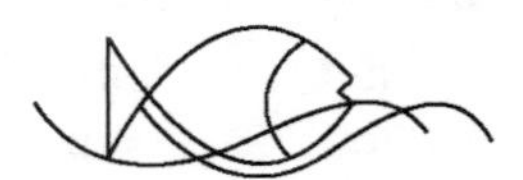

Significant advances have been made in the United States in recent years in the collection, management, and use of data related to marine fisheries. The National Marine Fisheries Service (NMFS) and the states have experimented with new technologies, such as vessel monitoring systems, electronic logbooks, acoustic fish detection, the Global Positioning System (GPS), electronic zebra code scanners, and modern tagging methods. Despite these advances, traditional fishery-independent surveys seem to be underfunded in some areas, commercial data are underused, and both may sometimes not be of high enough quality for their intended uses.

Congress has responded to the need for surveys by providing funds for new and more capable fishery research vessels, but budgets for using survey vessels are under continual pressure. Fisheries data collection, management, and use will suffer without provision of adequate, sustained funding from Congress that is applied appropriately by the Department of Commerce (DOC), the National Oceanic and Atmospheric Administration (NOAA), and NMFS. The committee did not analyze the fishery data collection budgets of the states or NMFS, so it cannot determine whether they need budget increases or could use their existing resources more efficiently. It does seem clear, however, that NMFS is being asked to provide more data to diverse groups for a variety of purposes.

Major advances in fisheries data management could be achieved by continuing the use of new computing and communication capabilities and increasing integration and standardization of data management on regional and national bases. NMFS and its state and regional partners on the U.S. Atlantic coast are implementing the Atlantic Cooperative Coastal Statistics Program (ACCSP), which seems to be a good model for regional data management systems because of its emphasis on data standardization and full information access by the program partners, including federal and state agencies, and participation by interstate commissions, commercial and recreational fishermen, and environmental advocates. At the direction of Congress, NMFS developed a plan for a nationwide Fisheries Information System, an umbrella data management system that will incorporate existing and planned regional systems and help them coordinate data standardization and access. Such new data management systems have become even more important as the most recent reauthorization of the Magnuson-Stevens Act added new requirements for identification of essential fish habitat.

The committee drew on input from its public meetings, published documents, and its own analyses and discussions to develop a series of findings and recommendations that (1) analyze the 1999 summer flounder stock assessment and provide advice to NMFS about ways to improve these assessments (see Appendix D) and (2) identify more general data issues needing attention. Many of the recommendations related to the summer flounder assessments should be applied more broadly in other U.S. fisheries; relevant examples will be provided later in this chapter. Improvements in the credibility of NMFS data collection and stock assessment procedures could be achieved if the following recommendations were implemented. Some of the major issues that emerged during the committee's work were the timing of data availability to managers and fishermen and the waste of the great potential of commercial and recreational fishermen as data collection partners.

Delay in the availability of data was a major theme of the committee's discussions. In particular, fishery-dependent data, especially commercial and recreational landings data, must be available on time scales that match needed management actions. The degree of timeliness needed varies depending on the type of data and the management system used. For example, for management based on total allowable catch, total catch data and data on discards in other fisheries should be available soon enough to allow closing a season early. Presently, most commercial data meet this criterion, but recreational catch data generally are not available on this time scale. Management with closed areas is best accomplished with real-time information about vessel location. The expansion of the recreational share of many fisheries has exacerbated the data timeliness problem because data collection systems for recreational fisheries are not designed to make data available in a usable form quickly, so that in-season[1] management of most recreational fisheries is presently not possible.

State and federal budget limitations will probably always constrain fishery-independent surveys. The portion of the NMFS budget for research and data collection has been diminished in recent years from budget cuts and "earmarks," without positive adjustments to base budget that would offset personnel costs and other costs increased by inflation. Commercial and recreational fishermen are a large potential source of data about the fish stocks they exploit. The committee believes, therefore, that it is imperative for NMFS and the councils to improve the quality of data available from commercial and recreational fisheries.

NMFS finds it difficult to use certain forms of commercial data, especially measures of fishing effort and CPUE, because of valid concerns about the data's usefulness; the committee believes these concerns could be better addressed through constructive engagement with the commercial sector. Improvements in data quality will occur only if NMFS and industry work cooperatively to create more effective and efficient data collection and management systems and create an environment that fosters the availability of accurate, precise data with adequate protection of privacy and confidentiality. Other stakeholders, such as environmental advocates and ceremonial and subsistence users should also participate in data collection and management so that their knowledge and interests are considered. Responsibility for the current failure to use certain kinds of fishery-dependent data (e.g., commercial catch rates and logbook data such as the landings by species and locations of catch) for stock assessments can be ascribed to both NMFS and the industry, and both need to make good-faith efforts to work together to improve data availability, thereby fostering improvements in management and sustained fisheries yields.

[1] In-season management refers to changes in the season length based on a sector's actual catch in relation to the total allowable catch allotted to that sector.

IMPROVING DATA COLLECTION

Matching Data Collection Costs to Benefits from Fisheries

Findings: In an ideal world, sufficient resources would always be available for data collection, but resources have never been adequate to meet this ideal and NMFS has not been able to obtain satisfactory increases in its data collection budget. The committee could find no existing analyses of the costs and benefits of data collection and management for specific fisheries, particularly of the ratio of marginal costs and marginal benefits for each additional dollar spent on data collection. The potential value of management advice (e.g., in terms of decreasing costs or increasing the long-term potential yield of fisheries) may not be related to the frequency and cost of surveys and other data collection and management systems (e.g., logbook programs, port sampling, observer programs). As a consequence, it is not clear whether allocation of data collection resources is based on objective analysis, or instead reflects regional tradition, bureaucratic inertia, or political incentives. Mismatches between costs and benefits may exist because social and economic factors are not accounted for properly. A comparison of data collection and management costs and the marginal benefit of such expenditures for fisheries management would allow NMFS to allocate its data collection effort more effectively within and across fisheries.

Data collection could be made more cost-effective through changes in the allocation of survey effort, collection of commercial and recreational catch and effort statistics, and optimization of regional monitoring through observer programs and vessel monitoring systems.

NOAA recognized the need for such a cost-benefit approach in its 1998 *Fisheries Data Acquisition Plan*:

> Because fiscal resources for fisheries management are not unlimited, a hierarchical system of priorities must be set to ensure that the most important data needs are met. First priority must go to endangered or threatened species. [examples given include marine mammals and overfished stocks]
>
> Another consideration when setting priorities is the value of the stock or species to the U.S. economy. However, the ecological importance of a species is also considered. Surveys may be conducted on a species of little economic value, but is an important forage fish for, or predator on other stocks, or is a bycatch species (NOAA, 1998).

The status of fisheries in the U.S. Exclusive Economic Zone as national public-trust resources suggests that a nationwide prioritization of expenditures would be desirable. Such an exercise would make costs and benefits more explicit and would give NMFS a more objective basis for changing data collection intensity or requesting additional resources. Implementation of a cost-benefit approach to data collection and management will be hindered if sufficient resources are not appropriated by Congress for necessary analysis and implementation.

Recommendations: NMFS should allocate its data collection resources by some rational plan between its own data collection efforts and the efforts of others. Data from non-NMFS sources will require special attention to such matters as data quality and coverage (by fishery, species, times of year, and location). Reallocation of data collection resources may not result in a net cost savings.

Congress should encourage NMFS to conduct a nationwide analysis of the costs and benefits of optimizing data collected for each fishery, including the value of fish stocks for commercial, recreational, and non-consumptive uses. Analyses should include appropriate multipliers to capture benefits of recreational and commercial fisheries to the broader economy (e.g., bait and tackle purchases, boat rentals, sales by fish dealers and retailers), as well as ecosystem benefits. For example, a table such as the following might be constructed:

Species or Species Complex	Potential Value of Harvest and Other Benefits	Spawning Stock Biomass	Long-Term Potential Yield	Importance of Species in Food Web/ Endangered Status	Precision Needed	Estimated Data Collection Cost (itemize by data source)

Such a table could be an extension of the information on recent average yield, current potential yield, and long-term potential yield given in the NMFS *Our Living Oceans* reports (e.g., NMFS, 1999). The primary intent of such an analysis would be to guide the federal investment in fisheries data collection and management. States could use a similar approach to evaluate the costs and benefits for individual fisheries at the state level, which may be only partially related to federal costs and benefits.

An interim measure could be to try the approach for one major fishery on each coast or within each council region. The key point is that benefits and costs of fisheries data collection and management need to be measured in common scales across fisheries, whether in dollars or in other scales that can be used to quantify non-economic costs and benefits. Quantifying environmental benefits of data collection may be difficult and proxy measures may be needed so that environmental benefits can be included in the total benefit of data collection. Examples of proxies might include the standing stock biomass and information about the importance of a species in the food web. Such an approach should consider the potential value of recovery of an overfished stock. This type of analysis would allow more informed decisions to be made about which fisheries merit increased funding for data collection and management.

As part of the recommended cost-benefit analysis, the precision of each data source should be determined, because it will not be cost-effective to measure one input (e.g., commercial or survey CPUE) with great precision, while only measuring another input approximately, because the precision of the output (e.g., estimated fishing mortality) and the management that is possible will often reflect the precision of the most imprecise input.

An analysis of the costs and benefits of data collection and management should also identify areas of research needed to make better cost-benefit decisions. The federal government may be subsidizing some fisheries by spending more on data collection and management than the fisheries are worth to the nation. When this is the case, the government could decide that such expenditures are not cost-effective. If this led to cutbacks in data collection and required more conservative management, industry should have the option to pay for data collection that might (or might not) allow higher TACs. Otherwise, where data collection and management costs exceed benefits to a specific fishery, less data-intensive management (reflecting perhaps more biologically conservative actions) should be considered by the regional fishery management councils. This review should be updated every 5 to 10 years to account for changes in the value of the fisheries, the development of more cost-effective techniques for data collection and management, and improved understanding of specific fisheries.

An important contribution to this cost-benefit analysis would be quantification of the costs and benefits of different data sources for different species. This information would allow determination of the appropriate mix of data types and sources, for example, sampling of landings, observers, vessel monitoring systems (VMSs), logbooks, and dealer reports. The appropriate mix may be different for each fishery, but VMSs may be particularly useful in fisheries managed with closed areas, observers are useful for all fisheries in which bycatch is a problem, and logbooks are useful in all areas.

The committee's recommendations assume that funding will constrain most fishery-independent surveys to groups of species, although additional funding would allow NMFS to examine how well its surveys characterize individual species within multispecies complexes. Surveys almost always focus on groups of species because NMFS does not have the resources (financial, personnel, ships) to optimize surveys for individual species, and the value of most individual species does not merit such an approach. This creates a dilemma: each survey will be sub-optimal for any particular species. How can this situation be remedied in the context of limited resources? If Congress, NMFS, the councils, industry, or environmental advocacy groups believe that better data are needed for any single species, (1) funding could be shifted from other sources or new funding appropriated, (2) fishermen who target the species could be asked to contribute to directed surveys of it (through taxes or in-kind contributions of services in helping with the surveys), (3) NMFS could find new ways to improve commercial and recreational data and use these data more extensively in stock assessments, and/or (4) new methods and technologies, such as acoustic surveys or adaptive sampling, could be implemented.

Greater Use of Fishery-Dependent Data

Findings: Data collected from scientifically designed surveys are often contrasted with that collected from fishery operations. In fact, both have benefits and drawbacks as data gathering mechanisms. Generally, greater control is exercised over the gathering of data from scientific surveys. For a scientific survey, a full statistical design can be implemented, accounting for potential sampling biases. Unfortunately, the cost per sample is large, often resulting in a low sample size, thereby reducing the precision. Loss of precision leads to uncertainty through greater variation in the estimates.

The Marine Recreational Fisheries Statistics Survey (MRFSS) is an example of a scientifically-designed survey that can provide sufficient precision for most modes at the regional level for the recreational component of many fisheries. However, it estimates only what is caught by anglers and does not monitor the fish population directly. Preferences of anglers will motivate them to fish in a manner that biases the catch toward their desired species and fish size.

Data from commercial fishery operations (e.g., from logbooks, observers, port sampling) do not represent random samples of the fish population. Rather, such data reflect the characteristics of those portions of the fish population that are subject to harvest. Cost per fish sampled is low relative to the cost of a survey and data available from intensive fisheries resemble a census more than a survey sample. Thus, the effective sample size for commercial data may be orders of magnitude larger than what may be available via surveys. The fact that it is a fishery-directed harvest in which fishermen are pursuing their own goals, however, means that only a portion of the fish population may be targeted and that targeting may change with time. As a consequence, data from fishery operations present a biased perspective of the population (e.g., fisheries target high concentrations of adult fish), and perhaps more importantly, this bias may change over time and not correlate well with actual fish abundance. Part of the bias arises from not understanding how social and economic factors affect when, where, and how fish are harvested. The lack of control over fishery targeting practices (i.e., lack of structured sampling design applied to the population) makes it necessary to account for potential biases when such data are used in stock assessment procedures.

The committee believes that it would be more cost-effective to find ways to improve the collection and use of data from commercial and recreational fisheries in stock assessments than for the government to conduct vastly increased surveys, although there *are* areas in which surveys need to be increased. Data obtained from commercial and recreational fisheries can be a valuable resource if their inherent biases can be understood and accounted for in the assessment process. The value of commercial data is not limited to population assessment alone. These data may also prove useful for tracking ecosystem changes or to monitor the effects of fishing regulations over time. Such effects are not always apparent from fishery-independent surveys, due to their directed objectives and cost restrictions.

Recommendations: NMFS and the councils should invest in finding ways to improve data from commercial and recreational fisheries to make these data more useful in stock assessments, rather than establishing new fishery-independent surveys. Existing surveys should be made more cost-effective by incorporating new technologies and management methods. In implementing this recommendation, NMFS will need to understand, account for, and reduce (if possible) the inherent biases in fishery-dependent data of different types. Use of fishery-dependent data also should be guided by the evaluation of costs and benefits recommended earlier.

Minimizing and Accounting for "Data Fouling"

Findings: Data fouling is a serious problem that can result from specific types of management. Also, any change in management regime can cause data fouling by changing the spatial and temporal extent of fishing discard rates, misreporting, and other factors. For example, trip limits established in the summer flounder fishery changed the areas targeted by commercial fishermen.

Recommendations: Assessments should take into account the effect of regulations on how fishermen conduct their operations and how this could change the composition of fish caught and data available for stock assessments and other purposes. More fundamentally, councils should explicitly consider the effects that proposed regulations and management regimes would have on data quality and should attempt to design systems that will achieve their management goals with as little data fouling as possible, for example, by implementing or expanding observer programs. NOAA (e.g., NMFS, Sea Grant) should support both internal and external research to identify and evaluate incentives for accurate reporting and disincentives for misreporting and to study the effects of regulations on the industry. In particular, NMFS should engage more social scientists to help build the knowledge base needed to move management beyond trial-and-error to a more predictive capability. Emphasis should be on the relationship between different types of regulatory approaches and fishermen's attitudes and behaviors toward fish harvesting and data reporting.

Fishery-Independent Surveys

Survey Design

Precision of Survey Data

Findings: Once in place, long-term fisheries survey programs are rarely evaluated to determine whether the survey design provides accurate and precise estimates of abundance or relative changes in abundance.

Precision is a function of survey design and sampling intensity and reflects the uncertainty associated with the survey indices. Precision can be evaluated for the NMFS surveys by examining the current allocations of survey sampling stations among statistically designed strata to determine whether they correspond to the current spatial distribution of the target species and to judge if the allocation of samples to strata increases the precision of the overall abundance estimates. Methods for evaluating the precision of stratified random surveys are available in the statistics literature and have been used to evaluate similar fishery surveys (Smith and Gavaris,

1993; Smith, 1996; see Appendix C for summer flounder example). Although this kind of evaluation may be complicated by the multispecies aspect of surveys such as those conducted for New England groundfish, it is useful to evaluate the survey design for each species to determine whether the existing design is a reasonable compromise for the mix of species.

Reallocation of survey effort among sampling strata is a useful method for refining surveys as priorities change. NMFS has employed this approach well in the past to improve the level of information gained from its surveys. However, many of the strata in the 1995 fall New England groundfish survey and the deep strata for the 1995 spring groundfish survey contain only one sample per stratum. When strata contain only one sample, variance is inestimable, except by extrapolation from other areas.

Recommendations: The statistical precision of each NMFS survey should be evaluated. Estimates of the maximum expected precision given the current number of tows should be calculated and used to evaluate the statistical power of the survey to detect changes in abundance over time. The gains, or potential gains, from better survey designs, or an increase in sampling intensity, should be evaluated in relation to the assessments in which the data are used. An increase in the precision of state surveys might also be achieved if similar analyses were applied to them.

NMFS should periodically review whether sampling effort should be reallocated, even when the overall objectives of the survey stay the same. For example, if the range of the species expands to offshore areas, NMFS may need to sample these areas more heavily. In the case of the summer flounder, this type of change was made, but user groups, particularly fishermen, did not seem to know the extent of the survey expansion and the results of the changes, both in terms of raw data and assessment results. In relative abundance plots, the symbols denoting abundance at each survey station should identify the location of survey stations where there was zero catch. Small changes like this could greatly enhance communication.

Appropriate statistical design should always include more than one tow per stratum. If this is not a feature of an existing sampling design or extra stations cannot be occupied during the execution of a survey, strata should be redefined before the survey or combined for analysis if the number of tows becomes limited during the survey.[2] In general, surveys should be cost effective, statistically well designed, conducted frequently enough to detect significant trends in abundance, and take into consideration the biology of the species. Standardization of survey methods within management units to meet these needs should increase economic efficiency, quality control, and comparability of data. Standardizing within management units may be necessary, but flexibility across species and management regions is needed to address the diversity of issues and environments observed in marine systems.

State surveys apply a great deal of sampling effort, which is often allocated in relatively few tows per stratum each month, several different months each year. For many states the monthly surveys seem individually inadequate to develop precise estimates of abundance. Combining these surveys across months may also be inappropriate if a species is seasonally migratory and will change distribution among strata over the several months of survey effort. States should determine whether consolidating all their survey effort into a single annual survey might yield much more information than monthly surveys. Such analysis and coordination could be undertaken under the auspices of the interstate marine fisheries commissions.

Accuracy of Survey Data: Frequency and Spatial Extent of Surveys

Findings: The frequency of NMFS surveys varies from stock to stock and region to region. For

[2] Strata may differ by species and year to year, so post hoc combination of strata is more appropriate than combining strata in the sampling phase.

example, there are two general trawl surveys per year, plus the winter flatfish survey, on the U.S. East Coast. Other surveys (e.g., West Coast Continental Shelf groundfish) are conducted once every three years (see Table 3-3). Many federal surveys (the Atlantic Coast is an exception) do not provide information about seasonal distribution because they are conducted only annually or less frequently. The seasonal distribution of a species may be important to monitor in terms of how it influences catch rate for both survey and fishing vessels and in terms of how vulnerable different components of the stock are at different times of the year. Infrequency of surveys may hinder management, especially for fully exploited fisheries and for fish stocks whose dynamics change significantly from year to year. NRC (1998a) found that the single most important factor in achieving accurate results using any stock assessment model is an accurate indicator of relative abundance over time. Where vessel time is limited, surveys with twice the effort every second year—or thrice the effort every third year—may lead to more reliable assessments than annual surveys. Sometimes annual surveys may be the best option, but not universally. It is possible that the biology of a species complex, the health of its stocks, or the management approaches being used would not require an annual survey, particularly if good fishery-dependent data were available and the fishery were not in a situation in which recruits make up a significant portion of the biomass.

Recommendations: The NMFS examination of the costs and benefits of data collection should include the frequency and timing of surveys in each region, with consideration of factors such as the biology of the managed species, state of the stocks, the current and potential economic value of the species, and the availability of other accurate indices of trend (e.g., commercial CPUE). NMFS should report their findings to Congress to help members of Congress understand subsequent realignments of survey activity that may be worked out between NMFS and the regional councils. In addition, NMFS should attempt to improve the quality of commercial and recreational data to the point that they could substitute for some survey data.

Findings: The accuracy of survey data is mainly a function of sampling over a stock's entire geographic range and adequate sampling of all age and size classes. Distributions of fish over time and space and at different life stages are important considerations in designing the spatial extent of surveys. As noted by NRC (1998a), surveys should consider that stocks may shift over time due to regime shifts and other environmental changes.

For some stocks, such as summer flounder, it is not clear in available documentation that the surveys cover the entire range of the stock or adequately account for seasonal movement. For example, catch numbers of summer flounder in the winter surveys increased all the way to the edge of the survey (at the continental shelf edge) in some areas. NMFS added stations along the shelf edge for the winter survey (as weather permits), in response to industry concerns. The assumption in the subsequent analysis was that there were no significant concentrations of fish beyond the survey zone. Another example of a survey that may not sample a stock's entire range is for Greenland halibut (turbot) in the Bering Sea. Turbot is primarily a deep-slope species, yet no information is available regarding its abundance in its prime habitat because the triennial slope survey does not sample deep enough.

The range of a stock can be monitored through the spatial distribution of abundance indices in the surveys and the locations of commercial or recreational catches. Using fishery activity to detect changes in a species range may not be effective, however, if management is changed in such a way that fishing time or place are restricted (e.g., trip limits for summer flounder reduce fishing activities far from port).

Recommendations: NMFS should ensure that the geographic ranges of its surveys cover the geographic ranges of each species managed. In particular, the spatial extent of surveys should be based on good evidence that the ranges of the target species included in the survey have been reached, using objective threshold criteria for deciding when the range has been covered sufficiently.

Information about the geographic extent of surveys and commercial catches should be shared with managers and stakeholders in an appropriate forum. The historic locations of summer flounder, as determined by catches from fishing activities or exploratory surveys in winter, need to be analyzed to evaluate the extent of the stock area with respect to existing strata boundaries. Although this information may already be available in various forms, it needs to be drawn together and presented in a form accessible to the public, to either correct misperceptions of the industry or initiate action to improve the sampling if industry concerns are warranted. NMFS should use information provided by commercial (e.g., logbooks) and recreational fishermen regarding geographic locations of stocks (e.g., summary plots of commercial or recreational harvest locations), keeping in mind that if regulations limit the places in which it is legal or cost effective for fishermen to fish, the spatial distribution of catches may not indicate where fish are actually to be found. NMFS can consult with fishermen by presenting planned survey locations to them. All user groups should watch for patterns in fisheries data that may appear to be associated with poor spatial coverage (i.e., seeing few large fish in the catch) and could indicate other issues of concern, such as underreporting of discards or unaccounted-for migrations.

The existence of separate stocks should be examined and, if found, their significance should be determined. Data to differentiate among stocks should include information collected from tagging studies, characterizations of parasitic fauna, genetic studies, recruitment patterns, and growth rates (NRC, 1994a). Stock identifications should recognize that unit stock concepts are defined not only by genetic uniqueness, but also by the degree to which stocks differ in local recruitment characteristics, feeding and spawning ground fidelity, growth patterns, and other life-history characteristics. Survey data, as well as fishery-dependent data, may need to be collected at a finer scale than presently to manage such stocks separately, but the added costs of more detailed data should be evaluated against the additional benefits that would result.

Essential Fish Habitat Data

Findings: A dramatic increase in information on essential fish habitats is likely to result from the emphasis on this matter in the Magnuson-Stevens Act. Although this information is being gathered principally to identify habitats critical to various life history stages of important marine species, the result will be a compendium of information that can be used more broadly to serve the needs of fisheries assessment and management. For example, knowledge of the distribution of favorable habitats for adult fish can be used to design more appropriate sampling strata, leading to increases in precision in population estimates. As another example, key habitats for eggs, larvae, and juveniles can be used to identify areas for more effectively monitoring year-class strength, leading to better short- and long-term projections of population trends.

Recommendations: Information on essential fish habitat should be collected and managed in the context of the broader needs of fisheries science, assessment, and management. Stock assessment scientists, fishery managers, and fishery ecologists should work together in setting objectives for collecting data related to essential fish habitat, and in establishing a means of managing and accessing these data. This latter goal should be developed in parallel with ongoing efforts that focus on more traditional data management. An important aspect of managing data related to essential fish habitat is to ensure that the format and content of the data conforms to other federal oceanographic data in a spatial and temporal framework.

Sampling Gear and Methods

Findings: The use of gear standardized over time is important for maintaining consistent measures of relative abundance. Surveys are generally designed to yield estimates of abundance that can be related to estimates based on catches with the same gear in previous years.

Commercial fishermen are constantly updating their gear and methods and try to fish in areas where they can maximize their profits. A common complaint by fishermen about surveys is that NMFS uses outdated gear, inefficient fishing methods, and random or set stations, which fishermen believe biases the species and size compositions of survey catches in ways that reduce the TAC. In some cases, commercial fishermen have been able to demonstrate that survey gear has not been used correctly, preventing the determination of even relative abundance, for example, in a West Coast fishery (Lauth et al., 1998).

Fishery-independent surveys for summer flounder and other groundfish in the Northeast region use 30-minute tows to increase the tows per station and increase precision, whereas commercial fishermen often tow their trawl nets for 3 hours or longer. Fishermen report that longer tows allow them to catch large fish that are hard to tire with shorter survey tows. Fishermen who attended the committee's meetings were unsure about what gear was used in Northeast Fishery Science Center's summer flounder surveys, but doubted whether the gear would sample summer flounder in an unbiased manner, given the tow duration.

Most U.S. surveys are conducted using bottom trawls, so mid-water and surface-dwelling fish species are undersampled. Hydroacoustic and other remote sensing methods are not commonly used for U.S. fishery assessments, with a prominent exception being the hydroacoustic surveys used to assess spawning aggregations of Alaskan pollock and Pacific whiting (e.g., Traynor, 1997). Acoustic methods do not work for flatfish because these methods have difficulty distinguishing fish from seafloor, but these methods can greatly extend survey capabilities for midwater fish.

Recommendations: When survey gear is outdated, has unstable performance or is hard to set up correctly, effort should be directed at improving the gear and providing some level of cross-calibration so the value of historic data is maintained. Key issues in gear selection include credibility of results as well as bias in sampling and sample-to-sample variability. If use of modern gear and methods could decrease bias and variability without seriously compromising survey data time series, NMFS should consider updating its gear and changing survey methods to make them more similar to current practice in the fishing industry. If changes are made, however, they should be done with appropriate parallel use of old and new gear for as long as necessary to provide adjustment factors for historic data.

NMFS should consider hiring commercial fishermen to participate in survey cruises to see how sampling gear is used and where surveys are conducted. Conversely, fishermen should acknowledge that even if survey gear is somewhat antiquated, as long as it operates as it was intended, it may still provide an adequate index of relative abundance.

NMFS should document and communicate information to stakeholders about the gear and methods (such as tow duration) used in its surveys, as well as an evaluation of the gear's efficiency at capturing different species. This is particularly important for summer flounder because the committee is unsure if roller gear, whether 6 inches or 36 inches, is appropriate for surveying this species. NMFS should endeavor to communicate to stakeholders why survey gear may be operating differently than commercial gear and how such differences may or may not affect assessments. In the case of summer flounder, the effect of tow duration on the size distribution of fish caught should be studied to understand differences in gear selectivity that may exist between surveys and commercial catch.

Gear and operating procedures used in all surveys should be evaluated on a regular basis (e.g., every 5-10 years and particularly at the time research vessels are changed) to ensure that they are still reasonable by industry standards. Modernizing gear and procedures is one way to increase the credibility of survey results to fishermen. Another means of boosting NMFS' credibility would be to run parallel surveys with commercial fishermen. This will help determine how catchabilities compare between the survey and the fishery and help improve communication between NMFS and commercial fishermen. It also can help in the interpretation of fishery-dependent data, for example, how commercial CPUE is changing over time relative to survey CPUE. Incorporating fishermen's knowledge into sampling design and analysis, especially about where and how long to fish, could improve sampling efficiency and NMFS' credibility with fishermen.

The interstate commissions should initiate or continue efforts to get states to standardize their surveys to improve the comparability of their survey data, decrease the frequency of their surveys to increase the stations on any specific survey, or in some cases, shift from doing surveys to other activities more useful for coastwide stock assessments.

New survey methods should be pursued. NMFS should increase its use of hydroacoustic and other developing methods to estimate the stocks of surface-dwelling, mid-water, and other vertically-oriented fish species (e.g., rockfish) that are not susceptible to bottom trawls usually used in NMFS surveys, where this would lead to a cost-effective gain in precision. Although acoustic methods don't work for flatfish, they may be useful for other species.

Survey Vessels

Findings: Fishery-independent surveys using calibrated vessels owned by NOAA provide vital and irreplaceable data for stock assessments and ecological monitoring and other fishery assessment purposes. Trawl surveys are especially susceptible to variability caused by differences among vessels, so calibrated vessels are a necessity. Many kinds of data can be collected only by using survey vessels, although NMFS charters 40 percent of its survey and research days at sea on commercial and university vessels annually for other purposes. NOAA survey vessels are aging and in need of replacement and it appears that the number of ships and ship time available for surveys is not adequate to meet all critical needs.

Recommendations: The committee endorses the efforts of Congress and NMFS to maintain a strong fleet of NOAA survey vessels, particularly for trawl and acoustic surveys, by replacing aging vessels with new, more capable, and more quiet ones. NMFS should continue to use charter and lease-back arrangements, where appropriate, even as the agency acquires new survey vessels. Congress should not only fund the construction of new vessels, but also should provide adequate funding for survey and research work performed by these vessels.

Adaptive Sampling

Findings: Adaptive sampling provides a means of increasing the precision of certain kinds of survey estimates, especially total biomass of some species, by using information obtained during the survey to determine where additional sampling should be done. In a sense, fishermen act as adaptive samplers when they locate and fish intensively in areas where the fish are present, but they do not sample in a way that is statistically valid. Although adaptive sampling is generally a single-species approach, sampling rules may be defined so that precision would be increased for groups of species that have similar distributions. Knowledge of habitat preferences can be used to design adaptive sampling schemes. Adaptive sampling is not appropriate for other purposes, such as mapping species distributions, monitoring recovery of depleted populations, or tracking migration, especially where the rarity or absence of some species is important information.

Recommendations: NMFS should explore alternative data collection and analysis designs, including using adaptive sampling, as an approach to fishery-independent surveys (e.g., as practiced in New Zealand and Canada). Adaptive sampling using commercial fishing vessels in conjunction with NMFS survey vessels also should be explored.

Data from Commercial Fisheries

Fishery-dependent data arise from the catch or catch rates of individual fishermen and therefore pertain to both the biological population of interest and the harvesters of that population. Because people harvest fish for a number of different reasons (commercial, recreational, ceremonial, subsistence) many different motivations influence the time, place, and gear employed in catching fish. These influences can obscure changes in the abundance of a stock, but stock abundance is still an important determinant of the catch. If confounding influences can be accounted for, fishery-dependent data can provide an important source of information regarding trends in fish populations and more generally trends in the fishery. Consequently, it is important to conduct research to understand the motivations of harvesters to enable accurate interpretation of fishery-dependent data. A necessary prerequisite for this research is the collection of data on the determinants of catch in commercial, recreational, and subsistence fisheries.

Standardized and Formalized Data Collection

Findings: NRC (1998a) recommended "that a standardized and formalized data collection protocol be established for commercial fisheries data nationwide." (p. 117). More specifically, "[s]ome regions, for example, the U.S. Northeast, are in the midst of the development and publication of new protocols, whereas other regions do not appear to use standardized methods. The lack of formalized, peer-reviewed data collection methods in commercial fisheries is worrisome. To the extent that formalized and standardized procedures are lacking, potential bias and improper survey conduct must exist, with unknown impact on data reliability" (pp. 116-117). The trend in new database management systems (e.g., ACCSP and the proposed Fisheries Information System) to standardize systems is an important development. The Fisheries Information System is intended to serve as a national umbrella of regional systems that meet regional needs. Nevertheless, the national system could fail to meet national needs if each region adopts different data standards.

Recommendations: As part of the Fisheries Information System, NMFS should work with regions to develop and publish a standardized data collection protocol and required data elements for commercial fisheries data (e.g., logbooks, observers, VMS) nationwide. A core set of data should be required from all commercial fisheries in all regions, with other data components being more specific to certain species, types of fisheries, or regional management needs. Standardization should take into account the cost-benefit analysis recommended earlier.

Incentives for Timely and Accurate Reporting

Finding: Commercial and recreational fishermen may misreport catch, bycatch and discards, or landings for a variety of reasons, some of which may be intentional (e.g., underreporting to avoid regulatory limitations and penalties or overreporting to increase standing for future allocation decisions based on historical catch levels). Reducing the incentives for misreporting (or increasing the incentives for accurate reporting) would ultimately improve the credibility and quality of fishery-dependent data and allow such data to be a better complement to fishery-independent surveys. Good evidence exists that improved cooperation among scientists, managers, and fishermen in designing and participating in a meaningful way in data collection can improve the quality of fishery-dependent data.

Recommendations: NMFS and fishery scientists should identify the most important incentives and disincentives that could be used to promote accurate reporting (see also our earlier recommendations on data fouling). NMFS and the regional councils should determine how to design regulations and management that provide incentives for fishermen to provide accurate and complete data, without compromising their personal economic welfare. It is important to encourage fishermen to follow regulations while simultaneously rewarding them for reporting the degree to which they did not. This objective can be achieved by using graduated penalties, so that minor infractions are more likely to be reported. Congress and NMFS should consider incentives to encourage commercial fishermen to invest in technologies that will provide data to make fisheries management more timely and effective.

Collectively, studies related to co-management, cooperative relationships between industry and scientists, and incentives for misreporting catch data indicate that early and active incorporation of fishermen's viewpoints into research and regulatory processes could improve fisheries management. Communication among groups is essential, including explanations of how decisions were reached.

As fishermen develop business relationships with commercial data management companies, it is important for NMFS to work with commercial fishermen and their third-party data providers to get the most useful data into the NMFS data system. Great care should be taken to avoid devaluing useful and innovative commercial products and processes by government regulations and bureaucracy.

Incentives for Commercial Assistance in Data Collection

Findings: Commercial fishermen are businessmen who must devote a great deal of time and effort to keeping their businesses solvent. Any time that they are not tending to their business creates an opportunity cost that must be balanced against the benefit of the alternative activity. Thus, the benefits of cooperation with NMFS and state fishery agencies will be judged against the costs of not fishing and fishermen's other business activities. Cooperation will not develop if this tradeoff does not result in a net benefit to fishermen. The Magnuson-Stevens Act allows councils to set aside part of the allowable biological catch for a fishery to compensate cooperating fishermen. In a recent example, scallop fishermen were allowed to land a larger catch to offset observer costs in scallop surveys and bycatch studies on closed areas of Georges Bank. Another mechanism used in New England fisheries is to link availability of disaster relief funds to fishermen's willingness to volunteer for joint NMFS-industry research activities.

Recommendations: NMFS and the regional councils should explore ways to offer incentives or to enhance existing incentives to enlist the help of commercial fishermen, such as charter arrangements and allowing fishermen to keep and sell catch obtained while assisting NMFS in research and surveys ("fish for research").

Observer Data

Findings: Observers are an essential source of some kinds of information needed to validate or adjust self-reported commercial data. From individual hauls, observers record catch data such as bycatch, discards, and interactions with prohibited species (e.g., sea birds, marine mammals, marine turtles, salmon, halibut, and crab). Observers can also record information about the area traveled by the boat, the depth of fishing, types of gear, changes in gear during a trip, catches by catcher-processor vessels, and size and age of catch as it relates to location. Some of this information is sometimes available from logbooks, but much of it can only be gathered on a routine basis from observers.

In some areas, such as the North Pacific region, where significant amounts of fish are processed at sea, observer data are crucial for managing quotas for target species and prohibited species catches on a real-time basis. Certain types of management systems (e.g., trip limits and some types of individual quota systems) may increase discarding and thus may be priorities for observer coverage. Regulatory discards seem to be a particular problem in the commercial summer flounder fishery, but fishermen are reluctant to report the full extent of discards because they fear that the information will be used to decrease their allowable catches or used to generate enforcement measures (i.e., fines) against boats. Although fishery-wide bycatch and discards may not be fully monitored by observers on all vessels, they can be estimated by comparing the composition of catches on vessels with and without observers, if appropriate assumptions are used. The same care and statistical principles are necessary in designing an observer program as in designing a survey.

Information gathered by skippers informally from observers and shared with cooperating vessels via a commercial data management firm has yielded a more efficient fishery with lower bycatch rates in the North Pacific region. Cooperation works in these fisheries because there is a fleet-wide maximum allowed bycatch of prohibited species, so it is in everyone's interest to notify competing fishermen of high-bycatch areas. The potential sharing of a good fishing location with competitors is an acceptable cost if it results in better fishing practices that lengthen the season and sustain fishery resources. The timeliness of this information sharing is an important factor and this information would be useless if available more than a few hours or days after observations were made. This system is possible only because the industry provides observer data voluntarily to a third party. Confidentiality problems would hinder NMFS from serving the same function.

Recommendations: NMFS should evaluate current levels of observer coverage, and means for increasing the effectiveness of observers and deployment designs. The effectiveness of observer coverage can be judged in terms of the needed precision from observer data compared with the precision available for a given level of observer coverage (see Figure 3-2). Minimal acceptable coverage depends on the objective of the management methods used, but care should be taken to ensure that sampled vessels are representative of the fishery as a whole. Figure 3-2 demonstrates that observer coverage for some fisheries must be increased well above present summer flounder coverage (0.6-0.8 percent), particularly for bycatch monitoring. NMFS should consider increasing observer coverage to at least 25 percent for commercial summer flounder trips over several years to obtain a better estimate of bycatch discards and misreporting in the summer flounder fishery.

Some information can be obtained only by at-sea observer programs, but at-sea observers are extremely expensive. Because observer programs are expensive to either NMFS or industry (in the case of Alaskan fisheries), careful consideration should be given to how much NMFS should pay and how much specific fleets should pay, for example, through a fee-based observer program. However, many fisheries do not generate the level of revenue seen in Alaskan fisheries and therefore could not support observers directly. The approach of allowing extra catch for vessels carrying observers, to pay for observer coverage, should be tested broadly. It is important that all observer programs that use less than 100 percent coverage institute statistical designs that apportion observer coverage in a statistically valid manner (e.g., see NOAA, 1999), with special attention to measurement of possible effects on catches of having an observer onboard. The goals of each observer program need to be specified very carefully and other, less expensive, means of obtaining the needed data should be examined. In some cases, the cost of obtaining good data on discarding and misreporting may be greater than the benefits of reducing these problems. Despite the costs and other problems, observers should be used when it is clear that observer coverage can provide necessary scientific or management information and would be cost-effective in relation to other data collection techniques.

NMFS should pursue efforts to develop a national database of observer data in its Fisheries Information System. To ensure that data within such a database are comparable among regions, standardized data collection forms and training for observers is needed. These national efforts would not preclude the necessary regional implementation of observer programs at the discretion of regional councils. As for other fisheries data, NMFS should find ways to release aggregated observer data in a timely manner, and sunset periods should be set for observer data.

Real-time availability of bycatch data can be particularly useful in helping fishermen avoid high bycatch areas. NMFS and Congress should find ways to encourage (in legislation and regulations) such voluntary, real-time sharing of bycatch information, including encouraging commercial and non-profit mechanisms to accomplish this goal.

Logbook Data

Findings: Logbooks compiled by fishermen and fish dealers can be effective and cost-effective sources of data concerning abundance and other characteristics of fish stocks and fisheries. But, if fishermen and fish dealers do not believe that stock assessments and fisheries management are responsive to their needs or concerned with their survival, they will not willingly submit accurate and complete logbook data. If forced to participate, they may provide biased information to try to influence the outcome of the assessment or because they do not believe it is worth their effort (or may even damage their interests) to provide accurate data. Logbook data are underutilized in many U.S. fisheries because of concerns about their validity and the cost and time required to get logbook data into electronic form. There can be a significant lag in the collection of logbooks and the availability of the data for management (e.g., California is a year behind in reporting logbook data to the PacFIN system). Such lags decrease the usefulness of logbook data for stock assessments and other purposes. A large proportion of fishing vessels now have computers onboard, making electronic logbooks feasible. Electronic logbooks may provide more accurate and timely spatial data that may eventually be critical for management of local stocks and for evaluation of essential habitat. However, doubts about the legal validity of electronic submissions have caused fishermen in some areas to provide data both in electronic form and as signed hardcopy forms in the federal format.

Recommendations: Logbooks should be required from fishermen and dealers. Dealer reports, observers, and other methods should be use to validate estimated weights in logbooks and provide additional economic information not contained in logbooks.

Commercial logbooks will be more useful for stock assessments if a standardized format is designed and required. The standards may differ by type of vessel, gear, stock, and fishery but there should be a minimum set of information collected from all fisheries and all vessels, including starting and ending times and coordinates of each tow (for trawl fisheries), or soak time and coordinates (for longline, trap, and pot fisheries), species and amounts caught, and species and amounts discarded. Congress should also consider mandating logbooks for charter/party vessel fisheries in the recreational sector because of the imprecision of measurements of catch by charter/party vessel fisheries available from the Marine Recreational Fisheries Statistical Survey (MRFSS). Finally, it should be recognized that if logbooks are required by more than one level of government (state, regional, federal, and international), agencies should coordinate their efforts to reduce the burden of duplicative reporting (which may involve more than merely multiple logbooks). To make logbook programs more effective, they should be made less onerous for fishermen to fill out accurately and submit in a timely manner. This goal can be achieved, in part, by involving fishermen in the design of logbook forms, and using focus groups and field testing to ensure that the forms actually provide the information they are designed to provide. The use of logbook data is particularly important for fisheries in which fishery-independent surveys are conducted infrequently or not at all.

Electronic submission of data should be developed in ways that reduce the errors that occur with data entry and transfer, and improve spatial data for management purposes. Fishermen should be encouraged to report by electronic means and NMFS should continue to promote commercial development of electronic logbooks. NMFS might continue development of electronic logbook systems via Department of Commerce Requests for Proposals for its Small Business Innovation Research grant program.

Fishermen will use electronic logbooks if they find that they have advantages over paper logbooks. One of the greatest incentives would be for NMFS to invest in the data processing personnel needed to take advantage of electronic logbook information. NMFS should also take whatever steps are needed to remove all doubts about whether logbooks meet legal requirements. NMFS should provide various value-added data products to fishermen based on their logbook data, for example, timely port- or regionally-averaged CPUE values, and reports to individual fishermen both on-line (password protected) and in written form on a monthly basis, to verify the fisherman's data. The International Pacific Halibut Commission sends logbook data to fishermen to verify their catch information. The Fisheries Information System is proposing to issue monthly statements to fishermen. Logbook data should be available to fishermen and others as soon as possible after they are collected.

The introduction of electronic logbooks will move the bottleneck in data collection and management to the landing ticket data. The delay in entry of landing ticket data can be up to six months. Such data consist of weight and value information collected by fish dealers and provided to the states. Landing ticket data are compared with logbook data before logbook data are used for stock assessments. To make efficient use of electronic logbooks to speed the availability of logbook data for stock assessments, it will be necessary to take the next step, getting landing ticket data into electronic format as quickly as possible. This situation illustrates that the entire fishery data system must be examined to eliminate bottlenecks that cause lags in data availability.

Historic information should also be converted into electronic form, if this would be cost effective. It is important that individual records remain confidential until sunset periods have passed, so the level of data aggregation will remain an issue.

Vessel Monitoring System Data

Findings: Vessel monitoring systems (VMSs) have been tested in several U.S. fisheries, demonstrating their usefulness in tracking the locations of fishing vessels and their usefulness to enforcement personnel, managers, and to the fishermen themselves. VMS data have numerous potential applications beyond monitoring of time and area closures and other enforcement uses. Spatial data from VMS units can be used to verify positions in both fishing logs and observer logs and generally to assist interpretation of fisheries data. VMSs may also prove to be valuable in putting catch data into a spatial format that can be linked to electronic logbooks. In some fisheries, however, the estimates of tow-by-tow catches and discards are so inaccurate that knowing the location of such retained catches and discards might not justify the cost of VMS systems for non-enforcement purposes. Fishermen in some fisheries have also used the ship-to-shore communication capabilities of some VMSs for personal and business purposes. The cost of VMS units has limited the interest of small boat owners, although mass production could reduce the unit cost of VMS devices.

Recommendations: NMFS should consider implementation of VMS programs in all U.S. fisheries for which such programs would be cost effective. Before implementing a VMS in any specific fishery for non-enforcement purposes, however, NMFS should determine whether tow-by-tow data on catch amount, discards, and species composition are accurate enough to justify the expense necessary to determine the location of each tow using a VMS unit. The costs and benefits in terms of data availability could be determined using computer simulations.

NMFS and councils should consider the usefulness of position data for all managed fisheries, whether for monitoring and enforcing management based on closed areas, providing spatial data to improve understanding of changes in distribution and abundance of fish stocks, or for other purposes. NMFS should ensure that VMS data are linked to other data sources and are made available to stock assessment experts (see also recommendations on data confidentiality). NMFS should recognize that VMSs will be willingly and rapidly adapted by fishermen who perceive that VMSs provide them with value-added information worth more than the cost of the unit. Development of the VMS approach should include system integration across technologies (i.e., VMS, FIS, electronic logbooks). NMFS and the regional councils should consider other mechanisms to encourage VMS use, such as:

- leasing units to fishermen,
- liberalizing management regimes when compliance can be monitored by VMS,
- making the units useful to fishermen in management of their businesses through a variety of communication capabilities,
- making it possible for fishermen to reach their individual or fishery quotas and stay on the fishing grounds until the last minute, and
- instituting NMFS "fleetnet" broadcasts of bycatch data to vessels on the grounds.

VMS data should not be confidential indefinitely, but should be available freely after a reasonable period.

Data from Recreational Fisheries

Marine Recreational Fisheries Statistical Survey (MRFSS)

Findings: Unlike most commercial fishery operations, in which a small number of vessels land large volumes of fish in a highly regulated manner at designated ports, most recreational fisheries tend to have a great number of individual fishermen who are highly dispersed in where they fish and how they land fish, operating in a system that is not uniformly regulated or licensed. MRFSS is used to estimate the number of fish removed by recreational fisheries. MRFSS uses phone surveys of households in coastal counties to assess effort and on-site intercept surveys to assess CPUE and species composition of recreational landings and bycatch. This is relatively expensive and inefficient because of the large reference frame of coastal households surveyed by random digit dialing, with a relatively small proportion being marine anglers, and the large number of recreational fishing sites. In many regions, phone surveys and on-site interviews conducted by MRFSS appear to provide the necessary information for estimating recreational fisheries catch and effort to be used in subsequent years' assessments, but MRFSS is not designed to provide information timely enough for in-season management. There is a three- to four-month delay in providing recreational catch estimates. This was not a great problem when MRFSS was developed, but the increasing proportion of catch taken by recreational fishermen is creating management problems in some fisheries. Overruns can be caused by recreational fisheries, putting councils in a position of having to compensate for overruns in the following year. For example, it has been estimated that the summer flounder recreational fishery exceeded its share of the TAC by 40 percent in 1998.

MRFSS serves as a de facto national standard for recreational data, insofar as individual states participate, although some states (e.g., Texas) have chosen not to participate because they fund their own extensive data collection activities. Other methods have been used to estimate recreational catch for particularly valuable fish species (e.g., Pacific halibut) and species monitored by measuring escapement (e.g., Pacific salmon). The devolution of MRFSS activities in the Gulf of Mexico region to regional control could hinder necessary standardization and quality control. MRFSS has a systematic plan for addressing known sampling problems, but seems to lack the personnel and financial resources to make rapid progress in updating its procedures.

Recommendations: In the short term, MRFSS should be extended to all coastal states with significant marine recreational fisheries that have requested inclusion. Additionally, methods to improve precision, such as longitudinal sampling (a high priority to MRFSS), should be included. In longitudinal sampling, households that report angling or a likelihood of future angling are recontacted in subsequent sampling waves. The retention of identified angling households, at least for the next sampling wave, would increase survey efficiency. Efficiency could also be increased if sampling strata were built up of the phone numbers of individuals who fish regularly, ones who fish occasionally, and ones who fish rarely, and sampling effort is allocated optimally among these strata. However, a problem that can arise with such an approach that samples avid anglers on an ongoing basis is that individual anglers eventually refuse to cooperate. This problem can be avoided by using longitudinal sampling in which any specific angler is contacted only a few times. NMFS should recognize that needs for recreational data may differ somewhat among regions and NMFS should work with regional councils and interstate commissions to identify region-specific recreational data needs, while maintaining an adequate level of nationwide standardization.

MRFSS continually identifies statistical and other issues that need to be addressed to improve its surveys. The committee has prioritized a set of questions and issues (some already identified by MRFSS) that NMFS should consider in future improvements to its recreational sampling programs.

1. How valid are the MRFSS assumptions (1) about the ability of the intercept survey to equally survey coastal and noncoastal residents who engage in each fishing mode, two-month sampling wave, and state; and (2) that fishing patterns with respect to types of fish caught are the same within these categories for residents of coastal and noncoastal counties? It is not economically feasible to sample persons in all noncoastal counties; however, a study designed to compare fishing patterns of residents of coastal and noncoastal areas may be warranted. A feasible first step might be to compare the fishing patterns of residents of coastal counties, already surveyed, in relation to their distance from the water (or perhaps their landing site). For the telephone survey, why is each county considered separately—thereby acting as individual de facto strata? It might be more efficient statistically to group counties and to stratify on region within the state. NMFS should test the efficiency of this alternative stratification scheme. NMFS should continue its comparisons of data quality and costs of using sampling frames based on salt-water fishing licenses versus frames based on random-digit dialing.

2. Cluster effects among fishermen on a single charterboat may arise from either the fact that all persons on a boat are fishing in the same area, or the fact that members of the group may share personal features such as skill in fishing or interest in a particular species. The MRFSS document does indicate that a cluster effect occurs in charter vessel fisheries due to sampling

all anglers who have similar demographic characteristics (ASMFC, 1994, page 1-10). How was this adjusted for in the analysis or possibly in the sampling effort? Adjustments on a "case-by-case basis" seems vague and should be better documented with objective criteria.

3. How does MRFSS deal with missing and unusual values? For example, when a fisherman is not home, MRFSS calculations assume that the missing person has the same patterns of fishing as those already surveyed. This seems highly unlikely; for example, persons not at home because they are working may fish much less than persons who are retired. If a fisherman has an unusually high catch exceeding the 95th percentile, the value is reduced to the 95th percentile (ASMFC, 1994, page 1-9). Are such outliers true large values or spurious? The impact of these outliers should be smaller if fewer, but still homogenous, strata are used. Additional study may be required to validate this practice. MRFSS might also explore ways to estimate the true catch at that upper end and add the estimate to the total below the 95th percentile.

4. How valid is the assumption in roving surveys that if a trip is half over, one can double the number of fish caught in the first half of the day (particularly if weather changes during the day) to estimate the day's total catch? It is unclear how time of day impacts survey outcome. For example, do early morning catch composition and numbers differ from afternoon catch? Is sampling across time of day conducted in an unbiased fashion? How interviewers attempt to get a cross-section of fisherman within each mode should be made clear in MRFSS documentation.

Some of these issues can be addressed using existing data, but new data may be needed to assess the validity of the assumptions made in the survey. Even though these effects may not markedly affect total catch estimates, it is important to determine their impact.

In addition to improving MRFSS, other options for improving recreational data include mandatory logbooks for all charterboat and party boat fisheries (as discussed earlier, these provide more precise catch data) and mandatory marine recreational fishery licenses nationwide. Another option would be to use a dual-frame approach, in which an incomplete recreational license frame could be augmented with some level of random-digit dialing. Another means to improve charter boat/party boat data would be to develop and use a complete list of vessels, require them to maintain logbooks to verify catch rates, and use the lists for telephone surveys of this fishing sector.

In-Season Monitoring of Recreational Fisheries Catch

Findings: MRFSS was designed to monitor recreational catch and effort each year to use in stock assessments run in subsequent years. MRFSS provides a relatively long time series of such data. It was not designed to track catch during a season for the purpose of monitoring recreational catch against the TAC.

Recreational fishing harvest cannot be kept within its portion of the TAC unless data on the catch are timely and accurate enough to allow in-season closures. For example, salmon and halibut fisheries on the West Coast use in-season management measures to allow early closures or extensions of the seasons, depending on how fast the TAC is approached. It is difficult for MRFSS in its present configuration to monitor in-season catch because of the nature of recreational fisheries. Unlike most commercial fisheries, in which relatively few individuals participate in the fishery, hold licenses, and can be contacted directly, marine recreational fishing is undertaken by many individuals, typically unlicensed, at multiple and diffuse access points, who are less accessible to gather data from than are commercial fishermen.

Each sampling wave lasts two months and post-wave processing may require an additional two months. Because MRFSS is a sampling survey, its data must undergo statistical procedures to estimate total catch by combining the telephone effort estimate with catch rate estimates derived from on-site intercept surveys. Effort data reported at the end of the process are two to four months old. Thus, a catch limit reached early in a given sampling wave would not be detected until two to four months later. MRFSS, as currently conducted, would have difficulty providing catch estimates timely enough to be used for closure of recreational fisheries with seasons shorter than four months. This can be contrasted with catch data from commercial fisheries, which are reported on a weekly to monthly basis and are thus amenable to tracking against the TAC.

In-season monitoring of recreational catch is as important as monitoring commercial catch in heavily exploited fisheries with a substantial recreational component because both sectors can contribute to TAC overruns, which diminishes the effectiveness of management. As might be expected, attempts to use MRFSS for in-season monitoring (particularly at fine scales) have not usually been successful. This lack of success should not be taken as proof that MRFSS is ineffective nor does it mean that in-season monitoring is impossible. The only conclusion that can be drawn is that an enhanced MRFSS or an entirely different system will be necessary if it is determined that in-season monitoring of recreational catch and effort are worthy and cost-effective goals.

Recommendations: NMFS should poll the regional fishery management councils to determine if they could use more timely recreational data if they were available. If councils want this type of data, NMFS should seek funding to design and test both augmentations to MRFSS and entirely new systems, with the additional goal of determining the costs and benefits of in-season recreational data availability. The optimal direction for management of recreational fisheries may be clear after NMFS explores means of very rapid retrieval and analysis of recreational data.

MRFSS data might be made more timely by using shorter, more numerous sampling waves; increasing the number of phone calls in the telephone surveys; and/or using license list frames and longitudinal sampling. An alternate approach to obtain in-season measures would be to institute a survey to monitor specific recreational fisheries. Such a survey could contact anglers weekly or even more frequently during the season. However, such a survey would still have to undergo quality control and statistical expansion procedures to produce estimates of total effort (and potentially catch) and this would produce some lag time in these estimates, although possibly shorter than for MRFSS. Although such a targeted approach could make recreational data more timely, it still might be difficult to use in short-season fisheries and for seasonally migrating stocks. Each of these options would undoubtedly increase the cost of collecting recreational fisheries data.

Despite these hurdles, the value of in-season estimates of recreational catch for important fisheries such as summer flounder deserves further study. Implementation of in-season tracking of recreational catch data could revolutionize management of fisheries with significant recreational catch. Offsetting the cost of making MRFSS data more timely would be the benefit of reducing the likelihood that recreational fisheries (e.g., the summer flounder fishery) would overrun their portion of the overall TAC for a fishery.

NMFS should explore the possibility of modifying MRFSS survey contracts to require rapid feedback to fishery managers and others and to weigh the financial and precision costs of such estimates against the value of greater timeliness.

The committee recognizes that each of these approaches will face substantial opposition from anglers, but believes that drastic steps are necessary because the expanding recreational components in many fisheries are making the fisheries less manageable.

Auxiliary Information

Some information cannot now be input directly into an assessment model (and thus could be considered as "auxiliary" to biological data), but is relevant to the assessment or decision-making process to help explain why fish stocks fluctuate (physical and biological oceanographic data), to predict how stocks might fluctuate (climate forecasts), and to understand the outcomes of fishery regulations (economic and social data). In general, stock assessment and forecast models do not incorporate social, economic, and environmental data. Multispecies virtual population analysis models have been developed to take into account predator-prey relationships among commercial species, but these models still do not generally consider non-commercial species or ecosystem-level effects. Also, existing multispecies models are used primarily to estimate natural mortality rates (an important task in itself) but not for more general management.

Social and Economic Data

Findings: Many fishery problems that appear to result from the biology of fish stocks are actually rooted in the economics of a fishery. For example, low catches and catch rates, poor recruitment, excessive impacts of fishing gear on habitats, bycatch, and attacks on data and science are the expected result of the economic factors of overcapitalization and overharvest, which are the end results of an open-access fishery. Economic data are needed to document the extent of overcapitalization and to assist in designing mechanisms to bring fishing, economic stability, and sustainable yields into balance. Such information is important because the behavior of fishermen, whether recreational or commercial, is influenced by economic conditions. The ability of fishermen to switch from targeting one fish stock to another is impacted by prices paid by fish dealers and by management systems (e.g., licensing and single-species quotas). As one fisherman told the committee: "There is an absolute need to develop systematic socio-economic fisher and vessel data that doesn't rely on 10-page surveys administered every 5 years in the face of a major allocation decision. When the agency attempts to collect socio-economic data against the backdrop of a particular decision, it is inevitable that fishers will try to anticipate the use to which the information is being put and answer accordingly." (D. Fraser, commercial fisherman, personal communication, 1999)

Recommendations: Congress should authorize and support NMFS in the routine collection of economic data for commercial and recreational fisheries. Congress must first make such data collection legal (see also NRC, 1999b) by eliminating prohibitions on collecting economic and financial fisheries data in the Magnuson-Stevens Act (Sec. 303[b][7] and 402[a]) to make commercial fisheries and processor data more accessible. Other confidentiality provisions in the Magnuson-Stevens Act and state laws are adequate to protect any economic data collected.

Overfished fisheries that are also overcapitalized should be identified. Economic data collected should include characterizations of the distribution of benefits and costs associated with existing fishing practices and with proposed regulations—who receives or will receive financial or social benefits versus who may bear the associated costs. Councils and NMFS should identify economic impacts of regulations before and after they are enacted. The success of such assessments will depend on the results of yet-to-be-funded research on the types of regulatory frameworks and economic contexts that promote fishermen's behavior that is conducive to sustaining marine fisheries. Fishermen requested that regulations be maintained for long enough to assess their impacts; this is a good strategy, but may require larger, less frequent changes that might be resisted more by the industry.

NMFS also should support the construction of coupled biological-economic models that incorporate behavioral assumptions regarding fishermen so that the effects of economic behavior on fishing effort can be modeled. These models will need an increased quantity of social and economic data from fishermen. Prices paid to fishermen by dealers and to dealers by processors, and fishing cost surveys are needed for such a stock assessment process using bioeconomic models. This approach has the potential to improve both stock assessment and management and to help predict the impact of management changes on fish stocks, individual fishermen, and fishing communities.

Social data needed include information about employment, community structure, and how fishermen may respond to management (including shifts to other fisheries), as well as social and economic data for recreational fisheries. PFMC (1998a) noted that baseline economic and social data on the fishing industry and communities is needed to better predict potential impacts of fisheries management. Examples of baseline descriptive data include "vessel characteristics, fishing strategies, catch mixes, and vessel mobility for both commercial vessels and recreational charter vessels." (PFMC, 1998a).

Environmental and Ecosystem Data

A variety of data should be collected to monitor how the environment affects fish stocks and marine ecosystems, and how fishing affects marine ecosystems.

Effects of Fishing on Marine Ecosystems

Findings: In recent years, increased concern has been expressed about the effect of fishing on marine ecosystems, in addition to its direct effect on target species of fish. The Magnuson-Stevens Act requires that fishery management plans minimize, to the extent practicable, adverse effects of fishing on essential fish habitat. Regulations implementing the act require that fishery management plans contain assessments of the impacts of fishing gear. The information available on fishing effects is quite limited. The ecosystem effects of fishing activities and ecosystem-based fisheries management are thus important concerns of fisheries management worldwide (Ecosystem Principles Advisory Panel, 1999; Hall, 1999; Kaiser and deGroot, 1999; NRC, 1999a). Clearly, fisheries management and fisheries science must both consider a wider range of factors than has previously been the case, when data collected for fisheries management tended to focus on individual species for the purpose of stock assessments.

In the past, ecosystem data have been collected to provide information for scientific research projects, as a by-product of sampling target species. To some fishery agencies (and bodies that are responsible for appropriating funding to these agencies), long-term ecosystem monitoring may appear to be unimportant, so that with any cost cutting or shifting demands on scientific staff, these may be the first activities to be eliminated. This can be a short-sighted and risky strategy. NMFS must do its best, now, to determine what data from the present era will be needed five, ten, twenty, or more years in the future, and ensure that the data are collected. Similarly, it must determine what long-range research strategies must be instituted now, and ensure that the data are available to develop and defend these strategies.

The effects of fishing on the wider ecosystem will require using studies of the effects of different fishing gear and fishing methods on target species, non-target species, and marine ecosystems. Studies of the fishing mortality rate on vulnerable species are particularly important. Mortality might be better understood by investigations of the size and age structure of populations. The amounts and kinds of data that surveys collect and process already press the limit of available resources. Any expansion of monitoring will require additional funding and personnel.

Recommendations: Congress and NMFS should ensure that adequate funding is available to conduct monitoring of marine ecosystems and to study the effects of fishing on them. Variables that should be monitored include the biomass and recruitment of commercial and non-commercial species, levels of fishing activity, the quantity and quality of essential fish habitats, and the occurrence of other events and processes that can confound the effects of fishing on marine ecosystems. Fishery surveys may provide a cost-effective platform for the conduct of necessary research. Studies of the feeding habits and the distribution and types of prey and predators of important non-target species should also be conducted to understand the functioning of the marine ecosystems that could be affected by fishing activities. After ensuring that ecosystem-level monitoring is underway, the next most important activity is to understand how fishing affects marine ecosystems. The results of such studies may help managers reduce negative effects of fishing.

Effects of the Marine Environment on Fish Populations

Findings: It has been known for some time that environmental factors influence recruitment, as indicated by the poor relationship between the size of the spawning stock biomass and recruitment for many species. Evidence continues to grow in support of such factors as ocean temperature, salinity, dissolved oxygen concentration, phytoplankton type and abundance, and predation as having a major impact on recruitment and subsequent year-class strength (Murawski, 1993; Hixon and Carr, 1997). In addition, the relationships between environmental factors—particularly temperature—and maturity and growth may be more discernable than relationships between environmental factors and recruitment. Direct incorporation of environmental data in stock assessments is limited, but many analysts now recognize the importance of environmental effects and environmental forecasting. However, many stock assessment models and management control rules still assume that recruitment is proportional to spawning stock biomass and that the major dynamics of populations are driven by fishing-induced mortality.

Both environmental factors and spawning stock biomass are likely to be strong sources of variation in population abundance. The focus on extreme views in the debate has largely stalled the development and acceptance of assessment tools and management protocols that account for both factors. One reason for this is that the two sides of the debate are based on different philosophical paradigms. It seems best for management to argue the role of density-dependence, as this provides a strong and clear mechanism for optimal control over population abundance and yield. It seems best for fishermen to argue for density-independence, as this discounts human-induced causes of population change. Unfortunately, the density-dependent arguments ignore real regime shifts in these systems, whereas the density-independent arguments ignore the effects of human intervention.

Another important reason to be able to understand and predict effects of the environment on fish populations is the need to rebuild fish populations to historic high levels as required by the Magnuson-Stevens Act. Such levels may have occurred under more favorable environmental conditions of the past, but may be unattainable under current conditions. Repeated missing of rebuilding targets and consequent disruptions to management could result.

Recommendations: NMFS scientists and managers, in association with fishery stakeholders (especially fishermen), should move toward envisioning fisheries management in a more complete ecosystem context. This process has already begun (NRC, 1999a) and the Magnuson-Stevens Act now requires consideration of essential fish habitat, for example, in fishery management plans (Sec. 303[a][7]). Such activities require routine use of environmental data for stock assessments in addition to the more traditional population statistics. High-quality, long-term data sets will continue to be required to assess the influence of environmental conditions on fisheries, including both density-dependent and density-independent factors. NMFS should ensure the capability to link its proposed Fisheries Information System with environmental data from other sources. Geospatial data will become increasingly important as managers try to move towards characterizing environmental effects.

NMFS should review its stock assessment and forecasting methods, as well as the life-history characteristics used in developing management thresholds, to determine the need for more explicit inclusion of environmental information. Important examples of fish and shellfish species for which management thresholds should be reviewed are those that exhibit a small number of dominant year classes (e.g., surf clams, ocean quahogs, West Coast rockfish, Pacific whiting) or short life spans (e.g., shrimp). NMFS should encourage development of stock projection models that include species-specific environmental variables in addition to density-dependent mechanisms where appropriate. In addition to using the relationship between environmental factors and recruitment in assessments, quantifying the relationships between environmental factors and the distribution and growth of fish and using these in stock assessments could yield improvements in assessments more quickly. NMFS scientists should develop the scientific methods needed to facilitate recognition of regime shifts, and take steps towards incorporating such methods where necessary (e.g., incorporating realistic expected recruitment in short-term projections that consider existing environmental regimes). Furthermore, NMFS should establish the means to collect the data required by the new generation of models. Many kinds of environmental data are already being collected by other government agencies (e.g., other parts of NOAA, NASA, EPA) and are available from these sources. It will be necessary to have environmental data on a scale and reference grid that is compatible with needs for fisheries management. Continued research on the effects of the environment on fish population size, individual growth, survivorship, and spawning potential in conjunction with density-dependent effects is crucial for determining the kinds of data collection and stock assessments that should be used for any given species.

Cooperation and Communication

Cooperation With Industry

Findings: Some important forms of fishery-dependent data are either not collected (e.g., some economic and social data) or are underutilized because their accuracy is mistrusted. The accuracy of fishery-dependent data can be suspect because of incentives to misreport catch levels and locations and the effects of changes in management on catch and reporting. Despite these potential shortcomings, better use of commercial data could increase the ability of NMFS to sample the ocean and augment its surveys in a cost-effective manner. An important point that arose during the public meetings is that fishermen would be more interested in collaborating if the information they provide is used to help them and they are able to find out the results of the information and analyses more quickly.

Cooperation of NMFS with industry differs considerably among regions of the United States. Gear type (trawl versus fixed), past interactions, and agency use of funds play a role in determining the degree to which fishermen are (or are not) involved in data collection. Involving fishermen in data collection has an important sociological role as the fishermen gain a degree of "ownership" of the information process. But, a scientific role is met as well, by providing NMFS with data they could not otherwise obtain and a means for fishermen and scientists to communicate directly. Such cooperation could be extended to include funding provided by industry for directed research. When fishermen are involved in the data collection process, there are fewer complaints about the quality of the data, based on experiences of committee members.[3] Fishermen

[3] For example, Stephen Smith reports: "In meetings at which only government survey results were presented, survey procedures and results were a common target of industry complaints, whereas in meetings where industry surveys were one of, or the only, survey, very little criticism was leveled at the survey (although the results may be much discussed). All surveys are designed cooperatively by scientists of the Canadian Department of Fisheries and Oceans and the fishing industry, and conducted on fishing industry vessels with both fishery and scientific personnel participating. The following references provide examples of industry-based surveys being used. In the first two (Hurley et al., 1998; Mohn et al., 1998), the industry survey is used in conjunction with the government surveys while in the latter two (Koeller et al., 1998; Roberts et al., 1998) the industry survey is the only one available."

should more often accept opportunities to participate in joint exercises.

An external report assessing NOAA fisheries research vessels also recommended that NMFS, nationwide, put more emphasis on developing "best practices" for interactions with fishermen, indicating that "NMFS as a whole needs to devote very serious effort and experimentation to the sociological aspects of its operations. They could use professional assistance in this area, as part of the legally-mandated increased emphasis on socioeconomic considerations in ecosystem-based management.... NMFS must partner with a much broader segment of the interested community—fishermen, academics, industrialists—to a much greater degree than it ever has, in all aspects of its fisheries oceanography and fisheries monitoring and survey efforts" (Dorman, 1998).

Recommendations: NMFS should identify approaches that maintain the statistical rigor needed for long-term fishery assessments, while making the best use of local knowledge among commercial fishermen with expertise about specific stocks and gear types that are efficient at catching targeted stocks. Canada, New Zealand, and other nations provide models for positive interactions with industry.

NMFS should consider hiring commercial fisherman to participate in surveys (in addition to opportunities for unpaid participation) to see how sampling gear is used and where the surveys are conducted. NMFS also should carry out some joint sampling cruises using NMFS and commercial vessels, with exchanges of crew.

Harms and Sylvia (1999) suggested that collaborative research between fishermen and scientists should be "undertaken only if there is (1) equal partnership in planning and implementation, (2) adequate funding, (3) competent management, and (4) commitment to begin small and build on success." If cooperative research is to be adopted on a broader basis, institutional changes in both industry and government agencies will be needed, and the recommendations of Harms and Sylvia should be considered.

Cooperation with Recreational Fishermen

Findings: The lack of a national program for saltwater fishing licenses greatly complicates estimation of recreational catch and effort. Such a requirement is controversial because many states do not presently require licenses, and anglers in those states do not want to face additional regulations and a perceived intrusion of government into their private lives. However, requiring licenses for marine recreational fishing (even free ones) could improve data collection efforts by providing a comprehensive sampling frame and eliminating the inefficient random-digit dialing surveys (but not the expensive intercept surveys). License frames are of greatest value when they obtain uniform information and are coordinated among the states. In theory, recreational effort assessments could be less costly (in terms of time, money, and staff) if based on license sample frames, because anglers would be identifiable and sampled more easily than with current methods. However, some states' requirements for saltwater licenses exempt certain classes of anglers (see Table 3-9), which would complicate attempts to use saltwater recreational fishing licenses.

Recommendations: NFMS should increase its dialogue with recreational fishermen to jointly develop and implement improved data collection for recreational fisheries, through MRFSS, specially designed and coordinated tagging studies, and alternatives to MRFSS that allow in-season adjustments to recreational quotas. MRFSS should continue to evaluate whether saltwater fishing licenses and longitudinal sampling would provide cost-effective alternatives to random-digit dialing, keeping in mind that merely using license frames may not make MRFSS more timely or cost-efficient if different states use vastly different licensing programs.

Review

Findings: The review of data collection procedures is usually handled tangentially in the course of reviewing stock assessments, but few reviews focus on data collection.

Recommendations: NMFS, in conjunction with the regional councils, should review all aspects of its data collection activities, on a fixed, publicly-announced schedule including all types of fishery-dependent and fishery-independent data. Such reviews should include both a scientific peer review and a stakeholder review. As part of this process, commercial fishermen and other stakeholders should participate in actual data collection exercises.

IMPROVING DATA MANAGEMENT

Defining User Groups and User Needs

Findings: Many types of stakeholders could be users of new fishery databases. Different user groups (federal and state agencies, fishermen, scientists, managers, environmental advocacy groups, consumers, local communities) have different needs and concerns about fish stocks and marine ecosystems. The primary focus of most existing data collection and management activities is on collecting sufficient data to conduct an accurate and precise scientific assessment of biomass relative to previous years. However, other potential users have different questions. For example, fishermen may want to know how the catch of different species correlates with environmental variables and how their business compares against port and fleet average performance. Community leaders may want to know how fishing is expected to fare in future years, whether fishing activity is shifting to or from their port, and the value of the local multiplier effect for different kinds of fishing-related businesses. Environmental advocates might want to know how fishing is affecting marine ecosystems and trends in bycatch. NMFS seems to be giving a low priority to putting fisheries data into the same geospatial format used for data from other agencies and working with other agencies to conform to the same data standards.

Recommendations: Research on stakeholder and user concerns and needs should be conducted and used to improve the outcome of any new fisheries data management system. NMFS and others responsible for data collection should recognize that the data they collect may be used in a broader context. Stakeholders should be enlisted to identify broader uses for both traditional and new data. NMFS, interstate commissions, and states should involve expected and potential data users and providers in designing their fishery data management systems. Mechanisms for involvement should identify and address the concerns and needs of data users and providers in relation to fishery databases. The kind of involvement used in design of the Atlantic Cooperative Coastal Statistics Program (ACCSP) may serve as a useful model.

Databases and Data Management Systems

Findings: ACCSP has been a good model of a regional data management system up to this point in its development. For example, it has identified a core data set and requires inclusion of all necessary data in a single database available to all partners. Less desirable are systems that allow one partner—either states, commissions, or the federal government—to have absolute control of data.

Recommendations: Regional databases should standardize their data collection, management, and quality control activities. The committee agrees with the directive of Congress in requesting a plan for a nationwide Fisheries Information System (FIS). The FIS design (based on coordinated regional systems) is good and its reliance on national standards is a positive feature. The FIS is ambitious, however, and for it to be successful, (1) Congress must provide adequate funding and (2) cooperation and balance among regions must be ensured. ACCSP, other regional databases, and the FIS should specify the national and international standards to which the system will comply. NMFS should:

- continue its plans to implement national data standards that are designed to promote data exchange,
- include metadata with all data sets to ensure that data can be understood properly,
- promote the wide distribution of shareable data and metadata,
- create the ability to cross-validate logbook, observer, dealer, and VMS records to assess and improve the quality of fishery-dependent data, and
- work with the U.S. Fish and Wildlife Service (FWS) to prevent confusion between the NMFS Fisheries Information System and the FWS Fishery Information System, a data system for freshwater migratory fish.

Findings: Estimating the costs and benefits of existing and planned fisheries data management systems is an important task. Some analyses of costs and benefits of regulatory impact are required as part of fishery management plans (FMPs; MSFCMA Sec. 303[a][9]), although FMPs do not require analysis of data management as a separate phase. Analysis of the actual costs and benefits of an operational FMP often include factors that are difficult to quantify in economic terms, such as better information availability and streamlining an activity. The committee agrees with NMFS that developing an FIS is critical for fulfilling its responsibilities pursuant to the Government Performance and Results Act (GPRA) of 1993 in terms of collecting relevant data, tracking progress in achieving performance goals, and ensuring integrity and accountability.

Recommendations: NMFS should continue to attempt to find ways to contain costs and increase benefits from its fisheries data management activities. In part, this can be accomplished by continued cooperation with states and regions in data management and looking for opportunities to build on existing efforts (e.g., through FIS), rather than duplicate them. Another means to achieve this goal would be increased use of commercial data and better recreational data that could be available by following recommendations given earlier in this chapter.

Institutional Arrangements

Findings: Several different paths have been taken in different regions of the United States to manage fisheries data. ACCSP and FIS are providing (or plan to provide) centralized regional and national databases, respectively. In contrast, several commercial firms are providing value-added products to fishermen for operational purposes, including data for smaller areas and for specific fleets or fleet segments. Government data systems provide the potential benefits of access to all users, long-term storage, and nationwide standards for data collection, but are presently limited by the confidentiality of data and the often significant time lag between data collection and availability. Commercial systems can quickly provide very specific products and can establish whatever level of confidentiality is specified by a consortium of data contributors and financial supporters.

Recommendations: NMFS should encourage both centralized governmental and decentralized commercial data management, depending on the characteristics of the product needed and the capabilities of different sources to achieve the identified product characteristics. Centralized governmental data management can be achieved by continuing the development of the planned umbrella Fisheries Information System. At the same time, NMFS should identify which sources of data and information might best be managed by commercial firms and find ways to encourage such commercial development through Small Business Innovation Research grants, Cooperative Research And Development Agreements, and other programs.

Findings: Institutional arrangements for management of data are evolving toward systems of regional and national coordination and access. The situation in the Gulf of Mexico region, where funding was appropriated directly to the Gulf States Marine Fisheries Commission, with no assurances of partnerships in system development and data access, could hinder the implementation of the FIS and degrade data quality.

Recommendations: The Gulf of Mexico, Pacific, and other regions should follow the ACCSP lead in identifying needs and recognizing where efficiencies could be gained and quality could be improved by standardizing protocols and identifying regional concerns. Congress should ensure that systems are compatible and that the proper level of centralization or decentralization is achieved. The role of commissions in data management needs special attention. The overriding issue is whether a centralized, decentralized, or mixed approach to data management is most efficient and meets the needs of different levels of government. In relation to the commissions, NMFS should determine:

- What are the benefits and costs to federal management of having the commissions involved in collecting and managing data from the states?
- Would it be more cost-effective for the commissions, NMFS, or private contractors to manage state data?
- Does the imposition of an additional administrative layer between the states and NMFS hinder data flow and imposition of standards for data collection and management?

Implementing Standards and Improving Quality Control

Findings: Standards of potential importance to fisheries data management range from commonly accepted data and information standards such as SQL, TCP/IP, Z39.50 searchable indices, CORBA, and Government FIPS standards for federal systems, to Internet standards (TCP/IP, HTTP). Metadata standards of potential importance include the Federal Geographic Data Committee (FGDC) Content Standard[4] for Digital Geospatial Metadata and the Biological Data Profile of the Content Standard of Digital Geospatial Metadata. The U.S. Fish and Wildlife Service (FWS) has established and promoted data standards to ensure compatibility of data management and use throughout the agency.

Recommendations: NMFS should follow standards set by the Federal Geographic Data Committee, so that its data are compatible with data from other agencies. This is of more than bureaucratic interest because several agencies collect data that could be useful to fisheries management if NMFS data and data from other agencies were in compatible formats. NMFS should follow the FWS example in the attention given to stewardship, promotion of, and accessibility to data standards through Web-based information sources. Any standards specific to fisheries should be set cooperatively among stakeholders, as practiced in other industries such as electronics and computers (e.g., by the American National Standards Institute). Computer hardware to be used onboard fishing vessels should comply with the NMEA-0183 standard for marine interfaces. NMFS should attempt to implement nationwide standard error-checking procedures.

Improving Technologies

Findings: Data entry without verification and manual transfer of logbook data to electronic systems are opportunities for errors to enter databases. Electronic logbooks are a promising new development being pursued in a NMFS project, jointly with the industry, in the Pacific Northwest region. VMSs are either being used or will be used soon in many U.S. fisheries.

Recommendations: NMFS and the regional councils should require double-entry and other verification techniques and continue to pursue the possibilities for electronic submission of logbook information. NMFS should implement VMSs in such a way that their value is much greater than the added financial and bureaucratic costs to fishermen.

[4] Data content standards provide semantic definitions of a set of objects and the relationships among them (see http://www.fgdc.gov).

Review

Findings: Many of the regional data management systems seem to have arisen and accreted from within states or regions without much external review or even periodic internal review.

Recommendations: NMFS and other agencies with data management responsibilities should have their systems reviewed by outside experts, considering both how well they meet their intended purpose and how well they adhere to relevant national and international standards.

IMPROVING DATA USE

Data in Stock Assessments

Findings: Fishermen often believe that data are mishandled or misrepresented to support hidden agendas. This perception may arise because fishermen do not understand or agree with the significance or scientific merit of a particular data collection method. However, some suspicion may be understandable because it is not always clear why some data are used and other data are ignored. Changes in assessment methods appear to occur frequently and such changes often come as a surprise to fishermen. In the case of summer flounder, commercial fishermen believe that NMFS surveys do not cover the full range of the species, information from commercial logbooks is ignored, sportfishing information is lacking in timeliness and quality, and changes are made in management actions before the consequences of previous actions are fully realized.

The Food and Agriculture Organization's report *Precautionary Approach to Fisheries* recommended that stock assessment processes should include "a process for assessment analysis that is transparent. . . ." (FAO, 1995, p. 14), to improve the understanding and trust of stakeholders in assessments. *Improving Fish Stock Assessments* (NRC, 1998a) gave extensive advice for improving data use in terms of stock assessments:

- A variety of assessment models should be applied to the same data.
- Greater attention should be devoted to including independent estimates of natural mortality in assessment models.
- Fish stock assessments should include realistic measures of the uncertainty in the output variables whenever possible.
- New stock assessment techniques should be developed that can yield accurate and precise estimates, even though some of the data are incomplete, ambiguous, and variable. New techniques should also take into account the effects of environmental fluctuations on fish populations.

Recommendations: NMFS should make its stock assessment process more transparent and accessible to stakeholders. Prior to a major modification in any aspect of a stock assessment procedure, NMFS scientists should discuss proposed changes, and the reasons for them, with key user groups (i.e., fishermen, managers, and environmental advocacy groups). In general, NMFS should make the objectives of each survey clear, based on existing and anticipated management programs. NMFS should present assessment results with and without the proposed changes in stock assessment procedures, to show how the results differ and why the modifications make the assessment more realistic. When a survey design is modified, stakeholders should be informed of the changes and what they are intended to accomplish. Feedback from stakeholders should be acknowledged and addressed either by incorporating suggestions or by providing a reasonable rationale as to why the existing approach is most appropriate. Objectives of stock assessments should be communicated to stakeholders in a way that can be understood. NMFS should review the objectives of surveys periodically and publicly to ensure that surveys are designed correctly to meet these objectives.

The committee supports the recommendations of NRC (1998a) listed above. Institution of these measures would help fishery scientists and managers understand fisheries better and understand the level of uncertainty more fully.

Access to Fisheries Data

Findings: It is difficult for individuals outside NMFS to access many types of fisheries data held by the agency, although NMFS does provide some basic queries for both commercial and recreational data on its Web site. Even data that do not need to be confidential can be difficult to find and use because they are kept on many different servers, do not feature user-friendly interfaces, and/or require special permission to access. Furthermore, data that are available are not always in a user-friendly or even a computer-friendly format. The Open GIS Consortium is well advanced in developing Web-based access to geospatial data and provides a ready standard for government data systems.

Recommendations: NMFS should develop and publicize (including on their Web site) a data access policy and instructions for accessing data, which may need to be different for different users because of confidentiality concerns. Standardized formats or access through standardized query programs should be enhanced so that any qualified user can access and use the data readily to answer questions that arise across data sets, and over a variety of scales within data sets. NMFS should participate actively in the interagency Open GIS Consortium and adhere to consortium guidelines in creating its systems for data management and access, so that its data are compatible with data from other federal agencies and accessible through the same means.

Fishery scientists in NMFS and academia should continue to have access to data from logbooks and observers without aggregation, if they agree to respect confidentiality of the data when reporting the results of their research. Access to data and metadata both within and outside NMFS should be encouraged because involving new people in data analysis can provide new perspectives about the data and a greater number and variety of users is more likely to reveal problems and errors in data sets, improving quality control.

Confidentiality

Findings: Confidentiality of fisheries data is restrictive to the point of hindering both research and management. State and federal restrictions to free access to data can hinder development of the kind of socioeconomic models that enable scientists and managers to determine whether regulations and management measures have been effective or to predict whether potential measures are likely to be effective. These policies neglect the rights of the public to have greater information about the use of public-trust resources. The privilege to exploit marine fish resources should carry some obligation on the part of fishermen, balancing reasonable protection of proprietary information against the large need of managers to be well informed and able to manage the fishery. For example, fishing in some areas may be detrimental to the environment. If information on fishing areas is confidential, interested stakeholders would have a difficult time determining how much fishing is being conducted in sensitive areas. Conversely, some level of confidentiality may be necessary to allow fishermen to maintain their businesses and to promote reporting of high-quality information about location, landings, and bycatch in some fisheries (e.g., reporting of halibut fishing information to the International Pacific Halibut Commission), information that might not be as accurate if it were not confidential. NMFS and its partners in development of new fishery data management systems (e.g., the ACCSP) have taken the approach of creating cooperative systems managed independently of NMFS and with access from all contributing partners. Because of such independence and broad participation, such systems may be more credible with all stakeholders than the current generation of systems that are less accessible to all stakeholders.

Coastal states (except Alaska) share the same data confidentiality standards, with the rule of thumb being that there must be three or more reporting entities (e.g., vessels or seafood dealers) at any level of data summary before the data are not confidential and can be available to the public. Alaska requires a minimum of four or more reporting entities. On the U.S. West Coast, states have authority to collect logbook information and they set policies on confidentiality. States use a common (NMFS-designed) logbook and store the data with NMFS, but NMFS must respect state data policies. For some states, NMFS is not allowed full access to all state data; for example, Alaska restricts access to confidential data, except to selected NMFS personnel for stock assessment and law enforcement purposes.

Two recommendations of NMFS (undated) that are particularly relevant to this report are that the next reauthorization of the Magnuson-Stevens Act should (1) establish a sunset period on confidentiality of fisheries data and (2) eliminate existing prohibitions on collecting economic and financial statistics from marine fisheries.

Recommendations: Congress and the states should re-evaluate their existing policies on data confidentiality, while respecting the rights of fishermen as small business owners as much as possible. The assertion that fisheries information has intrinsic value proprietary to the fisherman should be examined in light of the mandate to NMFS and the regional councils to manage fisheries as public-trust resources. Specific consideration should be given to establishing sunset periods for data confidentiality for all fisheries data, when the information no longer retains significant proprietary value. The proprietary periods should be determined for each fishery in a public forum including scientists, managers, fishermen, and environmental advocates, and should be included in the fishery management plan for each fishery, as each fishery is unique. Sunset provisions for data confidentiality should be developed by government (states and federal) working with data providers. The effects of the loss of confidentiality on precision and bias in logbook and other reporting should be considered in setting the proprietary period for each type of data.

Matching Management to Data Available

Findings: The effectiveness of fisheries management depends on the use of timely data of suitable accuracy and precision to provide answers to questions about a stock's current status, desired future status, and actions needed to achieve the desired status. Greater precision and accuracy often require greater resources for sampling and analysis. Moreover, some sources of inaccuracy and imprecision may be difficult to eradicate even with unlimited expenditures. Thus, the question for management should become: What level of inaccuracy and imprecision in advice and hence in management is tolerable if a particular management regime is to achieve its goals? The answer to this question clearly depends on the management regime chosen and its objectives. If managers are not prepared to pay for the needed precision or if that precision may not be achievable at any price, they may have to modify either their management objectives or management tools. As demonstrated in Chapter 3, management and data needs are closely related.

Recommendations: The regional fishery management councils and NMFS should work together to match management to data that are available at a reasonable cost. Such an analysis will depend on completion of the review of the costs and benefits of fisheries and data collection recommended earlier in this chapter.

Cooperation and Communication

Findings: Part of the image problem shared by NMFS and the regional councils is lack of communication on a level that is informative and appealing to stakeholders. Many stakeholder groups are affected by the collection and use of fisheries data, including commercial and recreational fishermen, and environmental advocacy groups. Few stakeholder groups have a good understanding of why fisheries data are collected and used in certain ways. Greater outreach to these audiences to improve their understanding and the perceived credibility of fisheries data is needed. User groups need to know about the information contained in a data set, even though the analysis for which it is used may be complex.

It is important for scientists and managers to improve their communication of the data available and to make such data available to stakeholders more readily and in a user-friendly form. When this is not achieved, a lack of trust develops between those who control access to data and those who cannot gain access. In many cases, disagreement of fishermen with the results of stock assessments can be traced to NMFS not explaining the sources of variability in the data and the uncertainty of the models being used.

In the current fisheries management system, several activities occur sequentially:

- Data are collected.
- Stock assessments are conducted.
- Management recommendations are made.
- Fish are allocated among user groups.
- Fishing regulations are designed and implemented.

For individuals, this process determines either their opportunity to make a living (commercial and charter sectors) or their ability to engage in their recreational activities. When their income or their favorite pastime is threatened, people respond by attempting to manipulate the management process at every stage. No amount of communication or transparency of the process will eliminate these conflicts among users. But a more open and innovative assessment process could improve the credibility of the resulting assessments and perhaps move harvester conflicts away from the stock assessments to other stages.

Recommendations: NMFS should improve its outreach to specific stakeholder audiences by seeking (1) perspectives from stakeholders regarding what information they would find useful and how best to get it to them and (2) perspectives from data gatherers and stock assessment scientists regarding what is important for user groups to know about the quality and limitations of data. NMFS and councils should ensure that stakeholders feel they are getting all the information they need to make decisions and understand how councils use data to make management decisions. Such outreach could take place in conjunction with meetings of the regional councils and/or at special meetings convened by NMFS in port areas. Outreach events would be one way to communicate information (and uncertainty) to target audiences in person. Another complementary approach would be to make data more accessible through Web-page queries and more sophisticated forms of graphical data presentation. Several efforts by NMFS and other parts of NOAA—providing the ability to query aggregated commercial and recreational data through the NMFS Web site and data visualizations provided through the Pacific Marine Environmental Laboratory Web site—are a first step in the next generation of data access. NMFS could use its new communication experts for this purpose, remembering that communication is a two-way process, not merely a one-way dissemination of information. Increased cooperation and communication between NMFS and industry is crucial for improving the collection and understanding of fisheries data. However, it is important that NMFS conduct all activities with industry in an open manner to avoid the appearance of collusion.

NMFS should improve its use of data visualization to communicate scientific information to the public, moving away from display in tabular form to more graphic displays, including plots, maps, and pictures, without lowering the quality of its data or presentations. The accuracy of the information available needs to be conveyed (e.g., maps are not always equally accurate over the entire range displayed). The methods and assumptions used for data summarization or analysis should be included with the data and analyses.

In communicating data and results of analyses, NMFS should tailor its approaches to different audiences that may require different levels of detail. It would be useful for the public to see what the different data sources (commercial, recreational, and survey) indicate about a fish stock, including overlays on maps to illustrate geographic coverage. Such communication should avoid statistical and other jargon so that non-specialists, fishermen, and the public can fully understand the significance of the information. NMFS should consider using state Sea Grant Marine Advisory Service units more often to help with outreach.

One innovative and useful approach is to conduct fishery assessment and management simulations with real fisheries data in a workshop setting to explore with commercial fishermen, recreational fishermen, environmental advocates, and others how assessments are developed and how management decisions are based on assessments. This approach has been used by several fishery scientists to provide opportunities to focus attention of stakeholders on the models and data, rather than on each other (Holling, 1978; Walters, 1986, 1994). In using this approach, it is important that the objectives of the workshops be very specific and that they be conducted outside NMFS, to provide an objective mediator. Simulation workshops should explore issues such as the following:

- What are the management implications of using random stratified methods for surveys versus going where fishermen know there are concentrations of fish?
- Why are survey gear and methods kept constant over time?
- Why is it difficult to use short-term charter vessels for trawl surveys?
- Does it make a difference in setting TACs whether a decline in recruitment, biomass, or catches are caused by fishing versus climate, habitat loss, pollution, or other environmental factors?
- How do unreported landings and bycatch affect assessments and subsequent management?
- How would observations of fishermen change the analyses if included? What kind of observations could and could not be included?

In each case, the goal should not be to justify NMFS procedures, but to expose the assumptions and procedures of modeling and TAC setting to the stakeholder communities and explore together what might be the consequences of changing assumptions. Such an activity might bring new insights to NMFS scientists, council members, commercial and recreational fishermen, environmental advocates, and others, and could indicate to NMFS changes needed in its assumptions or procedures. Preparation for simulation workshops should include thoughtful analysis and development of techniques and software. NMFS should seek assistance for such an effort from other parts of NOAA (e.g., the Joint Research Institutes and National Sea Grant College Program) and other sources.

NMFS should also create, and make available to the public, manuals describing survey operations that specify survey design, tow time, gear used, and other important factors, such as is done for West Coast fisheries (Anonymous, undated; Munro and Hoff, 1995; AFSC, 1998; Wilkins et al., 1998). NMFS should identify which data sources (including which individual surveys) are included in each stock assessment and what weight is given to each. If the weighting of data sources changes over time, NMFS should communicate what criteria are used to change weightings.

Uncertainty in Data, Models, and Model Outputs

Findings: Despite the significant effort that goes into designing and implementing fishery surveys to sample individual fish to characterize populations, survey data still usually enter assessments as point values with no corresponding estimate of uncertainty, such as variance. The result is an assessment that gives each observation equal value, whereas in reality some reflect more precise and/or more accurate information than others. This approach decreases incentives for improving data quality. One reason for not including uncertainty in models may be that it is difficult to distinguish among the different types of uncertainty entering an assessment model. Uncertainty can enter assessments through observation error, natural variation in the environment (process uncertainty), model misspecification, and uncertainty that arises in management implementation and institutional interactions. Another reason that uncertainty is sometimes not acknowledged is that managers and/or scientists are concerned that fishermen might use uncertainty as justification for less conservative regulations.

Recommendations: NMFS should take stronger steps to characterize uncertainty in their assessments and present uncertainty to fishery managers. This may involve an expansion of existing approaches, especially for key species or species complexes, to include a better representation of the level of each kind of uncertainty listed above. Recognizing that the added computational effort may slow the assessment process, the committee recommends that NMFS prioritize the types of uncertainty to address first, moving on to lower priority uncertainties as time and resources permit. Some sources of uncertainty might be determined less frequently and during model development, whereas others should be included with every new assessment. Approximations (e.g., Gaussian approximation using the Hessian approach) might be used for annual updates, whereas better and more complete estimates (e.g., Monte Carlo, Markov Chain, or full bootstrap procedures) might be carried out as assessments are reviewed or when new assessments are developed. NMFS must also decide how it wants to evaluate and present information on variability and uncertainty in a way that helps and does not hinder management. It is important that a case be made for why significant uncertainty should compel managers to be cautious, rather than giving a signal to stakeholders that stock assessment scientists do not know the condition of the stock and thus become an excuse for inaction or for selection of TAC levels that are too high. Presentations should include information about the risk involved in different TAC levels that could be set.

Review

Findings: The information content of an assessment, namely the data and the model structure and assumptions, should be subject to periodic peer review. This serves two purposes. The first is to have independent scientists scrutinize the process and the product to give public assurance of its scientific integrity, or to identify problems in the analyses. The second is to inject new ideas and research directions into the process that reflect the current state of the art. NMFS and the councils have experimented with different methods of peer review. Currently NMFS, in cooperation with the councils, has several ad hoc committees that review completed assessments prior to submission to a council. Two such committees are the Stock Assessment Review Committee (SARC), which reviews NMFS' East Coast assessments, and the STock Assessment Review (STAR) panel, which reviews NMFS' West Coast assessments. A standing body of advisory scientists called the Scientific and Statistical Committee (SSC) also exists for each council. In addition, several plan teams exist for each council, but their primarily role is to provide management guidance rather than provide a review per se. This peer review system, as it currently operates, has some shortcomings. The SARC and STAR processes, although comprehensive, are not truly independent because the review scientists are typically NMFS employees from other regions and state scientists from the region of concern. The SSC, on the other hand, although generally showing greater independence, at least in terms of composition (although NMFS scientists are on these committees also), does not have the opportunity to examine the issues in as fine a detail as the SARC/STAR panels or as, for example, the committee has in the course of its review. Together, the SSCs and SARC/STAR groups represent the thoroughness and independence needed in a review, but deeper consideration of this process might lead to a more constructive framework. NMFS is experimenting with a Center for Independent Experts to conduct independent reviews, but this approach most often has used single external reviewers rather than teams.

Recommendations: Congress should recognize the need for both a scientific assessment and evaluation process, and a thorough and objective review system. The committee acknowledges that NMFS has taken great effort to provide the means for establishing the scientific integrity of its work, but the committee also recognizes that it is too much to ask NMFS to objectively evaluate that same work. A greater degree of independence in the peer-review process is needed in order to maintain the integrity and scientific credibility of the NMFS assessments. To do this, the scientific products, such as the assessments produced by NMFS and others, should be peer reviewed by scientists who are not directly involved in the assessments or work directly for NMFS (scientists from state fishery agencies could be involved). Because of the limited number of stock assessment scientists outside NMFS and the large number of assessments produced each year, not every assessment should be reviewed each year, but every assessment should be externally reviewed on a regular basis, for example, every three to five years. The SSC could be considered the appropriate independent body for this review, if the SSC is made up of informed but otherwise independent scientists, but time is often a limiting factor for these volunteers in their deliberations leading to council advice. The committee therefore suggests that regional councils consider supporting a stock assessment scientist on staff. Such a scientist could be assigned the task of organizing assessment peer reviews while highlighting issues of scientific and managerial concern under the direction of the SSC and the council executive director. In this way, NMFS could present and defend their work in a public forum and the councils would be able to review this work in an objective fashion. Another vehicle that could be used for such independent reviews is the Center for Independent Experts that is funded by NMFS but which operates independently from NMFS.

RESEARCH NEEDS

Finding: Stock assessment science and fisheries management are still developing fields. Improvements in each are still needed and will be fueled by continued research and development.

Recommendations: Congress should support and NMFS should continue to fund research to improve our ability to characterize fish stocks quantitatively and manage them in the context of the important but sometimes conflicting goals of the Magnuson-Stevens Act and its National Standards. NMFS should fund both internal and external research (biological, economic, and social) relating to:

- developing methods for evaluating ecological benefits of fish stocks and fisheries;
- developing new methods for stock assessment;
- minimizing data fouling and misreporting;
- testing adaptive sampling for surveys, including both NMFS and industry data collection;
- testing electronic logbooks and VMSs that offer value-added features to fishermen;
- linking environmental, economic, and social data, and climate forecasts to stock assessments;
- studying the feeding habits and the distribution and types of prey and predators of important non-commercial species, to understand the functioning of the marine ecosystems affected by fishing activities;
- understanding the economic and social motivations of harvesters so that greater use can be made of fishery-dependent data;
- improving design of recreational fishing surveys; and
- conducting stock assessments combining recreational and commercial data with very different error and uncertainty structures.

Emphasis should be given to research exploring the relationships between different types of regulatory approaches and fishermen's attitudes and behaviors toward fish harvest and data reporting. Research should also identify the most important incentives and disincentives that could be used to promote accurate reporting.

References

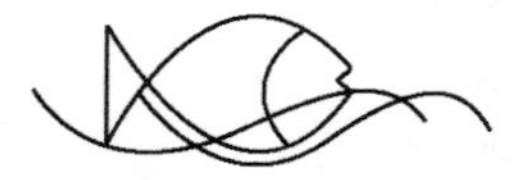

Able, K.W., and M.P. Fahay. 1998. *The First Year in the Life of Estuarine Fishes in the Middle Atlantic Bight.* Rutgers University Press, New Brunswick, New Jersey.

Able, K.W., and S.C. Kaiser. 1994. *Synthesis of Summer Flounder Habitat Parameters.* NOAA Coastal Ocean Program Decision Analysis Series No. 1, National Oceanic and Atmospheric Administration, Silver Spring, Maryland.

Alaska Fisheries Science Center (AFSC). 1998. *Report to the Fishing Industry on the Results of the 1998 Eastern Bering Sea Groundfish Survey.* AFSC Processed Report 98-09, Seattle, Washington.

Alverson, D.L., M.H. Freeman, S.A. Murawski, and J.G. Pope. 1994. *A Global Assessment of Fisheries Bycatch and Discards.* FAO Fisheries Technical Paper, Rome.

Anonymous. Undated. *Operation Plan and Sampling Manual: 1998 West Coast Triennial Bottom Trawl Survey.* Publisher unknown.

Anonymous. 1995. *Incorporating Uncertainty Into Stock Projections: Report of the Scientific Meeting.* April 3-7. Hobart, Tasmania, CCSBT Document SBFWS/95/inf-3.

Anonymous. 1999. Georgia Standard Operating Procedures (SOP) Manual. Unpublished.

Anthony, V. 1982. The calculation of $F_{0.1}$: A plea for standardization. Northwest Atlantic Fisheries Organization. Serial Document SCR 82/VI/64, Halifax, Canada.

Atkinson, D.B. 1989. Diel movements of beaked redfish and the implications of these for stratified random bottom trawl estimates of biomass and abundance. *North American Journal of Fisheries Management* 9:163-170.

Atlantic Coastal Cooperative Statistics Program (ACCSP). Undated a. *ACCSP: Good Data, Good Decisions for Fisheries Management* Brochure. Atlantic Coastal Cooperative Statistics Program, Washington, D.C.

Atlantic Coastal Cooperative Statistics Program (ACCSP). Undated b. Version 1.5 of ACCSP Program Design. Atlantic Coastal Cooperative Statistics Program, Washington, D.C.

Atlantic Coastal Cooperative Statistics Program (ACCSP) Computer Steering Committee. 1997. *Project Plan for the Development of an Atlantic Coastal Cooperative Statistics Program Information System.* Atlantic Coastal Cooperative Statistics Program, Washington, D.C.

Atlantic Coastal Cooperative Statistics Program (ACCSP) Coordinating Council. 1999. *Proposals Funded ACCSP Funding Cycle.* Atlantic Coastal Cooperative Statistics Program, Washington, D.C. Unpublished document.

Atlantic States Marine Fisheries Commission (ASMFC). 1994. *MRFSS User's Manual: A Guide to Use of the National Marine Fisheries Service Marine Recreational Fisheries Statistics Survey Database.* Special Report of the Atlantic States Marine Fishery Commission, Washington, D.C. Report No. 37.

Atlantic States Marine Fisheries Commission (ASMFC). 1996. *Maintaining Current and Future Fisheries Resource Survey Capabilities: Workshop Findings and Recommendations.* Washington, D.C.

Azarovitz, T.R. 1994. Northeast Fisheries Science Center Bottom Trawl Surveys. Pp. 4-7 in T. Berger (ed.), *Proceedings of the Workshop on the Collection and Use of Trawl Survey Data for Fisheries Management.* Special Report of the Atlantic States Marine Fisheries Commission, Washington, D.C. Report No. 35.

Benaka, L.R. (ed.). 1999. *Fish Habitat: Essential Fish Habitat and Rehabilitiation.* American Fisheries Society Symposium 22. American Fisheries Society, Bethesda, Maryland.

Berrien, P., and J. Sibunka. 1999. *Distribution Patterns of Fish Eggs in the U.S. Northeast Continental Shelf Ecosystem, 1977-1987.* NOAA Technical Report, NMFS 145, Woods Hole, Massachusetts.

Block, B.A., H. Dewar, C. Farwell, and E.D. Prince. 1998. A new satellite technology for tracking the movements of Atlantic bluefin tuna. *Proceedings of the National Academy of Sciences U.S.A.* 95:9384-9389.

Bolstein, R. 1992. *Review of Current Expansion Routines.* Final Report to NMFS, MRFSS (to John Witzig), dated August 4, 1992. Unpublished manuscript.

Bolstein, R. 1993. Develop Formulations for Adjusting Intercept Survey Data Analysis to Account for Weekend/Weekday Stratification. Final Report to NMFS, MRFSS (to John Witzig), dated July 15, 1993. Unpublished manuscript.

Bolz, G., R. Monaghan, K. Lang, R. Gregory, and J. Burnett. In press. *Proceedings of the Summer Flounder Ageing Workshop,* February 1-2, 1999. NOAA Technical Memorandum, Woods Hole, Massachusetts.

Brooks, E.N., K.H. Pollock, J.M. Hoenig, and W.S. Hearn. 1998. Estimation of fishing and natural mortality from tagging studies on fisheries with two user groups. *Canadian Journal of Fisheries and Aquatic Sciences* 55:2001-2010.

Burns, T.S., R. Schultz, and B.L. Brown. 1983. The commercial catch sampling program in the northeastern United States. Pp. 82-95 in W.G. Doubleday and D. Rivard (eds.), *Sampling Commercial Catches of Marine Fish and Invertebrates.* Canadian Special Publication in Fisheries and Aquatic Sciences 66.

Byrne, D. 1994. Stock assessment of New Jersey's nearshore recreational fisheries resources. Pp. 36-42 in *Proceedings of the Workshop on the Collection and Use of Trawl Survey Data for Fisheries Management,* T. Berger (ed.). Special Report of the Atlantic States Marine Fisheries Commission, Washington, D.C. Report No. 35.

Byrne, C.J., and J.R.S. Forrester. 1991. Survey vessels and gear modifications and their possible effects on assessment analyses. Pp. 20-27 in *Report of the Twelfth Northeast Regional Stock Assessment Workshop (12th SAW), Spring 1991.* NEFSC Ref. Doc. 91-03.

Campana, S.E., and C.M. Jones. 1998. Radiocarbon from nuclear testing applied to age validation of black drum *Pagonias cromies. Fisheries Bulletin* 96:185-192.

Campana, S.E., K.C.T. Zwanenburg, and J.N. Smith. 1990. ^{210}Pb/^{226}Ra determination of longevity in redfish. *Canadian Journal of Fisheries and Aquatic Sciences* 47:163-165.

Chambers, S. 1998. Catch accounting: Is it good enough? *Pacific Fishing* Nov. 1998:77- 80, 115.

Chao, C., and S.K. Thompson. 1997. Optimal sampling design under a spatial model. Technical Report 97–11, Department of Statistics, Pennsylvania State University, State College, Pennsylvania.

Cochran, W.G. 1977. *Sampling Techniques,* 3rd ed. John Wiley & Sons, New York.

Conser, R., and J. Powers. 1989. Extensions of the ADAPT VPA tuning method designed to facilitate assessment work on tuna and swordfish stocks. International Commission for the Conservation of Atlantic Tunas (ICCAT), Madrid, Spain. Working Document SCRS/89/43.

Consortium for Oceanographic Research and Education (CORE). 2000. *The Use of Scientific Information in Fisheries Management.* Consortium for Oceanographic Research and Education, Washington, D.C. http://core.ssc.erc.msstate.edu/fishmanage.html

Correia, J. 1994. Massachusetts inshore bottom trawl survey. Pp. 22-23 in *Proceedings of the Workshop on the Collection and Use of Trawl Survey Data for Fisheries Management,* T. Berger (ed). Special Report of the Atlantic States Marine Fisheries Commission 35.

Cressie, N.A.C. 1993. *Statistics for Spatial Data.* Revised Edition. John Wiley & Sons, New York.

Darby, C.D., and S. Flatman. 1994. *Virtual Population Analysis: Version 3.1 (Windows/DOS) Users Guide.* MAFF Information Technology Series No 1, Directorate of Fisheries Research, Lowestoft, U.K.

Daspit, W., P. Crone, and D. Sampson. 1997. Pacific Fisheries Information Network. Chapter 6 in D. Sampson and P. Crone (eds.), *Commercial Fisheries Data Collection Procedures for U.S. Pacific Coast Groundfish.* NOAA Technical Memorandum NMFS-NWFSC-31. National Oceanic and Atmospheric Administration, Silver Spring, Maryland.

Department of Fisheries and Oceans and Pacific Blackcod Fishermen's Association (DFO and PBFA). 1998. *Pacific Sablefish Fishery: Presentation for Panel Studying Partnering in Canada's Fishing Industry*, October 14, 1998. Seattle, Washington.

Deriso, R.B., T.J. Quinn II, and P.R. Neal. 1985. Catch-at-age analysis with auxiliary information. *Canadian Journal of Fisheries and Aquatic Sciences* 42:815-824.

Desfosse, J.C. 1995. *Movements and Ecology of Summer Flounder,* Paralichthys dentatus, *Tagged in the Southern Mid-Atlantic Bight.* Ph.D. Dissertation, The College of William and Mary, Williamsburg, Virginia.

Dorman, C.E. 1998. Memorandum to Sarah Laskin, Office of Management and Budget. NOAA NMFS Fisheries Research Vessels. Unpublished.

Dorn, M., J. Ianelli, and S. Gaichas. 1997. Uncertainty in estimates of total catch for target bycatch species at varying observer coverage levels in the Alaska groundfish fisheries. Unpublished manuscript.

Ecosystem Principles Advisory Panel. 1999. *Ecosystem-Based Fishery Management.* National Marine Fisheries Service, Silver Spring, Maryland.

Engås, A., and A. Soldal. 1989. Diurnal variations in bottom trawl catch rates of cod and haddock and their influence on abundance indices. *ICES Journal of Marine Sciences* 49:89-95.

Environmental Protection Agency (EPA), Quality Assurance Management Staff (QAMS) within the Office of Modeling, Monitoring Systems, and Quality Assurance in the EPA's Office of Research and Development. Undated. Glossary of Quality Assurance Terms. Available online at: http://www.epa.gov/emap/html/pubs/docs/resdocs/qa_terms.html, accessed 06/12/00.

Executive Order 12906. 1994. Coordinating geographic data acquisition and access: The National Spatial Data Infrastructure. *Federal Register* 59(71):17671-17674.

Felt, L.F. 1994. Two tales of a fish: The social construction of indigenous knowledge among Atlantic Canadian salmon fishers. Pp. 251-286 in C.L. Dyer and J. R. McGoodwin, (eds.). *Folk Management in the World's Fisheries*, University of Colorado Press, Niwot.

Fogarty, M.J., G. DeLaney, J.W. Gilliton, Jr., J.C. Poole, D.E. Ralph, P.G. Scarlett, R.W. Smith, and S.J. Wilk. 1983. *Stock Discrimination of Summer Flounder (*Paralicthys dentatus*) in the Middle and South Atlantic Bights: Results of a Workshop.* NOAA Technical Memorandum NMFS-F/NEC-18. National Marine Fisheries Service, Woods Hole, Massachusetts.

Fogarty, M.J., and S.A. Murawski. 1998. Large-scale disturbance and the structure of marine ecosystems: Fishery impacts on Georges Bank. *Ecological Applications* 8(1):S6-S22.

Food and Agriculture Organization (FAO). 1995. *Precautionary Approach to Fisheries.* FAO Fisheries Technical Report 350. United Nations, Rome.

Fournier, D. 1996. *An Introduction to AD Model Builder For Use in Nonlinear Modeling and Statistics.* Otter Research Ltd., Nanaimo, B.C., Canada.

Fournier, D., J. Hampton, and J.R. Sibert. 1998. MULTIFAN-CL: A length-based, age-structured model for fisheries stock assessment, with application to the south Pacific albacore, *Thunnus alalunga*. *Canadian Journal of Fisheries and Aquatic Sciences* 55: 2105-2116.

Fox, D.S., and Starr, R.M. 1996. Comparison of commercial fishery and research catch data. *Canadian Journal of Fisheries and Aquatic Sciences* 53:2681-2694.

Francis, R.C., and S.R. Hare. 1994. Decadal-scale regime shifts in the large marine ecosystems of the Northeast Pacific: A case for historical science. *Fisheries Oceanography* 3:279-291.

Francis, R.C., S.R. Hare, A.B. Hollowed, and W.S. Wooster. 1998. Effects of interdecadal climate variability on the oceanic ecosystems of the Northeast Pacific. *Fisheries Oceanography* 7:1-21.

Francis, R.I.C.C. 1984. An adaptive strategy for stratified random trawl surveys. *New Zealand Journal of Marine and Freshwater Research* 18:59-71.

Gallagher, D.W. 1987. The commercial fishing industry's involvement in data acquisition and marine biological research. Pp. 289-311 in J.L. Bubier and A. Rieser (eds.). *East Coast Fisheries Law and Policy.* Marine Law Institute, Portland, Maine.

Gavaris, S. 1988. An adaptive framework for the estimation of population size. Canadian Atlantic Fisheries Science Advisory Committee (CAFSAC). Research Document 88/29.

Geer, P.J. 1994. Virginia Institute of Marine Science's trawl survey. Pp. 61-79 in *Proceedings of the Workshop on the Collection and Use of Trawl Survey Data for Fisheries Management*, T. Berger (ed.). Special Report of the Atlantic States Marine Fisheries Commission, Washington, D.C. Report No. 35.

General Accounting Office (GAO). 1986. *Deactivating Research Vessels: National Oceanic and Atmospheric Administration's Use of Private Ships.* General Accounting Office, Washington, D.C.

General Accounting Office (GAO). 2000. *Problems Remain With National Marine Fisheries Service's Implementation of the Magnuson-Stevens Act.* GAO/RCED-00-69. General Accounting Office, Washington, D.C.

Gray, G. 1997. Telephone surveys for recreational fishing effort: Random digit dialing versus an angler license frame. Unpublished manuscript.

Gulf States Marine Fisheries Commission (GSMFC). 1996. *Framework Plan: Fisheries Information Network for the Southeastern United States.* Gulf States Marine Fisheries Commission, Ocean Springs, Mississippi.

Gulland, J.A. 1955. Estimation of growth and mortality in commercial fish populations. *MAFF Fish. Invest. Ser.* 2:18.

Gulland, J.A. 1969. *Manual of Methods of Fish Stock Assessment.* FAO Manual of Fisheries Science 4. Food and Agriculture Organization of the United Nations, Rome.

Hall, S.J. 1999. *The Effects of Fishing on Marine Ecosystems and Communities.* Blackwell Science, Ltd., Oxford, U.K.

Hampton, J. 1991. Estimation of southern bluefin tuna *Thunnus maccoyii* natural mortality rates from tagging experiments. *Fishery Bulletin* 89:591-610.

Hampton, J., and J. Gunn. 1998. Exploitation and movement of yellowfin tuna (*Thunnus albacares*) and bigeye tuna (*Thunnus obesus*) tagged in the northwestern Coral Sea. *Marine and Freshwater Research* 49:465-489.

Hamre, J. 1980. Biology, exploitation and management of the northeast Atlantic mackerel. *Rapp. Procès-Verb. Réun. Cons. Int. Explor. Mer.* 177:212-242.

Hanna, S.S., and C.L. Smith. 1993. Attitudes of trawl vessel captains about work, resource use, and fishery management. *North American Journal of Fisheries Management* 13:367-375.

Hanselman, D.H., T.J. Quinn II, J. Heifetz, D. Clausen, and C. Lunsford. In press. Spatial inference from adaptive cluster sampling of Gulf of Alaska rockfish. 17th Lowell Wakefield Fisheries Symposium on Spatial Processes and Management of Fish Populations. Alaska Sea Grant College Program, Anchorage, Alaska.

Hare, S.R., N.J. Mantua, and R.C. Francis. 1999. Inverse production regimes: Alaska and west coast Pacific salmon. *Fisheries* 24:6-14.

Harms, J., and G. Sylvia. 1999. Scientists, industry share more than they know. *Pacific Fishing* March:41-43.

Harrison, G. 1997. *Oracle SQL High Performance Tuning.* Prentice-Hall, Upper Saddle River, New Jersey.

Hayne, D.W. 1977. Report to W.E. Swingle, Executive Director, Gulf of Mexico Fishery Management Council. Official memorandum, dated December 20, 1977.

Hersoug, B., and S.A. Ranes. 1997. What is good for the fishermen is good for the nation: Co-management in the Norwegian fishing industry in the 1990s. *Ocean and Coastal Management* 35:157-172.

Hilborn, R., and C.J. Walters. 1992. *Quantitative Fisheries Stock Assessment: Choice Dynamics and Uncertainty.* Routledge, Chapman & Hall, New York.

Hixon, M.A., and M.H. Carr. 1997. Synergistic predation, density dependence, and population regulation in marine fish. *Science* 277:946-949.

Hoenig, J.M. 1983. Empirical use of longevity data to estimate mortality rates. *Fisheries Bulletin U.S.* 81:898-902.

Hoenig, J.M., C.M. Jones, K.H. Pollock, D.S. Robson, and D.L. Wade. 1997. Calculation of catch rates in roving creel surveys of anglers. *Biometrics* 53:306-317.

Hofmann, E.E., and T.M. Powell. 1998. Environmental variability effects on marine fisheries: Four case studies. *Ecological Applications* 8(1):S23-S32.

Holling, C.S. 1978. *Adaptive Environmental Assessment and Management.* John Wiley & Sons, New York.

Horwood, J. 1993. The Bristol Channel sole (*Solea solea* [L]): A fisheries case study. *Advances in Marine Biology* 29:215-367.

Hurley, P.C.F., G.A.P. Black, P.A. Comeau, R.K. Mohn, and K. Zwanenburg. 1998. Assessment of 4X haddock in 1997 and the first half of 1998. DFO Canadian Stock Assessment Secretariat Research Document. 98/136.

Hutchings, J., and R.A. Myers. 1994. What can be learned from the collapse of a renewable resource? Atlantic cod, *Gadus morhua*, of Newfoundland and Labrador. *Canadian Journal of Fisheries and Aquatic Sciences* 51:2126-2146.

International Council for the Exploration of the Sea (ICES). 1995. *Underwater Noise of Research Vessels: Review and Recommendation.* ICES Cooperative Research Report No. 209. Copenhagen, Denmark.

International Council for the Exploration of the Sea (ICES). 1997. Multispecies Working Group Report, ICES Headquarters 11-19 August 1997 ICES CM 1997/Assess:16. Copenhagen, Denmark.

International Council for the Exploration of the Sea (ICES). 1998. *Report of the Working Group on the Assessment of Northern Shelf Demersal Stocks*, ICES CM 1998/Assess:1. Copenhagen, Denmark.

Itano, D., and K. Holland. 2000. Tags and FADs—Tuna movement and vulnerability in relation to FADs and natural aggregation areas in Hawaii. *Aquatic Living Resources,* in press.

Jakobsen, T. 1992. Biological reference points for North-East Arctic cod and haddock. *ICES Journal of Marine Sciences* 49:155-166.

Jentoft, S. 1989. Fisheries co-management: Delegating government responsibility to fishermen's organizations. *Marine Policy* 13:137-154.

Jentoft, S., and B. McCay. 1995. User participation in fisheries management: Lessons drawn from international experience. *Marine Policy* 19:227-246.

Jentoft, S., B.J. McCay, and D.C. Wilson. 1998. Social theory and fisheries co-management. *Marine Policy* 22:423-436.

Johnston, R.S. 1992. Fisheries development, fisheries management, and externalities. World Bank Discussion Paper No. 165. The World Bank, Washington, D.C.

Jolly, G.M., and I. Hampton. 1990. A stratified random transect design for acoustic surveys of fish stocks. *Canadian Journal of Fisheries and Aquatic Sciences* 47:1282-1291.

Jones, B.W., and J.G. Pope. 1973. A groundfish survey of Faroe Bank. *Research Bulletin of the International Commission of Northwest Atlantic Fisheries* 10:53-61.

Jones, C.M., D.S. Robson, H.D. Lakkis, and J. Kressel. 1995. Properties of catch rates used in analysis of angler surveys. *Transactions of the American Fisheries Society* 124:911-928.

Jones, C.M., D. Wade, W. Check, K.H. Pollock, R.S. Cone, J.M. Hoenig, and D.S. Robson. 1990. Evaluation of recreational angler survey methods for the Chesapeake Bay. Final Report to the Chesapeake Bay Stock Assessment Committee.

Jones, W.J., and J.M. Quattro. 1999. Genetic structure of summer flounder (*Paralichthys dentatus*) populations north and south of Cape Hatteras. *Marine Biology* 133:129-135.

Kaiser, M., and S.J. deGroot (eds.). 1999. *Effects of Fishing on Non-Target Species and Habitats: Biological, Conservation, and Socioeconomic Issues.* Blackwell Science, Ltd., Oxford, U.K.

Kaltongga, B. 1998. Regional tuna tagging project: Data summary. Oceanic Fisheries Programme, Technical Report 35, Secretariat of the Pacific Community, Noumea, New Caledonia. (Web site: http://www.spc.org.nc/oceanfish/)

Kaplan, I.M. 1998. Regulation and compliance in the New England conch fishery: A case for co-management. *Marine Policy* 22:327-335.

Karp, W.A., and H. McElderry. 2000. Catch monitoring by fisheries observers in the United States and Canada. Pp. 261-284 in *Proceedings of the 1999 FAO International Conference on Integrated Fisheries Monitoring.* Food and Agriculture Organization of the United Nations, Rome, Italy.

Kennelly, S.J., and M.K. Broadhurst. 1996. Fishermen and scientists solving bycatch problems: Examples from Australia and possibilities for the Northeastern United States. Pp. 121-128 in *Proceedings of the International Conference on Solving Bycatch: Considerations for Today and Tomorrow.* Seattle, Washington.

King, P.A., S.G. Elsworth, and R.F. Baker. 1994. Partnerships—the route to better communication. Pp. 596-611 in P.G. Wells and P.J. Ricketts (eds.). *Coastal Zone Canada '94, Cooperation in the Coastal Zone.* Conference Proceedings. Volume 2. Coastal Zone Canada Association, Bedford Institute of Oceanography, Dartmouth, Nova Scotia.

Koeller, P., M. Covey, M. King, and S.J. Smith. 1998. The Scotian shelf shrimp (*Pandalus borealis*) fishery in 1998. DFO Canadian Stock Assessment Secretariat Research Document 98/150. Department of Fisheries and Oceans, Halifax, Nova Scotia.

Korsbrekke, K., and O. Nakken. 1999. Length and species-dependent diurnal variation of catch rates in the Norwegian Barents Sea bottom-trawl surveys. *ICES Journal of Marine Sciences* 56:284-291.

Laurec, A., and J.G. Shepherd. 1983. On the analysis of catch and effort data. *Journal du Conseil International pour l'Exploration de la Mer* 41:81-84.

Lauth, R.R., S.E. Syrjala, and S.W. McEntire. 1998. Effects of gear modifications on the trawl performance and catching efficiency of the West Coast Upper Continental Slope Groundfish Survey Trawl. *Marine Fisheries Review* 60(1):1-26.

Lynch, T. 1985. Rhode Island. Pp. 39-40 in *Proceedings of a Workshop on Bottom Trawl Surveys,* T.R. Azarovitz, J. McGurrin and R. Seagraves (eds). Special Report of the Atlantic States Marine Fisheries Commission. 17.

MacLennan, D.N., and E.J. Simmonds. 1992. *Fisheries Acoustics.* Chapman and Hall, London.

Marshall, C.T., N.A. Yaragina, Y. Lambert, and O.S. Kjesbu. 1999. Total lipid energy as a proxy for total egg production by fish stocks. *Nature* 402:288-290.

Maurstad, A., and J.H. Sundet. 1994. Improving the link between science and management: Drawing upon local fishers' experience. ICES C.M., pp. 1-6.

McCay, B.J., and S. Jentoft. 1996. From the bottom up: Participatory issues in fisheries management. *Society and Natural Resources* 9(3):237-250.

Megrey, B.A. 1989. Review and comparison of age-structured stock assessment models from theoretical and applied points of view. *American Fisheries Society Science Symposium* 6:8-48.

Methot, R.D. 1989. Synthetic estimates of historical abundance and mortality in northern anchovy. *American Fisheries Society Science Symposium* 6:66-82.

Methot, R.D. 1990. Synthesis model: An adaptable framework for analysis of diverse stock assessment data. *International North Pacific Fishery Commission Bulletin* 50:259-277.

Mohn, R.K. 1993. Bootstrap estimates of ADAPT parameters, their projection in risk analysis and their retrospective patterns. Pp. 173-184 in S.J. Smith, J.J. Hunt, and D. Rivard (eds.), *Risk Evaluation and Biological Reference Points for Fisheries Management.* Canadian Special Publication of Fisheries and Aquatic Sciences 120.

Mohn, R.K., L.P. Fanning, and W.J. MacEachern. 1998. Assessment of 4VsW cod in 1997 incorporating additional sources of mortality. DFO Canadian Stock Assessment Secretariat Research Document 98/78. Department of Fisheries and Oceans, Halifax, Nova Scotia.

Mowrer, J. 1994. Trawl survey summary for estuarine fisheries program in Maryland. Pp. 47-60 in *Proceedings of the Workshop on the Collection and Use of Trawl Survey Data for Fisheries Management*, T. Berger (ed.). Special Report of the Atlantic States Marine Fisheries Commission, Washington, D.C. Report No. 35.

Munro, P.T., and R.Z. Hoff. 1995. *Two Demersal Trawl Surveys in the Gulf of Alaska: Implications of Survey Design and Methods.* NOAA Technical Memorandum NMFS-AFSC-50. National Marine Fisheries Service, Seattle, Washington.

Murawski, S.A. 1993. Climate change and marine fish distributions: Forecasting from historical analogy. *Transactions of the American Fisheries Society* 122:647-658.

Myers, R.A., N.J. Barrowman, J.A. Hutchings, and A.A. Rosenberg. 1995. Population dynamics of exploited fish stocks at low population levels. *Science* 269:1106-1108.

Nandram, B., J. Sedransk, and S.J. Smith. 1997. Order restricted Bayesian estimation of the age composition of a population of Atlantic cod. *Journal of the American Statistical Association* 92:33-40.

National Marine Fisheries Service (NMFS). Undated. *Proposed Implementation of a Fishing Vessel Registration and Fisheries Information System. Report to Congress.* National Oceanic and Atmospheric Administration, Washington, D.C.

National Marine Fisheries Service (NMFS). 1981. Proceedings of the Summer Flounder (*Paralichthys dentatus*). Age and Growth Workshop, 20-21 May 1980, Northeast Fishery Science Center, Woods Hole, Mass. NOAA Technical Memorandum NMFS-F/NEC-11.

National Marine Fisheries Service (NMFS). 1995. *Fisheries Statistics of the United States, 1994.* Current Fishery Statistics No. 9400. U.S. Department of Commerce, Washington, D.C.

National Marine Fisheries Service (NMFS). 1996. *Our Living Oceans. Report of the Status of U.S. Living Marine Resources, 1995.* NOAA Technical Memorandum NMFS-F/SPO-19. U.S. Department of Commerce, Silver Spring, Maryland.

National Marine Fisheries Service (NMFS). 1999. *Our Living Oceans. Report on the Status of U.S. Living Marine Resources, 1999.* U.S. Department of Commerce, NOAA Technical Memorandum NMFS-F/SPO-41. Online version at http://spo.nwr.noaa.gov/olo99.htm

National Oceanic and Atmospheric Administration (NOAA). 1992. *Proceedings of the NEFC/ASMFC Summer Flounder,* Paralichthys dentatus, *Aging Workshop.* NOAA Technical Memorandum NMFS-F/NEC-89, Woods Hole, Massachusetts.

National Oceanic and Atmospheric Administration (NOAA). 1993. *Report to Congress: Satellite Capabilities for Fisheries Enforcement.* Submitted to Committee on Merchant Marine and Fisheries of the House of Representatives and the Committee on Commerce, Science, and Transportation of the Senate. NOAA, Department of Commerce, Washington, D.C.

National Oceanic and Atmospheric Administration (NOAA). 1998. *NOAA Fisheries Data Acquisition Plan.* NOAA, Washington, D.C.

National Oceanic and Atmospheric Administration (NOAA). 1999. *Development of a Process for Long-Term Monitoring of MMPA Category I and II Commercial Fisheries: Proceedings of a Workshop held in Silver Spring, Maryland, June 15-16, 1998.* Edited by Aloysius J. Didier, Jr. and Victoria R. Cornish, in collaboration with workshop participants. NOAA Technical Memorandum NMFS-OPR-14. National Marine Fisheries Service, Washington, D.C.

National Research Council (NRC). 1994a. *An Assessment of Atlantic Bluefin Tuna.* National Academy Press, Washington, D.C.

National Research Council (NRC). 1994b. *Review of NOAA's Fleet Replacement and Modernization Plan.* National Academy Press, Washington, D.C.

National Research Council (NRC). 1998a. *Improving Fish Stock Assessments.* National Academy Press, Washington, D.C.

National Research Council (NRC). 1998b. *Review of Northeast Fishery Stock Assessments.* National Academy Press, Washington, D.C.

National Research Council (NRC). 1999a. *Sustaining Marine Fisheries.* National Academy Press, Washington, D.C.

National Research Council (NRC). 1999b. *Sharing the Fish: Toward a National Policy on Individual Fishing Quotas.* National Academy Press, Washington, D.C.

National Research Council (NRC). 2000. *Marine Protected Areas*. National Academy Press, Washington, D.C.

Northeast Fisheries Science Center (NEFSC). 1997. *Report of the 25th Northeast Regional Stock Assessment Workshop (25th SAW): Stock Assessment Review Committee (SARC) Consensus Summary of Assessments*. Northeast Fisheries Science Center Reference Document 97-14. National Marine Fisheries Service, Woods Hole, Massachusetts.

O'Boyle, R. 1993. Fisheries management organizations: A study in uncertainty. Pp. 423-436 in S. Smith, J.J. Hunt, and D. Rivard (eds.), *Risk Evaluation and Biological Reference Points for Fisheries Management*. Canadian Special Publications in Fisheries and Aquatic Sciences 120.

O'Boyle, R.N., D. Beanlands, P. Fanning, J.J. Hunt, P.C.F. Hurley, T. Lambert, J. Simon, and K.C.T. Zwanenburg. 1995. An overview of joint science/industry surveys on the Scotian Shelf, Bay of Fundy, and Georges Bank. DFO Atlantic Fisheries, Research Document 95/133. Department of Fisheries and Oceans, Halifax, Nova Scotia.

Open GIS Consortium. 1999. Interoperability Program—Web Mapping Testbed. Unpublished Brochure.

Osborn, M., and H. Lazauski. 1995. Proceedings of the Workshop on Marine Recreational Fisheries Statistics Collection in the Gulf of Mexico, held February 7-9, 1989.

Pacific Fishery Management Council (PFMC). 1998a. *Research and Data Needs 1998-2000*. Pacific Fishery Management Council, Portland, Oregon.

Pacific Fishery Management Council (PFMC). 1998b. *West Coast Fisheries Economic Data Plan*. Pacific Fishery Management Council, Corvallis, Oregon.

Packer, D.B., and P. Hoff. 1999. Life history, habitat parameters, and essential habitat of Mid-Atlantic summer flounder. *American Fisheries Society Symposium* 22:76-92.

Palmer, C.T., and P.R. Sinclair. 1996. Perceptions of a fishery in crisis: Dragger skippers on the Gulf of St. Lawrence cod moratorium. *Society and Natural Resources* 9:267-279.

Parma, A.M. 1993. Restrospective catch-at-age analysis of Pacific halibut: Implications on assessment of harvesting policies. Pp. 247-265 in G. Kruse, D.M. Eggers, R.J. Marasco, C. Pautzke, and T.J. Quinn II (eds.), *Proceedings of the International Symposium on Management Strategies for Exploited Fish Populations*. Alaska Sea Grant College Program Report No. 93-02. University of Alaska, Fairbanks.

Pauly, D. 1980. On the interrelationship between natural mortality, growth parameters, and mean environmental temperature in 175 fish stocks. *Journal du Conseil International pour l'Exploration de la Mer* 42:116-124.

Pelletier, D. 1998. Intercalibration of research survey vessels in fisheries: A review and an application. *Canadian Journal of Fisheries and Aquatic Sciences* 55:2672-2690.

Pinkerton, E. (ed.). 1989a. *Cooperative Management of Local Fisheries. New Directions for Improved Management and Community Development*. University of British Columbia Press, Vancouver.

Pinkerton, E. 1989b. Introduction: Attaining better fisheries management through co-management—Prospects, problems, and propositions. Pp. 3-33 in E. Pinkerton (ed.). *Cooperative Management of Local Fisheries*. University of British Columbia Press, Vancouver.

Pollock, K.H. J.M. Hoenig, and C.M. Jones. 1991. Estimation of fishing and natural mortality when a tagging study is combined with a creel survey or port sampling. *American Fisheries Society Symposium* 12:423-434.

Pollock, K.H., J.M. Hoenig, C.M. Jones, D.S. Robson, and C.J. Greene. 1997. Catch rate estimation for roving and access point surveys. *North American Journal of Fisheries Management* 17:11-19.

Pollock, K.H., C.M. Jones, and T.L. Brown. 1994. *Angler Survey Methods and Their Application in Fisheries Management*. American Fisheries Society Special Publication 25, Bethesda, Maryland.

Pope, J.G. 1972. An investigation of the accuracy of virtual population analysis using cohort analysis. *Research Bulletin of International Commissions of Northwest Atlantic Fisheries* 9:65-74.

Pope, J.G. 1983. Analogies to the status quo TACs: Their nature and variance. Pp. 99-113 in W.G. Doubleday and D. Rivard (eds.), *Sampling Commercial Marine Fish and Invertebrate Catches*. Canadian Special Publication of Fisheries and Aquatic Sciences 66.

Pope, J.G., and D. Gray. 1983. An investigation of the relationship between the precision of assessment data and precision of total allowable catch. Pp. 151-157 in W.G. Doubleday and D. Rivard (eds.), *Sampling Catches of Marine Fish Invertebrates*. Canadian Special Publication of Fisheries and Aquatic Science 66.

Pope, J.G., and J.G. Shepherd. 1982. A simple method for the consistent interpretation of catch-at-age data. *Journal du Conseil International pour l'Exploration de la Mer* 40:176-184.

Pope, J.G., and J.G. Shepherd. 1985. A comparison of the performance of various methods for tuning VPAs using effort data. *Journal du Conseil International pour l'Exploration de la Mer* 42:129-151.

Pope, J.G., and T.K. Stokes. 1989. Use of multiplicative models for separable virtual population analysis (VPA), integrated analysis, and the general VPA tuning problem. *American Fisheries Society Symposium* 6:92-101.

Quinn, T.J. II, and R.B. Deriso. 1999. *Quantitative Fish Dynamics*. Oxford Press, New York.

Quinn, T.J. II, D. Hanselman, D. Clausen, C. Lunsford, and J. Heifetz. 1999. Adaptive cluster sampling for rockfish populations. Pp. 11-20 in *Proceedings of the American Statistical Association.* 1999 Joint Statistical Meetings, Biometric Section. American Statistical Association, Alexandria, Virginia.

Rijnsdorp, A.D., and B. Ibelings. 1989. Sexual dimorphism in the energetics of reproduction and growth of North Sea plaice, *Pleuronectes platessa. London Journal of Fisheries Biology* 35:401-415.

Robert, G., M.A.E. Butler, and S.J. Smith. 1998. Georges Bank scallop stock assessment—1997. DFO Canadian Stock Assessment Secretariat Research Document 98/69. Department of Fisheries and Oceans, Halifax, Nova Scotia.

Roberts, K.J., J.W. Horst, J.E. Roussel, and J.A. Shephard. 1991. *Defining Fisheries: A User's Glossary.* Louisiana Sea Grant College Program, Louisiana State University, Baton Rouge. Available at http://nsgd.gso.uri.edu/source/lsuh91001.htm.

Rose, G.A., B. deYoung, D.W. Kulka, S.V. Goddard, and G.L. Flecher. 2000. Distribution shifts and overfishing the northern cod (*Gadus morhua*): A view from the ocean. *Canadian Journal of Fisheries and Aquatic Sciences* 57:644-663.

Rose, G.A., and D.W. Kulka. 1999. Hyper-aggregation of fish and fisheries: How catch-per-unit-effort increased as northern cod (*Gadus morhua*) declined. *Canadian Journal of Fisheries and Aquatic Sciences* 56(Supplement 1):118-127.

Rosenberg, A., and S. Brault. 1993. Choosing a management strategy for stock rebuilding when control is uncertain. Pp. 243-249 in S. Smith, J.J. Hunt, and D. Rivard (eds.), *Risk Evaluation and Biological Reference Points for Fisheries Management.* Canadian Special Publications in Fisheries and Aquatic Sciences 120.

Rosenberg, A., P. Mace, G. Thompson, G. Darcy, W. Clark, J. Collie, W. Gabriel, A. MacCall, R. Methot, J. Powers, V. Restrepo, T. Wainwright, L. Botsford, J. Hoenig, and K. Stakes. 1994. *Scientific Review of Definitions of Overfishing in U.S. Fishery Management Plans.* NOAA Technical Memorandum NMFS-F/SPO-17. National Oceanic and Atmospheric Administration, Washington, D.C.

Rountree, R.A. 1994. Broad-scale distribution patterns of summer flounder and their prey based on bottom trawl surveys collected from 1973-1992 between Cape Hatteras and the Scotian Shelf. Seminar presented at the 12th Annual Meeting of the American Fisheries Society, held 21-25 August 1994, Halifax, Nova Scotia.

Sampson, D.B., and P.R. Crone (ed.). 1997. *Commercial Fisheries Data Collection Procedures for the U.S. Pacific Coast Groundfish.* NOAA Technical Memorandum NMFS-NWFSC-31. National Marine Fisheries Service, Seattle, Washington.

Seber, G.A.F., and S.K. Thompson. 1993. Environmental adaptive sampling. Pp. 201–220 in *Handbook of Statistics*, vol. 12 (Environmental Statistics), G.P. Patil and C.R. Rao (eds.). North Holland/Elsevier Science Publishers, New York.

Seber, G.A.F., and C.J. Wild. 1989. *Nonlinear Regression.* John Wiley & Sons, New York.

Shepherd J.G., and M.D. Nicholson. 1991. Multiplicative modelling of catch-at-age data and its application to catch forecasts. *Journal du Conseil International pour l'Exploration de la Mer* 47:284-294.

Shepherd, J.G., and J.G. Pope. 1993. Alternative methods for estimation of immigration to the Icelandic cod stock. *Fisheries Oceanography* 2:254-259.

Sinclair, A., D. Gascon, R. O'Boyle, D. Rivard, and S. Gavaris. 1991. Consistency of some Northwest Atlantic groundfish stock assessments. *NAFO Scientific Council* 16:59-77.

Sissenwine, M.P., and A.A. Rosenberg. 1993. Marine fisheries at a critical juncture. *Fisheries* 18(10):6-14.

Sissenwine, M.P., and J.G. Shepherd. 1987. An alternative perspective on recruitment overfishing and biological reference points. *Canadian Journal of Fisheries and Aquatic Sciences* 44:91-918.

Smith, E.M. 1995. The nature of Nature: Conflict and consensus in fisheries management. *Aquatic Living Resources* 8(3):209-213.

Smith, S.J. 1996. Analysis of data from bottom trawl surveys. Pp. 25-53 in H. Lassen (ed.), *Assessment of Groundfish Stocks Based on Bottom Trawl Surveys.* NAFO Scientific Council Studies 28.

Smith, S.J., and S. Gavaris. 1993. Improving the precision of fish abundance estimates of Eastern Scotian Shelf cod from bottom trawl surveys. *North American Journal of Fisheries Management* 13:35-47.

Smith, S.J., M.J. Lundy, and R. Claytor. 1999. Scallop production in areas 4 and 5 of the Bay of Fundy: Stock status update for 1999. DFO Canadian Stock Assessment Secretariat Research Document 99/170. Department of Fisheries and Oceans, Halifax, Nova Scotia.

Sogard, S.M. 1997. Size-selective mortality in the juvenile stage of teleost fishes: A review. *Bulletin of Marine Sciences* 60(3):1129-1157.

Somerton, D.A., J. Ianelli, S.J. Walsh, S.J. Smith, O.R. Godø, and D. Ramm. 1999. Incorporating experiments derived estimates of trawl efficiency into the stock assessment process: A discussion. *ICES Journal of Marine Sciences* 56:299-302.

Squires, D., H. Campbell, S. Cunningham, C. Dewees, R.Q. Grafton, S.F. Herrick, Jr., J. Kirkley, S. Pascoe, K. Salvanes, B. Shallard, B. Turris, and N. Vestergaard. 1998. Individual transferable quotas in multispecies fisheries. *Marine Policy* 22(2):135-159.

Steele, J.H. 1998. Regime shifts in marine ecosystems. *Ecological Applications* 8(1):S33-S36.

Sylvia, G. and J. Harms. Undated. Review and analysis of industry-scientific cooperative fisheries research programs: Potential application to the West Coast groundfish fishery. Unpublished research proposal. Oregon State University, Hatfield Marine Science Center. Newport, Oregon

Tagart, J.V. 1997. Groundfish data collection in Washington. In D.B. Sampson and P.R. Crone (eds.), *Commercial Fisheries Data Collection Procedures for U.S. Pacific Coast Groundfish.* NOAA Technical Memorandum NMFS-NWFSC-31, Seattle, Washington. http://www.nwfsc.noaa.gov/pubs/tm/tm31/tm31.html

Terceiro, M. 1999. *Stock Assessment for Summer Flounder for 1999.* Northeast Fishery Science Center Reference Document 99-19. National Marine Fisheries Service, Woods Hole, Massachusetts.

Thompson, S.K. 1990. Adaptive cluster sampling. *Journal of the American Statistical Association* 85:1050-1059.

Thompson, S.K., F.L. Ramsey, and G.A.F. Seber. 1992. An adaptive procedure for sampling animal populations. *Biometrics* 48:1195-1199.

Thompson, S.K., and G.A.F. Seber. 1996. *Adaptive Sampling.* Wiley, New York.

Ticheler, H.J., J. Kolding, and B. Chanda. 1998. Participation of local fishermen in scientific fisheries data collection: A case study from the Bangweulu Swamps, Zambia. *Fisheries Management and Ecology* 5:81-92.

Traynor, J. 1997. Midwater fish surveys at the AFSC. *Alaska Fishery Science Center Quarterly Report* Jan.-March:1-9.

Turnock, J., and W.A. Karp. 1997. Estimation of salmon bycatch in the 1995 pollock fishery in the Bering Sea/ Aleutian Islands. Unpublished manuscript prepared for the North Pacific Fishery Management Council, Anchorage, Alaska.

Vølstad, J.H., W. Richkus, S. Gaurin, and R. Easton. 1997. Analytical and statistical review of procedures for collection and analysis of commercial fishery data used for management and assessment of groundfish stocks in the U.S. exclusive economic zone off Alaska. Project report prepared by Versar, Inc. for the National Marine Fisheries Service, (Available from William A. Karp, National Marine Fisheries Service, 7600 Sand Point Way NE, Seattle, WA 98115, USA).

Wainright, S.C., M. Fogarty, R. Greenfield, and B. Fry. 1993. Long-term changes in the Georges Bank foodweb: Trends in stable isotope composition of fish scales. *Marine Biology* 115:481-493.

Walsh, S.J. 1991. Diel variation in availability and vulnerability of fish to a survey trawl. *Journal of Ichthyology* 7:147-159.

Walters, C. 1986. *Adaptive Management of Renewable Resources.* McMillan, New York.

Walters, C. 1994. Use of gaming procedures in evaluation of management experiments. *Canadian Journal of Fisheries and Aquatic Sciences* 51:2705-2714.

Walters, C., and P.H. Pearse. 1996. Stock information requirements for quota management systems in commercial fisheries. *Review of Fish Biology and Fisheries* 6:21-42.

West Coast Groundfish Stock Assessment Review Panel. 1995. *West Coast Groundfish Assessment Review.* Submitted to the Pacific Fishery Management Council, Portland, Oregon.

West, K.H., and C.J. Wilson. 1994. NCDMF current trawl surveys. Pp. 97-115 in *Proceedings of the Workshop on the Collection and Use of Trawl Survey Data for Fisheries Management*, T. Berger (ed.). Special Report of the Atlantic States Marine Fisheries Commission, Washington, D.C. Report No. 35.

Wilkins, M.E., M. Zimmermann, and K.L. Weinberg. 1998. *The 1995 Pacific West Coast Bottom Trawl Survey of Groundfish Resources: Estimates of Distribution, Abundance, and Length and Age Composition.* NOAA Technical Memorandum NMFS-AFSC-89. National Marine Fisheries Service, Seattle, Washington.

Appendixes

APPENDIX A

Committee Biographies

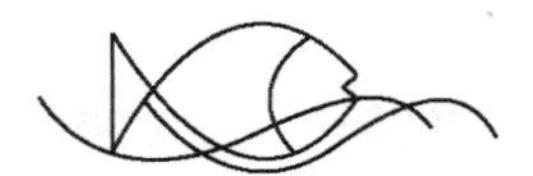

Patrick Sullivan chaired the committee. He earned a Ph.D. in biostatistics from the University of Washington in 1988. Dr. Sullivan is currently an assistant professor in the Department of Natural Resources at Cornell University. Prior to that, he served for ten years as a staff scientist for the International Pacific Halibut Commission. His research interests include the assessment and management of fisheries resources and the statistical modeling of biological systems.

Kenneth Able earned a Ph.D. in biological oceanography at the College of William and Mary in 1974. He is currently a professor at Rutgers University and director of Rutgers's Tuckerton Marine Field Station. Dr. Able's research interests include life history, ecology, and behavior of fishes; recruitment processes; and habitat structure and function. He is a pioneer in the study of the biology of summer flounder.

Cynthia Jones earned a Ph.D. in oceanography from University of Rhode Island in 1984. Dr. Jones is currently a professor of biology at Old Dominion University. Her main areas of research are fisheries and population ecology; she works extensively with recreational fisheries data.

Karen M. Kaye earned an M.S. in computer systems management from the University of Maryland in 1996. She has worked for the federal government and private industry for the past 20 years and is presently the information systems coordinator at the U.S. Geological Survey.

Barbara Knuth earned a Ph.D. in fisheries and wildlife sciences from Virginia Polytechnic Institute and State University in 1986. She is currently an associate professor and unit co-leader in the Human Dimensions Research Unit of Cornell University's Department of Natural Resources. Dr. Knuth's research focuses on integrating human dimensions information and techniques into natural resource decision-making processes, program evaluation, and fisheries management.

Brenda L. Norcross earned her Ph.D. in marine science from the College of William and Mary in 1983. She is an associate professor at the University of Alaska, Fairbanks. Dr. Norcross' primary areas of expertise include fisheries oceanography, fish habitat, and the influence of ocean conditions on distribution and recruitment of larval and juvenile fish, especially of flatfish species.

Estelle Russek-Cohen earned a Ph.D. in biostatistics from the University of Washington in 1979. She is currently a professor at the University of Maryland. Dr. Russek-Cohen's research interests include statistical methodology, analyzing experimental and survey data, and multivariate and bioassay methods.

John Sibert earned a Ph.D. in zoology from Columbia University in 1968. He is currently manager of the Pelagic Fisheries Research Program at the University of Hawaii, Honolulu. His research interests include fisheries oceanography, statistics, and the inclusion of spatial heterogeneity in population dynamics models.

Stephen Joseph Smith earned an M.Sc. degree in statistics from the University of Guelph, Canada, in 1979. He is currently a research scientist for the Canadian Department of Fisheries and Oceans at the Bedford Institute of Oceanography. His primary research interests are in the field of resource management and modeling of marine fisheries, with a concentration in statistics.

Steven K. Thompson earned a Ph.D. in statistics from Oregon State University in 1982. He is currently an associate professor at Pennsylvania State University. Dr. Thompson's research interests include sampling theory and methods, environmental statistics, statistics of hidden populations, adaptive sampling, and general issues in design and inference.

Richard D. Young earned a Ph.D. in economics from the University of California, Santa Barbara, in 1979. He participates in the Pacific Coast groundfish, crab, and shrimp fisheries as the owner and operator of the fishing vessel *City of Eureka* and the owner of the *Willola*, both based in Crescent City, California. Dr. Young has participated in a variety of research and management activities related to fisheries and is presently a member of the Scientific and Statistical Committee of the Pacific Fishery Management Council.

Consultant

John G. Pope earned a B.Sc. in mathematics from the University of London (U.K.) in 1962. He spent much of his career at the Lowestoft Fisheries Laboratory before taking early retirement in 1997. Mr. Pope has taken a leading role in the fisheries science of the International Council for the Exploration of the Sea. He is currently a Professor II at the Universitetet i Tromsø, Norway, and director of NRC (Europe), Ltd.

Appendix B

Acronyms

ACCSP	Atlantic Cooperative Coastal Statistics Program
ACFCMA	Atlantic Coast Fisheries Cooperative Management Act
AFMIS	Advanced Fisheries Management Information System
AGDB	Alaska Groundfish Data Bank
AKFIN	Alaska Fisheries Information Network
ANSI	American National Standards Institute
AOSN	Autonomous Oceanographic Sampling Network
ASCII	American Standard Code for Information Interchange
CalCOFI	California Cooperative Oceanic Fisheries Investigations
CPUE	catch per unit effort
CTD	conductivity-temperature-depth recorder
CTDEP	Connecticut Department of Environmental Protection
EFCL	Electronic Fish Catch Logbook
EPA	Environmental Protection Agency
F	fishing mortality rate
FGDC	Federal Geographic Data Committee
FIN	Fishery Information Network
FIPS	Federal Information Processing Standards
FIS	Fisheries Information System
FMP	fishery management plan
FOCI	Fisheries Oceanography Coordinated Investigations
FRCC	Fisheries Resource Conservation Council
FRV	fishery research vessel
FVR	fishing vessel registration
GIS	geographic information system
GLOBEC	Global Ocean Ecosystems Dynamics program
GPRA	Government Performance and Results Act
GPS	Global Positioning System
IFQ	individual fishing quota
ITA	information technology architecture
ITIS	Integrated Taxonomic Information System

ITS	Integrated Transportation System
LOOPS	Littoral Ocean Observing and Prediction System
MADMF	Massachusetts Department of Marine Fisheries
MSFCMA	Magnuson-Stevens Fishery Conservation and Management Act
MRFSS	Marine Recreational Fisheries Statistics Survey
NAS	National Academy of Sciences
NASA	National Aeronautics and Space Administration
NBII	National Biological Information Infrastructure
NEFSC	Northeast Fishery Science Center
NIST	National Institute of Standards and Technology
NJDF	New Jersey Department of Fisheries
NMFS	National Marine Fisheries Service
NRC	National Research Council
NSDI	National Spatial Data Infrastructure
NWFSC	Northwest Fishery Science Center
PacFIN	Pacific Fishery Information Network
PFEL	Pacific Fisheries Environment Laboratory
PSMFC	Pacific States Marine Fisheries Commission
RM-ODP	Reference Model of Open Distributed Processing
SABRE	South Atlantic Bight Recruitment Experiment
SARC	Stock Assessment Review Committee
SEAMAP	Southeast Assessment and Monitoring Program
SFA	Sustainable Fisheries Act
SQL	Structured Query Language
SSC	scientific and statistical committee (of regional fishery management councils)
STAR	STock Assessment Review panel
TAC	total allowable catch
USDA	United States Department of Agriculture
USGS	United States Geological Survey
VPA	virtual population analysis
VMS	vessel monitoring system
VRS	vessel registration system
VRML	Virtual Reality Modeling Language
VTR	vessel trip report
WMT	Web Mapping Testbed

APPENDIX C

Evaluation of Summer Flounder Surveys

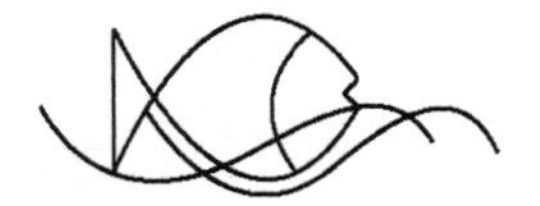

CATALOG OF SURVEY SERIES USED

The National Marine Fisheries Service (NMFS) conducts three seasonal surveys to monitor a large number of commercial fish species on the Atlantic Coast. These surveys cover a large number of strata, but not all strata (or stations) are used in constructing the abundance estimates for summer flounder. For example, the spring survey index uses a subset of 28 of the 56 offshore strata (Figure C-1) and using 1995 as an example, 94 stations were used to calculate the abundance estimate (Table C-1). This is equivalent to a sampling intensity of 1 station per 246 square nautical miles. Similar information on the strata used for the abundance indices for the fall and winter survey are presented in Tables C-2 and C-3, respectively. The selection of strata for each of the seasonal survey series appear to reflect general ideas about the distribution of summer flounder. In the fall, the strata used to construct the index are restricted to the inshore strata (see Figures C-2 to C-4) from Cape Hatteras up the coast, but not including the Gulf of Maine and the most inshore of the offshore strata (see Figure C-1) south of Cape Cod. The winter survey includes offshore strata on the eastern side of Georges Bank and all but the very deepest strata (deeper than 183 m or 100 fathoms) from Hudson Canyon to Cape Hatteras. In the spring survey, all offshore strata south of Georges Bank are used, including the very deepest offshore strata.

The 1996 stock assessment for summer flounder lists 9 survey series[1] for fish of 1 year or greater (used age by age) and 8 young-of-the-year[2] survey series used in tuning the ADAPT model for reconstructing the population (NEFSC, 1997). All series were reportedly used with equal weight, although the Northeast Fishery Science Center (NEFSC, 1997) mentions that series were eliminated from the tuning if the series trends did not match the virtual population analysis (VPA) estimated trends. No details are given by the NEFSC (1997) as to what series were actually eliminated.

Trends for ages 1+ for the nine adult surveys are presented in Figure C-5. The Northeast Fisheries Science Center (NEFSC) winter survey and

[1] A Delaware state survey was included in the 1999 assessment.

[2] Young-of-the-year are age-0 fish, less than 12 months old.

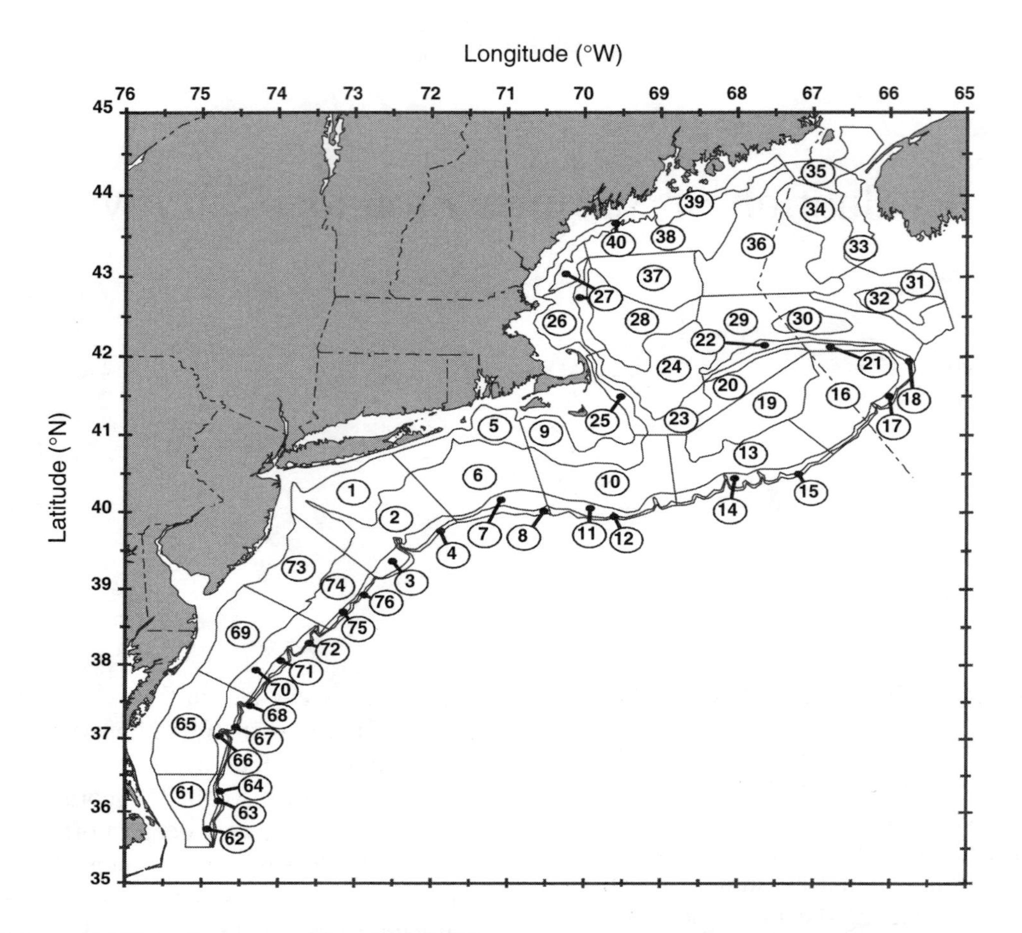

FIGURE C-1 Strata sampled on the NEFSC/NMFS offshore bottom trawl surveys. Depths range from 27 m to greater than 200 m.

the New Jersey (NJBMF) trawl survey show the largest changes in abundance over time. Similar plots are given for young-of-the-year (age-0) fish in Figure C-6, in which abundances are largest and most variable in the North Carolina and Maryland surveys.

NEFSC Surveys

The main features for each of the federal and state age-1+ surveys are presented in Table C-4. The first three series are the standard multi-species surveys conducted by the NEFSC. The fall and spring surveys are the longest running series, with the former starting in 1963 and the latter in 1967 (Azarovitz, 1994). These two series cover the continental shelf between Nova Scotia and Cape Hatteras. The winter series is the newest, having been initiated in 1992. It was designed primarily to monitor flatfish species in the mid-Atlantic and southern New England regions. Figures C-1 to C-4 show both the inshore and offshore strata sampled.

The spring and fall NEFSC surveys currently use a standardized #36 Yankee otter trawl rigged with 16-inch rollers on the sweep, 5-fathom trawl legs, and 1,000-pound Polyvalent doors (Azarovitz, 1994). From 1972 to 1981, the spring

TABLE C-1 Mean Catch (Number) of Summer Flounder by Stratum for NEFSC 1995 Spring Survey

Strata[a]	Stations	Proportion	Mean	Standard Deviation
01 01	8	0.075	1.62	3.46
01 01	7	0.090	0.14	0.38
01 02	7	0.075	0.14	0.38
01 03	2	0.020	0.50	0.71
01 04	**1**	**0.007**	**0.00**	**N/A**
01 05	5	0.053	0.20	0.45
01 06	8	0.092	1.00	1.41
01 07	2	0.018	2.50	3.54
01 08	**1**	**0.008**	**0.00**	**N/A**
01 09	5	0.055	0.00	0.00
01 10	8	0.098	0.25	0.46
01 11	2	0.022	0.00	0.00
01 12	**1**	**0.006**	**0.00**	**N/A**
01 61	3	0.047	7.67	6.11
01 62	2	0.009	4.00	5.66
01 63	2	0.003	0.00	0.00
01 64	**1**	**0.002**	**0.00**	**N/A**
01 65	7	0.102	1.71	3.30
01 66	3	0.020	1.33	1.53
01 67	**1**	**0.003**	**0.00**	**N/A**
01 68	**1**	**0.002**	**0.00**	**N/A**
01 69	6	0.087	1.17	1.17
01 70	4	0.037	2.00	1.63
01 71	2	0.010	0.00	0.00
01 72	**1**	**0.004**	**0.00**	**N/A**
01 73	5	0.077	0.20	0.45
01 74	4	0.046	2.00	1.41
01 75	2	0.005	0.00	0.00
01 76	**1**	**0.002**	**0.00**	**N/A**

NOTE: Stations = number of stations per stratum. Proportion = proportion of the total survey area in each stratum. N/A = standard deviation could not be calculated because only one station was fished in that stratum. **Bold** = strata with only one station.

[a] The NEFSC strata coding system is used here, with the first two digits indicating whether a stratum is offshore (01, Figure C-1) or inshore (03, Figures C-2, C-3, and C-4).

surveys used a larger #41 Yankee trawl before converting back to the #36 Yankee trawl. A modified version of the #36 Yankee is used for the winter survey. The main modifications of the winter gear include a rubber disk (4 inch)-cov-

TABLE C-2 Mean Catch (Number) of Summer Flounder by Stratum for NEFSC Fall Survey

Strata	Stations	Proportion	Mean	Standard Deviation
01 01	7	0.119	1.29	3.40
01 05	3	0.070	4.67	5.51
01 09	5	0.072	2.60	2.79
01 61	3	0.062	1.67	1.53
01 65	7	0.134	0.86	1.21
01 69	5	0.115	0.40	0.89
01 73	5	0.101	1.40	2.07
03 02	2	0.003	1.50	2.12
03 03	**1**	**0.001**	**4.00**	**N/A**
03 04	2	0.001	5.50	4.95
03 05	2	0.003	1.50	0.71
03 06	**1**	**0.001**	**13.00**	**N/A**
03 07	2	0.002	11.00	14.14
03 08	2	0.007	2.50	0.71
03 09	**1**	**0.002**	**30.00**	**N/A**
03 10	2	0.002	18.50	6.36
03 11	2	0.011	0.50	0.71
03 12	**1**	**0.002**	**3.00**	**N/A**
03 13	2	0.004	4.00	4.24
03 14	2	0.005	0.50	0.71
03 15	**1**	**0.001**	**35.00**	**N/A**
03 16	2	0.003	11.50	7.78
03 17	2	0.011	9.00	5.66
03 18	**1**	**0.005**	**27.00**	**N/A**
03 19	2	0.010	8.00	2.83
03 20	2	0.017	14.00	2.83
03 21	**1**	**0.001**	**11.00**	**N/A**
03 22	2	0.007	5.50	3.54
03 23	2	0.008	9.00	2.83
03 24	2	0.003	0.50	0.71
03 25	2	0.008	2.50	0.71
03 26	2	0.007	6.00	2.83
03 27	**1**	**0.002**	**6.00**	**N/A**
03 28	2	0.010	1.50	0.71
03 29	2	0.009	5.50	2.12
03 30	**1**	**0.004**	**0.00**	**N/A**
03 31	2	0.014	2.50	2.12
03 32	2	0.005	3.50	0.71
03 33	**1**	**0.004**	**6.00**	**N/A**
03 34	2	0.008	3.00	0.00
03 35	2	0.004	3.00	4.24
03 36	2	0.006	1.00	1.41
03 37	2	0.015	0.50	0.71
03 38	2	0.011	2.50	2.12
03 39	**1**	**0.002**	**0.00**	**N/A**

(continued)

TABLE C-2 Continued

Strata	Stations	Proportion	Mean	Standard Deviation
03 40	2	0.008	4.50	6.36
03 41	2	0.018	0.50	0.71
03 42	**1**	**0.002**	**1.00**	**N/A**
03 43	2	0.008	4.00	5.66
03 44	2	0.014	0.50	0.71
03 45	2	0.008	0.50	0.71
03 46	2	0.013	0.50	0.71
03 55	5	0.023	0.20	0.45
03 56	**1**	**0.003**	**0.00**	**N/A**
03 60	2	0.006	0.00	0.00
03 61	**1**	**0.006**	**0.00**	**N/A**

NOTE: Stations = number of stations per stratum. Proportion = proportion of the total survey area in each stratum. N/A = standard deviation could not be calculated because only one station was fished in that stratum. **Bold** = strata with only one station.

TABLE C-3 Mean Catch (Number) of Summer Flounder by Stratum for NEFSC 1995 Winter Survey

Strata	Stations	Proportion	Mean	Standard Deviation
01 01	8	0.075	1.62	3.46
01 02	7	0.062	10.14	10.88
01 03	2	0.017	18.00	16.97
01 05	5	0.044	1.00	1.22
01 06	9	0.077	3.00	3.46
01 07	2	0.015	18.00	7.07
01 09	5	0.046	0.00	0.00
01 10	8	0.082	5.88	8.51
01 11	2	0.019	6.50	7.78
01 13	9	0.071	2.44	2.55
01 14	3	0.020	2.00	1.00
01 16	9	0.089	0.00	0.00
01 17	**1**	**0.011**	**0.00**	**N/A**
01 61	4	0.040	47.00	32.63
01 62	2	0.007	3.00	1.41
01 63	**1**	**0.003**	**0.00**	**N/A**
01 65	8	0.085	35.38	26.20
01 66	4	0.017	61.00	53.65
01 67	2	0.003	6.50	0.71
01 69	8	0.073	12.00	14.07
01 70	4	0.031	31.75	22.95
01 71	2	0.008	9.50	4.95
01 73	5	0.064	2.60	3.44
01 74	4	0.038	12.25	5.12
01 75	2	0.004	6.00	4.24

NOTE: Stations = number of stations per stratum. Proportion = proportion of survey area in each stratum. N/A = standard deviation could not be calculated because only one station was fished in that stratum. **Bold** = strata with only one station.

ered chain sweep replacing the rollers and the addition of 30-fathom ground cables ahead of the net. All trawls are lined with a half-inch stretched mesh liner in the codend and upper belly to retain small fish. The spring and fall surveys used BMV Oval doors until 1985 when the manufacturer could no longer meet NEFSC specifications; these were then replaced with Polyvalent doors. The potential effects on species catchability of changing the doors was investigated by NEFSC and significant differences were found. Species-specific correction factors are used to convert results from surveys with BMV doors to Polyvalent door equivalents (Byrne and Forrester, 1991; Azarovitz, 1994). The survey abundance estimate for summer flounder does not include all the strata fished during the survey and the actual number of tows used to calculate abundance are also listed in Table C-4.

All of the NEFSC surveys use a stratified random design with bathymetric limits as the primary stratifying variable (<9 m, 9-18 m, 18.1-27 m, 27.1-55 m, 55.1-110 m, 110.1-185 m, and 185.1-365 m), with additional subdivisions introduced to spread sampling over the entire area. In the 1970s and 1980s sampling coverage was extended south of Cape Hatteras and inshore and as many as 450 stations were sampled in each survey. More recently, sampling south of Cape Hatteras and in Canadian waters has been eliminated and sampling has been reduced in depths <18 m and >110 m to save money and time. Stations are located randomly in each stratum and the number of stations allocated to each stratum is proportional to area. Azarovitz (1994) reports

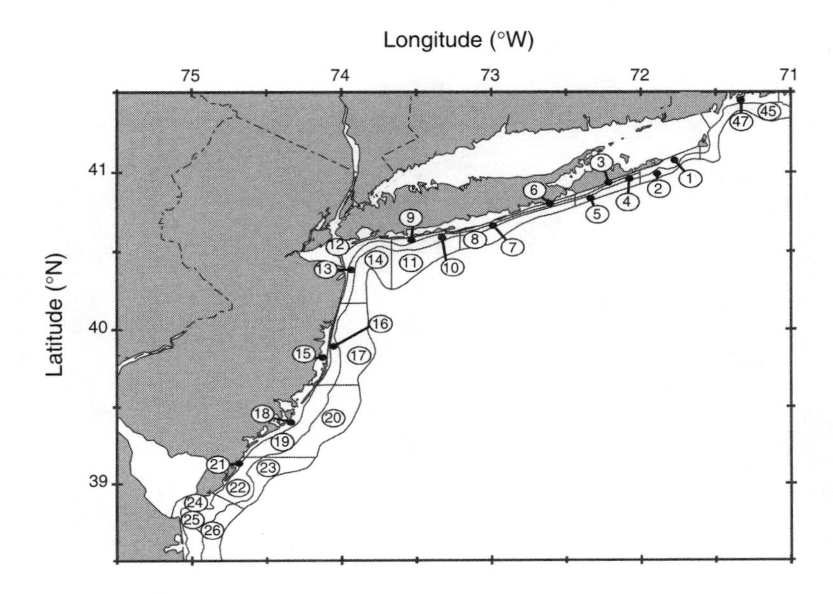

FIGURE C-2 Strata sampled on NEFSC inshore bottom trawl surveys from Buzzards Bay, Massachusetts, to Delaware Bay, Delaware. Depths range from 0 m to 27 m.

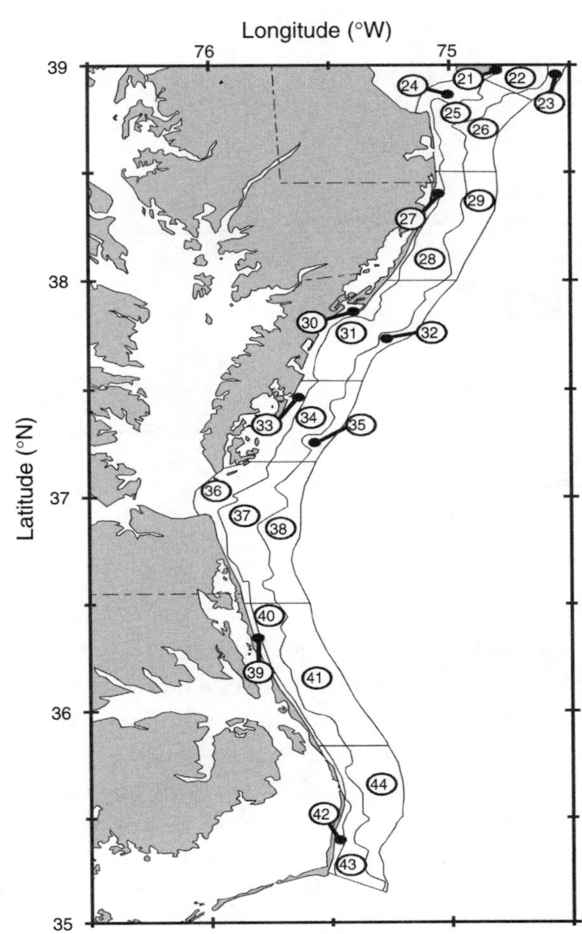

FIGURE C-3 Strata sampled on NEFSC inshore bottom trawl surveys from Delaware Bay, Delaware, to Cape Hatteras, North Carolina. Depths range from 0 m to 27 m.

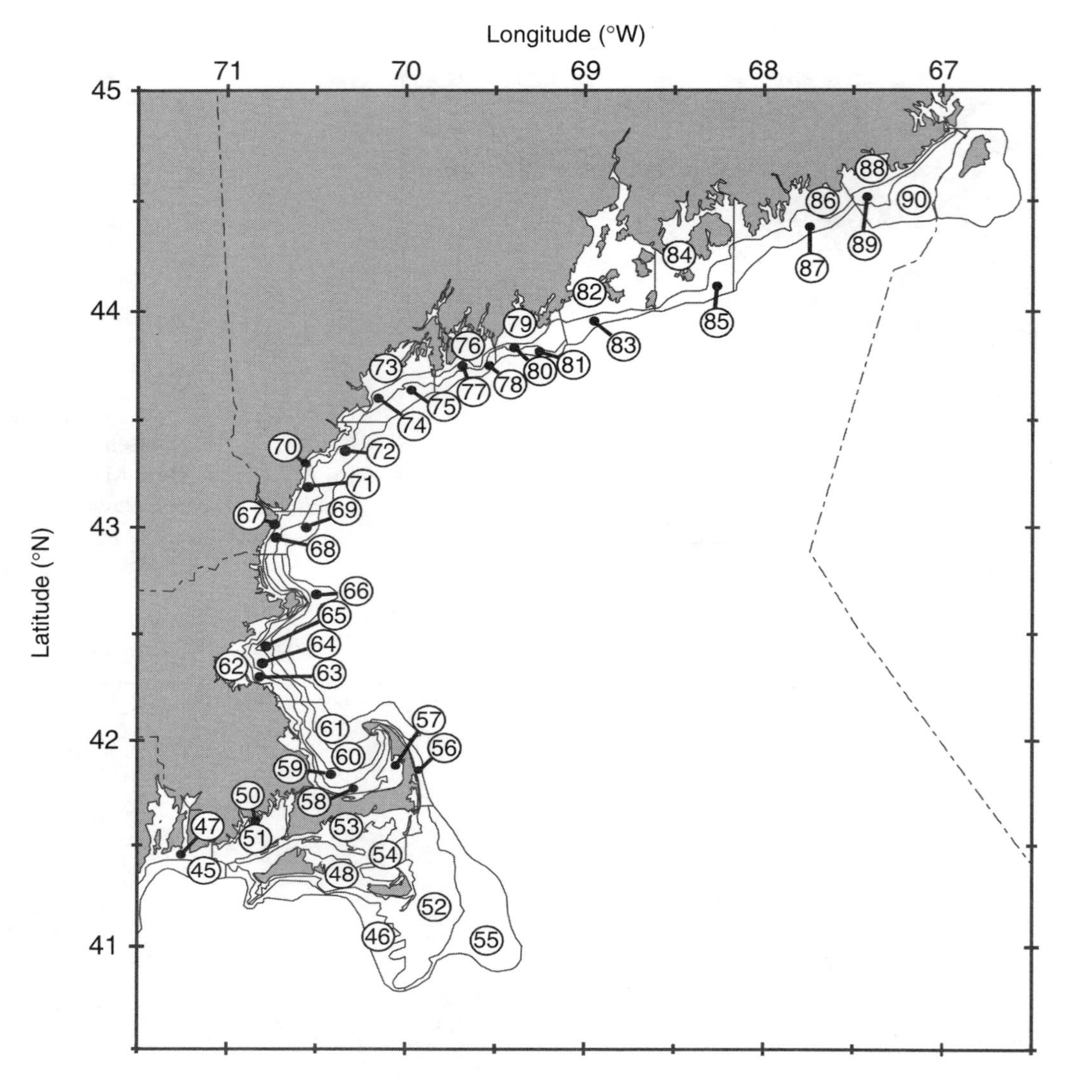

FIGURE C-4 Strata sampled on NEFSC inshore bottom trawl surveys from Eastport, Maine, to Buzzards Bay, Massachusetts. Depths range from 0 m to 54 m.

that 320 stations are sampled in each survey, which equates to one station every 885 square nautical miles. A 30-minute tow at 3.5 knots is made at each station. Stations are fished at all times of the day and night. The surveys are run at the same general time each year. No allowance is made for differences in catch between day and night, which could be quite important for some species.

State Surveys

The Rhode Island Department of Fish and Wildlife has conducted a trawl survey of its coastal waters since 1979 (Lynch, 1985). Originally, a stratified random design with 11 depth strata was used to monitor the waters in Rhode Island Sound, Block Island Sound, and Narragansett Bay. Since 1988 a fixed station de-

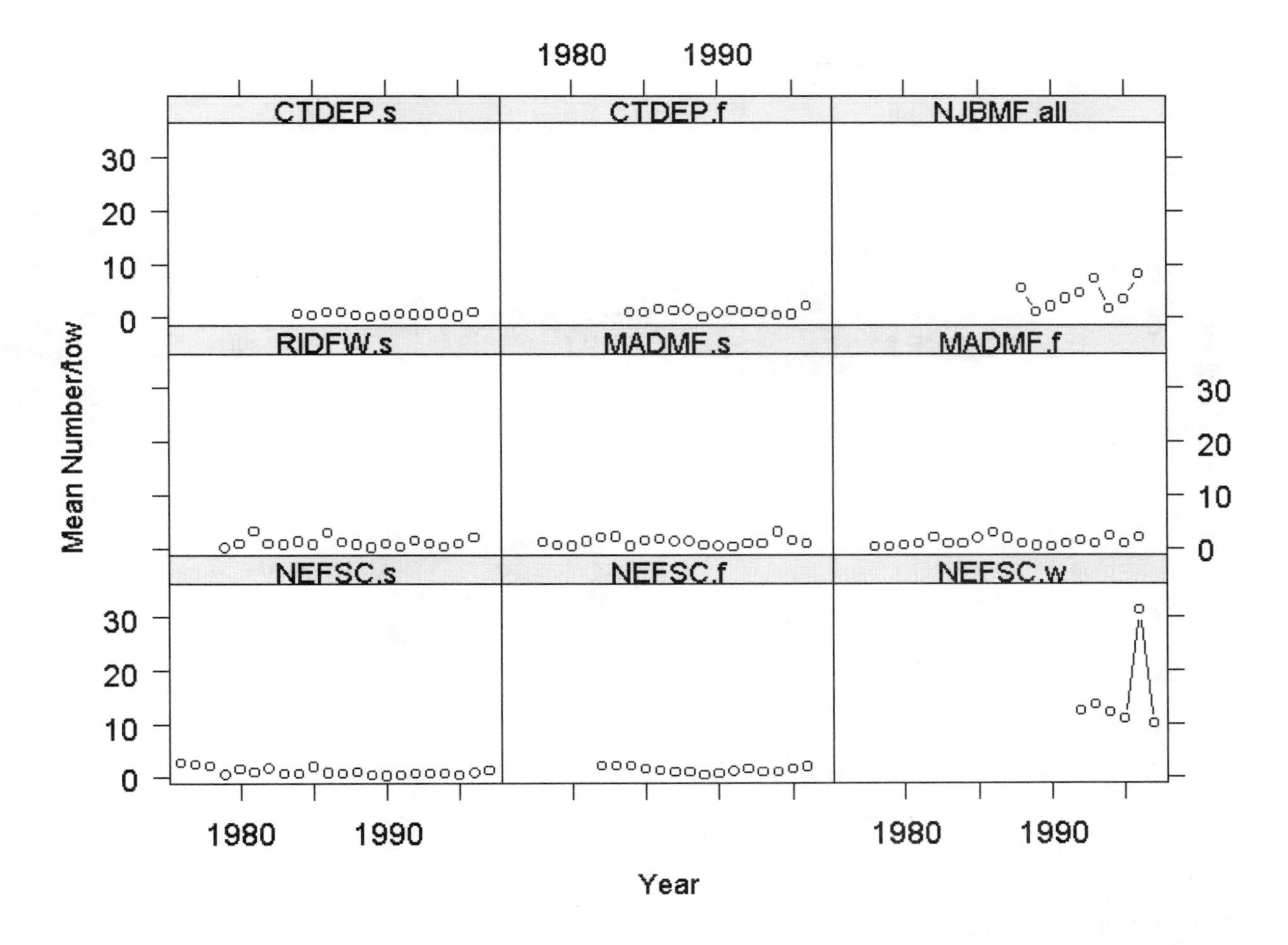

FIGURE C-5 Mean number of summer flounder per tow for estimates of age 1+ fish from surveys used in summer flounder assessment.

NOTE: CTDEP = Connecticut Department of Environmental Protection; MADMF = Massachusetts Department of Marine Fisheries; NEFSC = Northeast Fisheries Science Center; NJBMF = New Jersey Bureau of Marine Fisheries; RIDFW = Rhode Island Department of Fish and Wildlife. s = spring survey; f = fall survey; w = winter survey; and all = all surveys.

sign has been used in the first two areas, whereas the stratified random design has been retained for the latter area. This change was implemented because of problems of finding trawlable bottom when using random stations. A ¾-scale 340 High Rise bottom trawl is used at each station and towed for 20 minutes at an average speed of 2.5 knots.

The State of Massachusetts Department of Marine Fisheries has conducted a spring (May) and a fall (September) inshore survey since 1978 (Correia, 1994) in cooperation with the NEFSC. A total of 23 strata are defined over 6 depth zones and 5 regions along the coast. One hundred stations are allocated in proportion to strata area with a minimum of 2 stations per stratum, resulting in a sampling intensity of 1 station per 19 square nautical miles. Fishing occurs only during daylight hours and consists of a 20-minute tow at 2.5 knots at each station. Shorter tows are

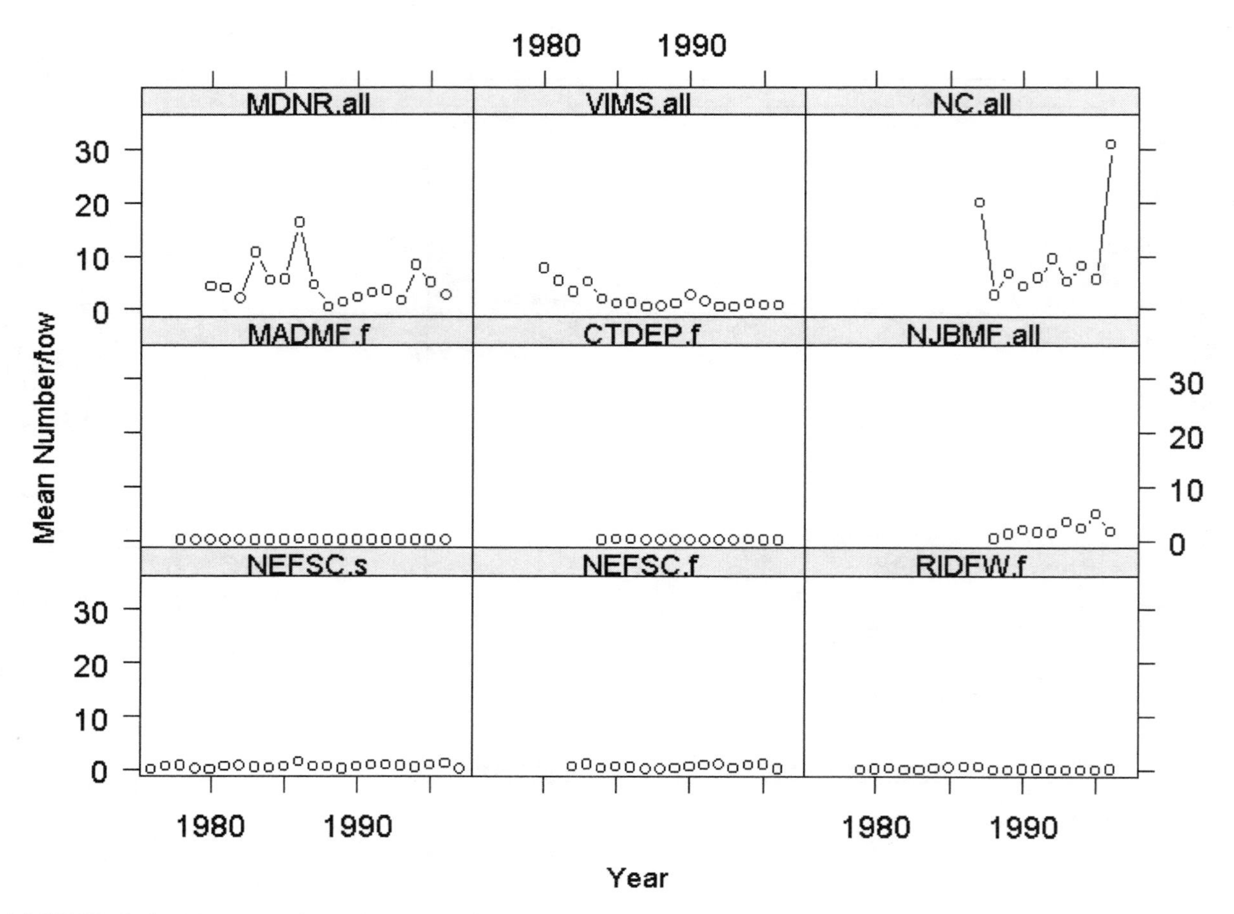

FIGURE C-6 Mean number of summer flounder per tow for young-of-the-year (age 0) from surveys used in summer flounder assessment.

NOTE: CTDEP = Connecticut Department of Environmental Protection; MADMF = Massachusetts Department of Marine Fisheries; MDNR = Maryland Department of Natural Resources survey; NC = North Carolina Pamlico Sound survey; NEFSC = Northeast Fisheries Science Center; NJBMF = New Jersey Bureau of Marine Fisheries; RIDFW = Rhode Island Department of Fish and Wildlife; and VIMS = Virginia Institute of Marine Sciences survey. s = spring survey; f = fall survey; w = winter survey; and all = all surveys.

made when dogfish are detected in the area. Numbers caught in tows of less than 20 minutes are linearly corrected from the actual time towed to the standard 20-minute tow. The fishing gear is a North Atlantic type 2-seam whiting trawl.

Connecticut's Department of Environmental Protection has surveyed its waters since 1984 (Johnson, 1994). Originally, 40 stations were occupied every month from April through November. In 1991 the series was changed to include a spring period of April through June and a fall period of September through November. Each month, 40 stations are allocated to strata defined by depth interval and bottom type for a sampling intensity of 1 station per 20 square nautical miles. At each station, a 30-minute tow at 3.5 knots is made with a 14 × 9.1 m High Rise otter trawl during daylight hours only.

A survey of New Jersey's coastal water has been conducted throughout the year since 1988 by the New Jersey Bureau of Marine Fisheries (Byrne, 1994). Initially, the survey was con-

TABLE C-4 Major Features of Surveys Used in Summer Flounder Assessments

Survey	Season	Time Period	Design	Number of Stations[a]	Gear	Notes
NMFS/NEFSC	Spring (March)	1976-1997	Stratified random	320 (116)	Yankee 36	Sampling and coverage reduced in recent times
	Fall (October)	1982-1996	Stratified random	320 (94)	Yankee 36	
	Winter (February)	1992-1997	Stratified random	320	Flounder trawl	
Rhode Island	April	1980-1996	Hybrid	84	¾ 340 Hi-Rise	
Massachusetts	May	1978-1996	Stratified random	100	North American type, 2-seam, whiting trawl	
	September			100		
Connecticut	April-June	1984-1996	Stratified random	120	Sweep net	Vessel change in 1990
	September-November			120		Temporal change in 1991
New Jersey	January	1991-1996	Stratified (constrained) random	39	Three-in-one trawl	Vessel change in 1991
	April, June, August, and October	1988-1996		39 per month	Three-in-one trawl	December and February dropped in 1990, 1991
North Carolina	June	1987-1996	Stratified random	100	2 30-ft. mongoose trawls	Change in area and time in 1990
	September		Stratified random	100		Optimal allocation
VIMS		1979-1996		2 to 4 per stratum per month	Semi-balloon otter trawl	
Maryland	April to October	1972-1996	Fixed station	20 per month	Semi-balloon otter trawl	

NOTE: NEFSC = Northeast Fisheries Science Center; VIMS = Virginia Institute of Marine Sciences.

[a] The figures in parentheses are the actual number of tows used for the summer flounder abundance estimates in 1995.

ducted in February, April, June, August, October, and December. In 1990-1991, the December and February surveys were replaced by a January survey, which has been continued to the present. The survey has 15 strata with latitudinal boundaries corresponding to adjacent NEFSC strata (except at the north and south state boundaries) and depth boundaries at the 30-, 60-, and 90-foot isobaths. Station allocation is semi-random, with the initial allocation of two stations per stratum such that one station is always assigned to the north and one to the south half of each stratum. A third station is randomly positioned in each of the nine larger strata. Sampling intensity is approximately 1 station per 46 square nautical miles. A 3-in-1 2-seam trawl is towed at each station during daylight hours for 20 minutes at an average speed of 2.8 knots.

The estimates of age-0 fish from the NEFSC spring survey and the fall surveys by NEFSC, Rhode Island, Massachusetts, Connecticut, and New Jersey, are supplemented by three young-of-the-year surveys of the summer flounder stock. The longest series has been conducted by the Maryland Department of Natural Resources (Mowrer, 1994). The current design (implemented in 1992) consists of 20 fixed trawl stations located from Assawoman Bay to Chincoteague Bay (close to the Virginia border). Six-minute tows are made at these stations each month from April to October using a 4.9-m semi-balloon otter trawl. This trawl is equipped with a 40-meter chain between the trawl doors, along with tickler chains connected to the doors. The abundance index is calculated as a form of the geometric mean.

The Virginia Institute of Marine Science juvenile fish trawl survey (Geer, 1994) has undergone many changes to its design and objectives since its inception. Since 1979 Virginia has conducted monthly monitoring of fixed stations in the main Virginia tributaries of Chesapeake Bay. The fixed sites are located in the river channels and spaced at approximately 5-mile intervals from the mouth of the river to the freshwater interface in each river. As of 1994 each river was divided into two equal distance strata for calculating abundance estimates. A pilot study using a stratified random design was initiated in 1991. A stratified random survey is made of the Chesapeake Bay on a monthly basis, with the exception since 1991, of January to March, during which there is only one cruise. Stratification in the bay is based on depth and latitudinal zones. Two to four trawling sites are chosen for each stratum each month. All stations are sampled using a 9.14-m lined semi-balloon otter trawl that is towed along the bottom for 5 minutes during daylight hours only. Catches are log-transformed (using ln[catch+1]) and the mean log catch is back-transformed to give a form of the geometric mean abundance per tow. Estimates of abundance are calculated using the re-transformed means of the log data and strata weighting as described by Cochran (1977).

The North Carolina Division of Marine Fisheries instituted a stratified random survey of Albemarle and Pamlico Sounds in 1987 (West and Wilson, 1994). The Albemarle strata were dropped from the survey beginning in spring 1990. At present, sampling is from the first to the third week every June and September. Seven strata are defined; the first three are rivers (Neuse, Pungo [not sampled through entire series], and Pamlico) and the remaining four are shallow (<12 ft) and deep strata (≥12 ft) in the eastern and western portions of Pamlico Sound. A minimum of three stations are allocated to each stratum, with the final allocation optimally allocated based on sampling in the same month in previous years. A total of 50 to 53 stations are sampled in each June and September. At each station, two 30-ft. mongoose trawls are towed for 20 minutes at 2.5 knots. Abundance estimates are presented as arithmetic means or "geometric" mean number of individuals per tow.

Allocation of Stations

The committee did not have the time or resources to conduct an evaluation of all of the different survey series or even all of the years within any one series. Instead, the three seasonal surveys conducted by NEFSC in 1995 were arbitrarily chosen to evaluate the survey design. Many strata in the NEFSC surveys have only 1 station (8 out of 28, Table C-1; 14 out of 56, Table C-2; and 2 out of 25, Table C-3). The spring and winter survey strata with only 1 station have never had any summer flounder in the survey catches, but these strata have had some of the largest catches in the fall survey. Within-stratum variance estimates cannot be calculated for strata with only 1 station. As a result, variance estimates for the stratified random mean or total abundance are probably underestimated.

The allocation of stations is approximately proportional to stratum area for the offshore (01) strata for each of the NEFSC surveys. Allocations for the inshore (03) strata in Table C-2 appear to have been made separately and at a minimal level. Proportional allocation is a reasonable compromise when trying to design an efficient survey of more than one species with very different spatial distributions. However, proportional allocation at the current sampling level will assign only one station to some offshore strata. Although it is possible that severe restrictions on the operation of the survey (i.e., time restrictions or demands for increased coverage) may have resulted in small overall sample sizes, it is difficult to understand why so many strata (between 8 and 29 percent) included only one station. Larger strata, with at least two stations in each, would provide a more sound statistical design without increasing total sample size.

It is common practice in groundfish surveys elsewhere to allocate a minimum of two stations per stratum no matter what overall allocation scheme is used. Some of the strata in the NEFSC survey, typically either shallow inshore or deep offshore, initially are allocated only 2 stations because of their relatively small areas. Both types of areas are subject to "bad tows": inshore areas usually because of untrawlable bottom, gear damage, and conflicts with inshore fixed gear; offshore usually because of the difficulty of towing at greater depth in stronger currents, which sometimes results in crossed trawl doors/tow wires, gear damage due to unseen uncharted bottom structure, and conflicts with offshore fixed gear. Although it is common practice on trawl surveys to have alternative stations available as a contingency to avoid the statistical problems of having only one station per stratum, NEFSC reports that it is not always possible to repeat bad tows within the scheduling constraints of their surveys.

TABLE C-5 Stratified Estimates for Summer Flounder from the 1995 NEFSC Surveys

Estimates	Winter	Spring	Fall
Stratified mean	10.93	1.09	2.40
Standard error	1.31	0.23	0.33
Relative error (percent)	12	21	14
Number of stations	116	94	122
95 percent confidence intervals			
Lower	8.36	0.63	1.76
Upper	13.50	1.54	3.04

The stratified estimates of mean number per tow, associated standard error, relative error, and confidence intervals for the 1995 surveys are presented in Table C-5. The NEFSC survey analysis program calculates confidence intervals as ±1.96*standard error. The relative error is calculated as the ratio of the standard error to its stratified mean. Although relative errors in the range of 10 to 20 percent, such as those listed in Table C-5, are not unusual for groundfish surveys, as noted above the variance was almost certainly underestimated because of a large proportion of strata having only one station.

Winter Flatfish Survey

NMFS describes the winter survey (Azarovitz, 1994) as a flounder survey. This survey is used mainly to monitor three species of flounder: summer, winter (*Pleuronectes americanus*), and yellowtail (*Pleuronectes ferrugineus*). The committee was supplied with data on catches for each of these three species from the 1996 winter survey. The catches for all three species were provided for those strata identified as being used for the summer flounder abundance estimate. Unlike the 1995 survey data discussed above, none of the strata provided had only one station. The winter groundfish survey series has not been used to estimate the abundance of winter flounder because the age structures have yet to be read (M. Terceiro, NMFS, personal communication, 1999). Abundance estimates for yellowtail have been developed from the southern New England stock, consisting of tows made in strata 5, 6, 9, and 10 (see Figure C-1).

The main focus of a fishery survey is to estimate as precisely as possible the mean, total, or some other aspect of fish abundance. One of the advantages offered by stratified random designs over such designs as simple random sampling is that the former designs provide a more precise estimate of the mean if the strata boundaries encompass similar densities of animals and higher sampling rates are used in the strata with the larger variances. How well each design achieves this goal of increased precision can be evaluated after the fact by comparing the variance of the estimate achieved to the variance that would have been obtained if a simple random sample design had been used. Given that the level of precision reflects the amount of information in the survey design, the simple random design is considered to be the minimal information case. In the case of stratified random designs, the difference between the simple random sampling variance and the stratified random sampling variance for the mean can be estimated from the data as (Smith and Gavaris, 1993),

$$\mathrm{Var}(\bar{y}_{SRS}) - \mathrm{Var}(\bar{y}_{STR}) = \sum_{h=1}^{L}\left(\frac{1}{n} - \frac{W_h}{n_h}\right)W_h s_h^2 - \left(\frac{N-n}{n(N-1)}\right)\left(\sum_{h=1}^{L}(\bar{y}_h - \bar{y}_{STR})^2 - \sum_{h=1}^{L} W_h(1-W_h)\frac{s_h^2}{n_h}\right) \qquad \text{(C-1)}$$

where,

n = total number of stations sampled in survey
n_h = total number of stations sampled in stratum h
W_h = proportion of the total area in stratum h
N = total possible number of stations in survey
$\bar{y}_h$ = mean in stratum h
s_h^2 = variance in stratum h
$\bar{y}_{STR}$ = stratified mean
$\bar{y}_{SRS}$ = simple random sampling mean

A positive difference indicates that the stratified random design resulted in a smaller variance and that the design added useful information about the population sampled. With a positive difference, the stratified design is referred to as more efficient than the simple random sample design because it provided a smaller variance for the same total sample size. The difference between the two variances consists of two components. The first term on the right-hand side of Equation C-1 is the allocation component and reflects the contribution of the sample-to-strata allocation scheme to the difference between the variances. Stations can be allocated to strata arbitrarily, proportional to the stratum area, proportional to some function of the stratum variance, or in other ways. The allocation term will be negative, zero, or positive for these three schemes, respectively.

The second term of the right-hand side of Equation C-1 is referred to as the strata component

TABLE C-6 Evaluation of Survey Design from the Winter 1996 Survey for Three Major Flounder Species

Statistics		Flounder Species: Summer	Winter SF strata	Winter WF strata	Yellow SF strata	Yellow YF Strata
Stratified mean		31.25	1.98	3.01	16.82	16.00
Standard error of the mean		7.55	0.56	0.86	3.12	5.17
Relative error (%)		24	28	29	19	32
95% confidence interval						
Lower		17.24	0.90	1.47	11.91	7.14
Upper		45.34	3.14	4.82	22.06	26.97
Efficiency	Total	39.07	21.41	14.88	9.09	2.75
	Allocation	–5.32	–11.84	–15.72	–59.46	–1.65
	Strata	45.05	33.25	30.60	68.55	4.39
	Maximum (%)	91.17	88.81	82.25	86.61	19.69
Minimum standard error		2.23	0.21	0.39	1.20	4.70

NOTE: SF strata refers to the group of strata used to construct the summer flounder survey index; WF strata refers to the group of strata used to construct the winter flounder survey index; and YF strata refers to the group of strata used to construct the yellowtail flounder survey index.

and measures the contribution of the strata to the reduction in variance. The purpose of stratification is to obtain homogeneous groupings such that the variance in a stratum is smaller than the variance over all strata. The larger this difference, and the smaller the within-stratum variance, the larger the difference between the simple random sample variance and the stratified random variance. In all of the efficiency analyses herein, the quantity in Equation C-1 will be presented as a percentage of the simple random sampling variance of the mean. The terms in Equation C-1 are based on random data, so the variance components are also random variables rather than precise measures.

Stratified estimates over all the strata provided were calculated for each of the three species from the 1996 winter survey (Table C-6). In addition, the stratified mean number per tow and associated precision estimates of the efficiency of the survey design for each species were calculated. The survey design provided more precise estimates of mean number per tow for all three species, with the design providing the greatest improvement for summer flounder (39.07 percent) and the least for yellowtail flounder (9.09 percent). In all three cases, the allocation of stations to strata was sub-optimal with the most severe case being for the yellowtail flounder (–59.46 percent). If only the yellowtail YF strata were used for the yellowtail abundance index, the total efficiency would be 2.75 percent with an allocation component of –1.65 percent.

The strata component was fairly substantial for all three species but the negative allocation component worked against any gains in precision offered by the strata boundaries for all three. If stations could be allocated in a optimal fashion (i.e., proportional to area and standard deviation of each stratum), the resultant estimated maximum efficiency along with the minimum standard error this would yield is provided in the table. In terms of standard error, the optimal allocation would result in substantial reductions of around 62 to 70 percent over simple random sampling.

The actual number of stations and the num-

TABLE C-7 Optimal Allocation Exercise for the 1996 NEFSC Winter Survey

Stratum	Actual	Summer	Winter	Yellowtail	Average
01 01	8	1	7	6	5
01 02	8	5	0	0	2
01 03	3	1	0	0	0
01 05	5	0	47	11	19
01 06	10	3	25	16	15
01 07	2	1	0	0	0
01 09	4	0	37	1	13
01 10	10	2	7	30	13
01 11	3	1	0	0	0
01 13	7	0	0	13	4
01 14	4	0	0	0	0
01 16	2	0	0	38	13
01 17	2	0	0	0	0
01 61	4	55	0	0	18
01 62	2	9	0	0	3
01 63	2	1	0	0	0
01 65	9	18	0	0	6
01 66	4	5	0	0	2
01 67	3	0	0	0	0
01 69	9	8	0	5	4
01 70	5	4	0	0	2
01 71	3	0	0	0	0
01 73	6	2	0	3	2
01 74	5	7	0	0	2
01 75	3	0	0	0	0
Total	123	123	123	123	123

NOTE: Because of the distribution of the stocks, there are some strata that should not be sampled. One constraint on the allocation optimization exercise should be that at least 2 stations are actually sampled in each sampled stratum. Some strata should have 0 stations because few flounder were observed in them in previous sampling. All the sampled strata have at least two stations per stratum in the average allocation.

ber of stations that would result from an optimal allocation for each species are presented in Table C-7. This table was constructed assuming that the observed sample variance was known before the survey and equal to the (unknown) population variance. The three species appear to have different spatial patterns of variability, so that no simple sampling scheme could be optimal for all three. There was some spatial overlap between winter and yellowtail flounders, but there was no overlap between these flounders and summer flounder. The first two species are found mainly in the northern group of strata in this survey (1-17, see Figure C-1), whereas summer flounder are found primarily in the southern group of strata (61-75). The actual allocation of stations used in 1996 appears to favor the distribution of winter and yellowtail over that of summer flounder. The last column in this table gives the number of stations per stratum averaged over the optimal allocations for each species. This allocation is presented as a simple compromise over all three species to determine if a single allocation pattern could be devised that

TABLE C-8 Results of Reallocation Exercise for Flounder Species Caught in the 1996 NEFSC Winter Survey

	Flounder Species		
Statistics	Summer	Winter	Yellow
Mean	29.97	1.98	16.82
Standard error of the mean	5.18	0.35	1.93
Relative error (%)	17	18	11
Efficiency			
Total	78.59	76.65	72.75
Allocation	29.39	38.27	3.22
Strata	49.20	38.38	69.53
Maximum (%)	92.33	89.69	87.02
Minimum standard error	2.80	0.21	1.20

TABLE C-9 Percentage of Total Number of Stations to be Allocated to Each Stratum Based on Optimal Allocation (Summer Flounder, 1995 and 1996 NEFSC Winter Surveys)

Strata	1995	1996
01 01	2.80	0.87
01 02	7.27	3.97
01 03	3.09	0.67
01 05	0.58	0.00
01 06	2.85	2.25
01 07	1.17	0.58
01 09	0.00	0.00
01 10	7.45	1.98
01 11	1.56	0.43
01 13	1.95	0.17
01 14	0.21	0.20
01 16	0.00	0.20
01 17	0.00	0.00
01 61	13.83	43.92
01 62	0.11	7.31
01 63	0.00	0.72
01 65	23.86	14.77
01 66	9.58	4.05
01 67	0.02	0.29
01 69	11.01	6.73
01 70	7.56	3.24
01 71	0.45	0.30
01 73	2.37	1.61
01 74	2.10	5.67
01 75	0.19	0.07

would improve the precision for all three species. The results for this allocation are presented in Table C-8, which assumes that the observed sample means and variances are reasonable estimates of the population values for the strata. In all cases (compare with Table C-6) there is a reduction in standard error as well as a substantial increase in efficiency. The allocation component is now positive for all three species.

Therefore, it appears that there is some potential for improving the allocation scheme for all three flounder species at the same time. We usually have only estimates from the previous years' data upon which to base the design of the current year survey. If there is some stability over time as to the relative variability of different strata, an allocation scheme in one year (or averaged over a number of years) could be used to design the allocation scheme in the next. The optimal allocations for summer flounder in 1995 and 1996 are similar in that the more and less variable strata are consistent over the two years (Table C-9). The actual rankings of strata by variability are not, however, exactly the same in the two years.

If the optimal allocation for summer flounder from the 1995 survey had been used to allocate stations in 1996 (again assuming all things equal), the predicted efficiency would have been 83 percent, with a positive allocation component. Unfortunately, the same allocation would have resulted in negative efficiencies for winter flounder and yellowtail in 1996. The committee did not have the data to try a compromise allocation from the three species in 1995 for the 1996 survey.

Finally, if the pattern persists of greater catches of yellowtail and winter flounder in the northern group of strata and greater catches of summer flounder in the south group of strata, the allocation of stations to strata might be a compromise allocation based on data from previous years. Thus, for yellowtail and winter flounder, stations would be allocated in the northern strata (1–17) and for summer flounder, stations would be allocated in the south (61–75). Although the optimal or compromise allocation calculations may result in no tows being allocated to some of the strata, it would be prudent to include at least two tows in each of these strata (even where "1" is indicated) in case the spatial distribution of the different species change. This will probably result in a loss of efficiency but this loss is unlikely to be large. If spatial patterns are not very persistent, an adaptive allocation scheme (Thompson and Seber, 1996) might be beneficial, again using some combination of catches of the flounder species in the current survey to allocate additional stations to the more variable strata.

Appendix D

Review of Summer Flounder Assessments

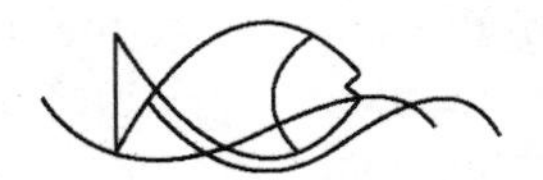

INTRODUCTION

The quality and appropriateness of a data set can sometimes best be judged from analyses and inferences derived from it. The strengths and shortcomings in the available information become apparent by examining results (e.g., parameter estimates), associated uncertainty, and sensitivity to assumptions. To evaluate summer flounder data, a number of standard assessment and data analysis techniques were applied to ascertain which data were informative and which data may be lacking, and to examine the level of information contained in model assumptions. The model structure and assumptions add information, but the added information may be wrong, depending on how well the model and assumptions match the actual situation.

EXPLORATORY ANALYSIS

The design of the NEFSC seasonal surveys is described in Appendix C. The committee analyses presented there indicate that some efficiency may be gained by reallocating sampling effort among sample strata. This gain in efficiency depends, of course, on the current state of the fish population and may have to be revised as sampling objectives change, depending on species-specific sampling and management priorities. Different and changing priorities in the data collection and fisheries management for specific species may lead to conflicting objectives. Managers, scientists, and stakeholders need to identify species or species complexes of major interest and specify the relative risk associated with the uncertainty for each species.

State agencies also collect data used in the National Marine Fisheries Service (NMFS) stock assessments (see Appendix C). There may be some redundancy in these data, but other reasons (e.g., political representation, verification) may exist for continuing them. Not all the survey information included in the SAW25 Report (NEFSC, 1997) was used in the 1996 NMFS assessment. Survey data on age classes 5 and older, for example, are typically excluded from the analyses for all but the Northeast Fisheries Science Center (NEFSC) winter survey. The NEFSC winter trawl survey includes information on 5+ age classes, but these data have been available only since 1992.

Table D-1 summarizes the number of surveys per age class and year in state and federal data sets. Coverage varies by age class and over time. Ages 1 and 5+ receive less coverage in general when compared with the middle three age classes, presumably due to the catchability characteristics of the gear employed. Some surveys are missing in the most recent years due to a time lag in availability. Information on age classes 0 and 2 through 4 will therefore have the greatest influence on the overall fit of the model because those data are the most abundant.

The log-transformed mean number per tow of summer flounder at age derived from the survey data is the relative abundance measure used by the assessment model. Trends in abundance can be seen by viewing a single age class across years as tracked by the individual surveys (NEFSC, 1997), but these trends can also be viewed in summary by examining the averaged log(CPUE+1) across surveys (see Figures D-1a-c). The log transformation is used to achieve stability in the variances and to facilitate contrast across the widely ranging abundance values. Such a transformation is commonly applied to fishery survey indices prior to inclusion in assessments because of the log-normal variation often seen in these indices; consequently, a log scale is the most appropriate scale on which to view these data in an exploratory analysis.

When survey-by-survey data are viewed (see Figures 4-8 in Terceiro, 1999), the NEFSC winter survey shows a much greater range in data values than the other surveys and thus must contribute a significant signal to the fit even though the length of the time series is limited. The second point, which can be seen more easily in the figures averaged across surveys (see Figures D-1a-c) is that the relative abundance indices may have peaked with the 1995 cohort. The trends indicated by the two observations available for the 1998 cohort go in opposite directions, creating a conflict in the data that may lead potentially to equally likely but conflicting trends in model estimates. These are important features to track in interpreting the assessment results below.

Another data set that goes into the assessment is the total catch at age (Figure D-2) representing the combined catch from the commercial and recreational fisheries. From the data available, we have no way of knowing how commercial and angling effort has changed in recent years, but the upward trend in ages 3 through 5, the slightly downward trends in age classes 1 and 2, and the significant drop in catch of age-0 fish should, in conjunction with the survey indices, determine the nature of the fit by the assessment models. Interpretation of these data is complicated by the 1992 changes under Amendment 2 to the summer flounder fishery management plan, which included an annual landings quota, a minimum size limit at 33 cm, and a minimum mesh size of 140 mm. These management actions probably influence what is seen in the combined commercial and recreational catch data. Shifts from commercial to recreational harvest also should influence the catch at age as each fishery exhibits different size and age selectivity patterns. Finally, weight at age goes into the assessment. The summer flounder weight at age has not shown any directional trends over time (Figure D-3). Thus, the dynamics exhibited in model estimates should reflect mainly the catch and relative abundance indices.

ASSESSMENT

A unified picture of the factors contributing to stock dynamics can now be developed using a system of equations linking the various components. This system of equations (a model) is derived from basic principles that are assumed to represent the system. These models add structure to the estimation process by characterizing relationships that exist among population variables (e.g., population size) and observations (e.g., catch and catch per unit effort). When using models for analysis, one should always keep in mind that (1) models are simplified representations of the system; (2) assumptions made in modeling impose a structure on the information that will influence assessment results; (3) no model works well with poor data (see NRC, 1998a); and (4) even if the data are high quality

TABLE D-1 Number of Surveys by Year and Age Class for Which Relative Abundance Measures are Available. (Surveys include NEFSC winter, spring, and autumn surveys, MADMF spring and fall surveys, CTDEP spring and fall surveys, RIDFW fall and fixed surveys, an NJDFW survey, and a DEDFW survey.)

Year	Age 0[a]	Age 1	Age 2	Age 3	Age 4	Age 5+
1982	4	1	2	3	3	0
1983	4	1	3	4	4	0
1984	4	1	4	5	5	0
1985	5	1	5	6	6	0
1986	5	1	5	6	6	0
1987	6	1	5	6	6	0
1988	7	2	6	6	6	0
1989	6	2	6	6	6	0
1990	7	3	7	6	6	0
1991	7	3	8	7	6	0
1992	7	4	9	8	7	1
1993	7	4	9	8	7	1
1994	7	4	9	8	7	1
1995	7	4	9	8	7	1
1996	7	4	9	8	7	1
1997	7	4	9	8	7	1
1998	6	4	9	8	7	1
1999	0	2	5	6	6	1

[a] Information on young-of-the-year fish (age 0) is available from a separate set of surveys and include data from Connecticut, Virginia, North Carolina, Maryland, New Jersey, Massachusetts, and the NEFSC.

NOTE: CTDEP = Connecticut Department of Environmental Protection; DEDFW = Delaware Department of Fish and Wildlife; MADMF = Massachusetts Department of Marine Fisheries; NEFSC = Northeast Fisheries Science Center; NJDFW = New Jersey Department of Fish and Wildlife; and RIDFW = Rhode Island Department of Fish and Wildlife.

and the assumptions are appropriate, there still might not be enough data to yield accurate and precise estimates.

Assessment Methodology

Three modeling approaches were applied to the summer flounder data to examine the influence of the data, model structure, and model assumptions on key assessment outputs. The approaches taken were based on the Laurec-Shepherd VPA method (Laurec and Shepherd, 1983; Darby and Flatman, 1994), the ADAPT method (Gavaris, 1988), and an AutoDifferentiation (AD) Model Builder implementation (Fournier, 1996) of a CAGEAN method (Deriso et al., 1985). Results from these methods were compared with results from the 1999 NMFS ADAPT assessment.

Laurec-Shepherd Virtual Population Analysis (VPA)

The Laurec-Shepherd method is an example of an ad hoc VPA tuning procedure. A number of these procedures are described in the literature and used in actual fisheries. The main characteristic of these procedures is that they accept the

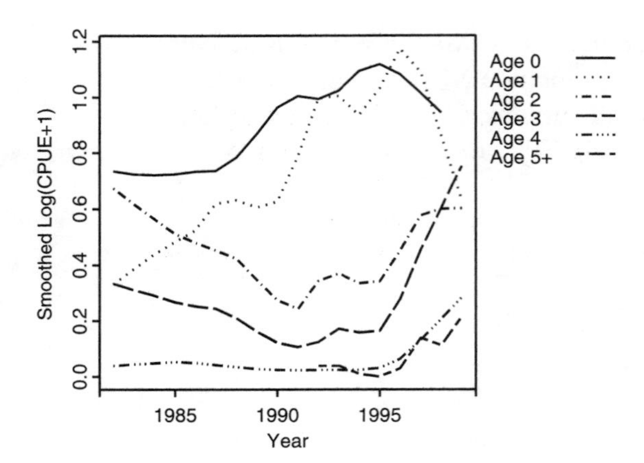

FIGURE D-1a Log (CPUE+1) averaged over surveys for each age class and smoothed over time.

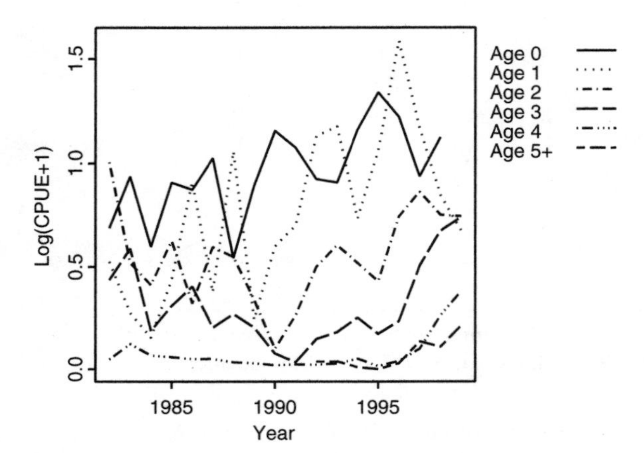

FIGURE D-1b Log(CPUE+1) averaged over surveys for each age class.

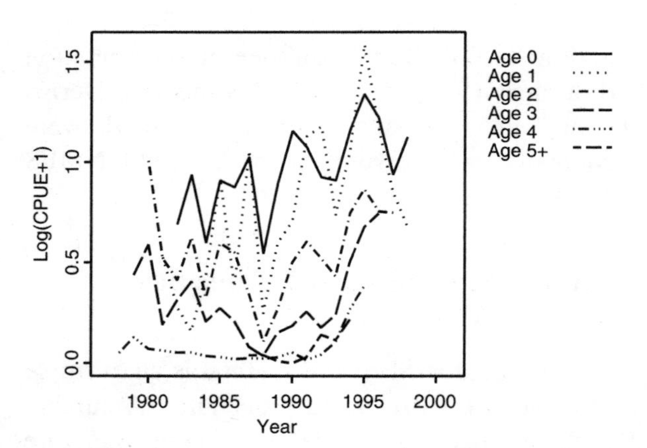

FIGURE D-1c Log(CPUE+1) averaged over surveys for each age class and lagged so that cohort indices coincide.

catch-at-age data as exact and then adapt the resulting range of possible population biomass estimates to fit auxiliary data depicting trends in the biomass. Auxiliary data can include commercial and survey catch rates. The influence of auxiliary data on the estimates is based on a standard goodness-of-fit criterion. The methodology used here is described in detail in Darby and Flatman (1994), and software (Lowestoft Tuning Package) is available to implement it.

The basic idea of the method is that an initial virtual population analysis is conducted from a reasonable but arbitrary starting point to arrive at numbers ($N_{y,a}$) at year (y) and age (a) for all years and ages:

$$N_{y,a} = N_{y+1,a+1}\, e^{M} + C_{y,a}\, e^{M/2}.$$

Here M represents an assumed instantaneous natural mortality rate. Under these values of population size and using observed catch rates ($CPUE_{f,y,a}$), estimates of catchability (q) are made for each survey or fleet (f) for each year and age of the analysis:

$$q_{fya} = \frac{CPUE_{f,y,a}}{N_{y,a}}\ .$$

For each survey-age combination the mean and standard error (SE) of log catchability are calculated across years. Each survey-age mean is then used to produce an estimate of fishing mortality in the most recent year of the survey. This fishing mortality estimate is then prorated up to a survey-based estimate of the fishing mortality generated by all vessels. An inverse variance weighted mean of the various survey results is then calculated. This provides an overall estimate of fishing mortality in the most recent year for each age of fish. Fishing mortality on the oldest age group is set as some specified ratio of the average fishing mortality on some specified range of younger ages. The resulting fishing mortalities for the last year and for the oldest true age in the analysis are then used to reinitiate the VPA.

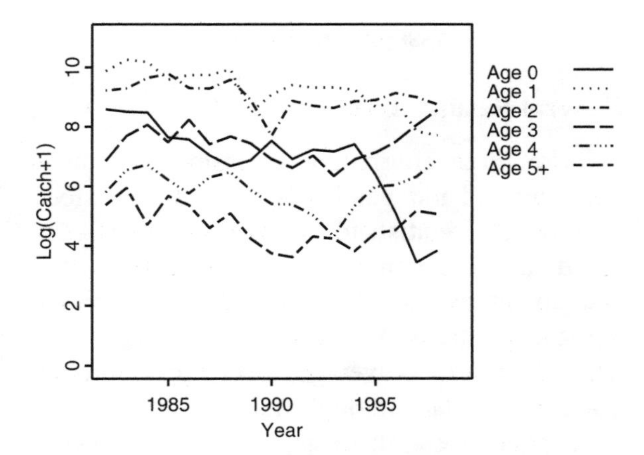

FIGURE D-2 Log(total catch+1) in thousands by age class.

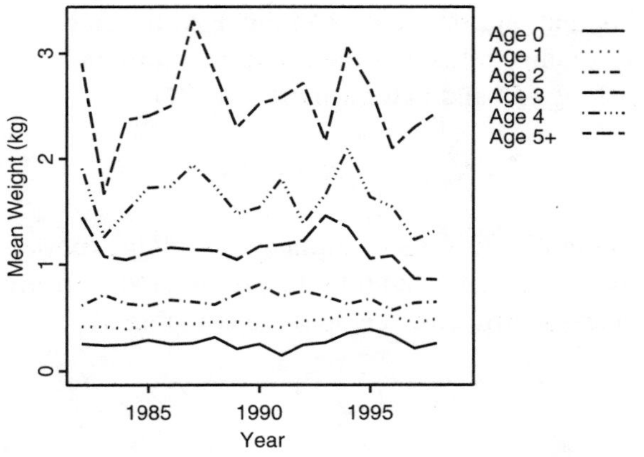

FIGURE D-3 Weight at age in kilograms.

This cycle continues until the process converges and results stabilize.

The method allows various decisions to be made about the weighting to be applied to past years' data and allows for some weight to be given to an assumption that the fishing mortality in the last year is the same as the average of a specified number of past years. This assumption of "shrinkage" to the mean fishing mortality is most useful when auxiliary tuning data are variable and when not too much trend has occurred in fishing mortality in recent years. The chief virtue of the Laurec-Shepherd method is its simplicity and consequent ease of understanding.

ACON ADAPT

The ADAPT analysis was performed by Robert Mohn of the Canadian Department of Fisheries and Oceans using ACON software.[1] The mathematical form of the model closely follows the one described above for the Laurec-Shepherd VPA procedure. The model fits the log of the survey observations at ages 0-4 in the terminal year using a non-linear least-squares algorithm. No "plus group" (i.e., cumulative ages 5 and older) is incorporated. For each year before the terminal year, the fishing mortality (F) at the oldest age (age 4) was estimated as half that of the ages 3, yielding a dome-shaped partial recruitment. Usually, the degree of domedness would be "tuned" by iteration to convergence for recent years; this was not done for this study. The software requires that the observation month for each survey be listed. Because the data were identified by season rather than month, the observation month was approximated by season (e.g., month 7 was assigned to the Northeast Fisheries Science Center summer series). The catchability coefficients (q) for each survey age were determined algebraically at each iteration of the non-linear least-squares algorithm. The error is assumed to be log-normal and the catch is assumed to be error free. Natural mortality was assumed to be 0.2 for all ages and years.

AD Model Builder CAGEAN

The CAGEAN framework differs from the VPA approaches mainly in that it assumes observation variability (ε) associated with the catch data:

$$C_{y,a} + N_{y,a} \frac{F_{y,a}}{F_{y,a} + M} \left(1 - e^{-(F_{y,a} + M)}\right) + \varepsilon.$$

[1] http://dfomr.dfo.ca/science/acon/index.html

Population abundance (N) changes through survivorship, reflecting the combined influences of fishing (F) and natural mortality (M)

$$N_{y+1,a+1} = N_{y,a}\, e^{-(F_{y,a}+M)}.$$

Generally, it is also assumed that fishing mortality can be separated into the product independent year-specific and age-specific components

$$F_{y,a} = f_y s_a.$$

(the so-called separable model assumption). This separable model assumption was relaxed in an effort to explore how it influences interpretations of recruitment. The analysis was conducted using the AD Model Builder software (Fournier, 1996) using a concentrated log-likelihood form for the optimization (Seber and Wild, 1989). Log differences in the sum of squares were applied to the catch and survey observations. Catchability was estimated for each survey data set. A constraint was imposed on the effort deviations reflecting a type of shrinkage assumption as discussed for the VPAs above.

Assessment Results

General Results Across Methods

The three alternative assessment methods were applied independently by different stock assessment scientists to the same summer flounder data and compared with the NMFS 1999 assessment results. The general trends in spawning stock biomass (Figure D-4) and fishing mortality (Figure D-5) averaged over ages 2 through 4 appear similar although there is not an exact correspondence. There appears to have been a general decline in the stock from 1982 through 1990, followed by a general increase in stock biomass through the 1990s. Complementary trends can be seen in the corresponding fishing mortality estimates, with an increase shown in the 1980s and a decrease in the 1990s.

Two things are worth noting in viewing these trends. First, the range of estimates varies substantially under the different model formulations, indicating that even with the same data, different models, employing different assumptions and implemented by different scientists, can produce different results. Second, the biomass trends in the last three years differ among methods, with certain estimates going in opposite directions. Although this is cer-

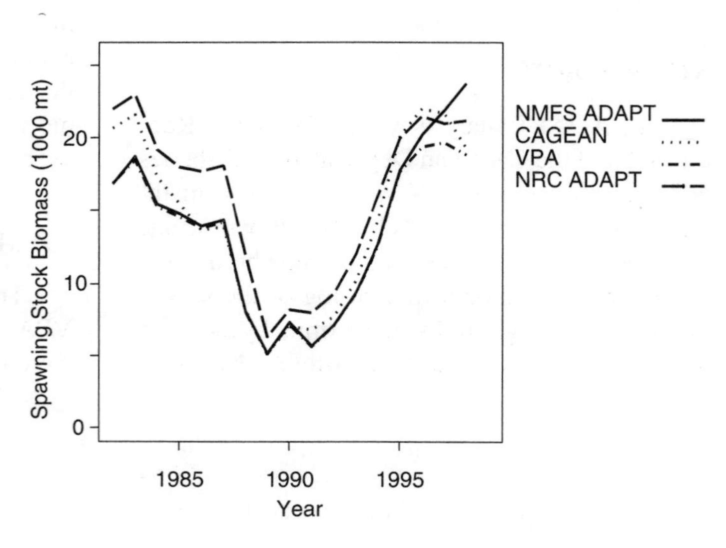

FIGURE D-4 Summer flounder spawning stock biomass as estimated by NMFS, as compared with three independently conducted model applications.

NOTE: NMFS = National Marine Fisheries Service; NRC = National Research Council; and VPA = virtual population analysis. ADAPT and CAGEAN are two stock assessment techniques, as described in the text.

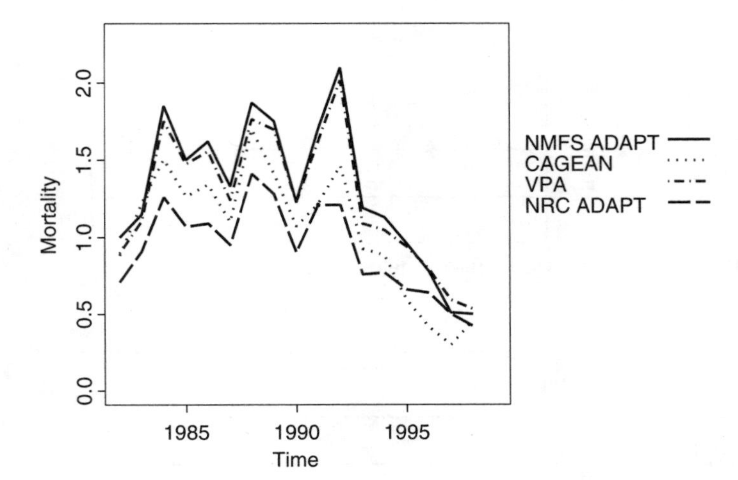

FIGURE D-5 Summer flounder fishing mortality for ages 2-4 as estimated by NMFS and compared with three independently conducted model applications.

tainly problematic for managers and stakeholders who are trying to make management decisions, it must be recognized that the variability among models in the last years is greater than the inter-annual variability of any specific model. Nonetheless, departures in the relative trends in abundance between models indicates that we should further explore how the models respond to the data.

Laurec-Shepherd VPA

Figure D-6 shows graphs of the residuals of Ln(q[f,*y*,*a*]) for each survey, with different symbols to show the ages. A number of points become apparent from these graphs.

1. Most series are quite variable. Residuals of greater than 1 (and less than –1) are common.
2. A number of the series may have some auto-correlation with long-term variations between higher and lower catchability.
3. These trends often seem similar between the various age classes represented by different surveys, but the trends vary among surveys. This might suggest variation in local availability to the various (mostly area-based) surveys rather than some change (e.g., mis-reporting errors on catch-at-age data) that would be reflected in all series.

The standard errors of the Ln(q[f,*y*,*a*]) calculated from the full 1998 SAW tuning data are shown in Table D-2. These include an input value of 0.5 for the shrinkage mean. In general the standard errors are quite high. Most are higher than 0.5, which corresponds approximately to a 50% coefficient of variation.

The standard errors in Table D-2 may be converted to the inverse variances used to combine estimates of fishing mortality in the last year (1998) of the analysis (results are shown in Table D-3). It is apparent that the Laurec-Shepherd method gives substantially more weight to the results for some surveys for some ages than others (e.g., on age 0, the New Jersey survey gets an 18 percent weighting, whereas Rhode Island fixed station survey gets only 3 percent). The Delaware 30-foot Trawl Survey gets a low weighting for young fish but a higher weighting for 2-and 3-year-old fish. The shrinkage mean for each of the 4 ages gets a weighting of about 13 to 16 percent.

The combined estimates of standard error of the terminal fishing mortality on ages 0-3 are 0.22, 0.27, 0.22, and 0.20, respectively. Thus, considerable variation might be expected in the various estimates of fishing mortality rate and population size at age.

A further diagnostic given by the Lowestoft

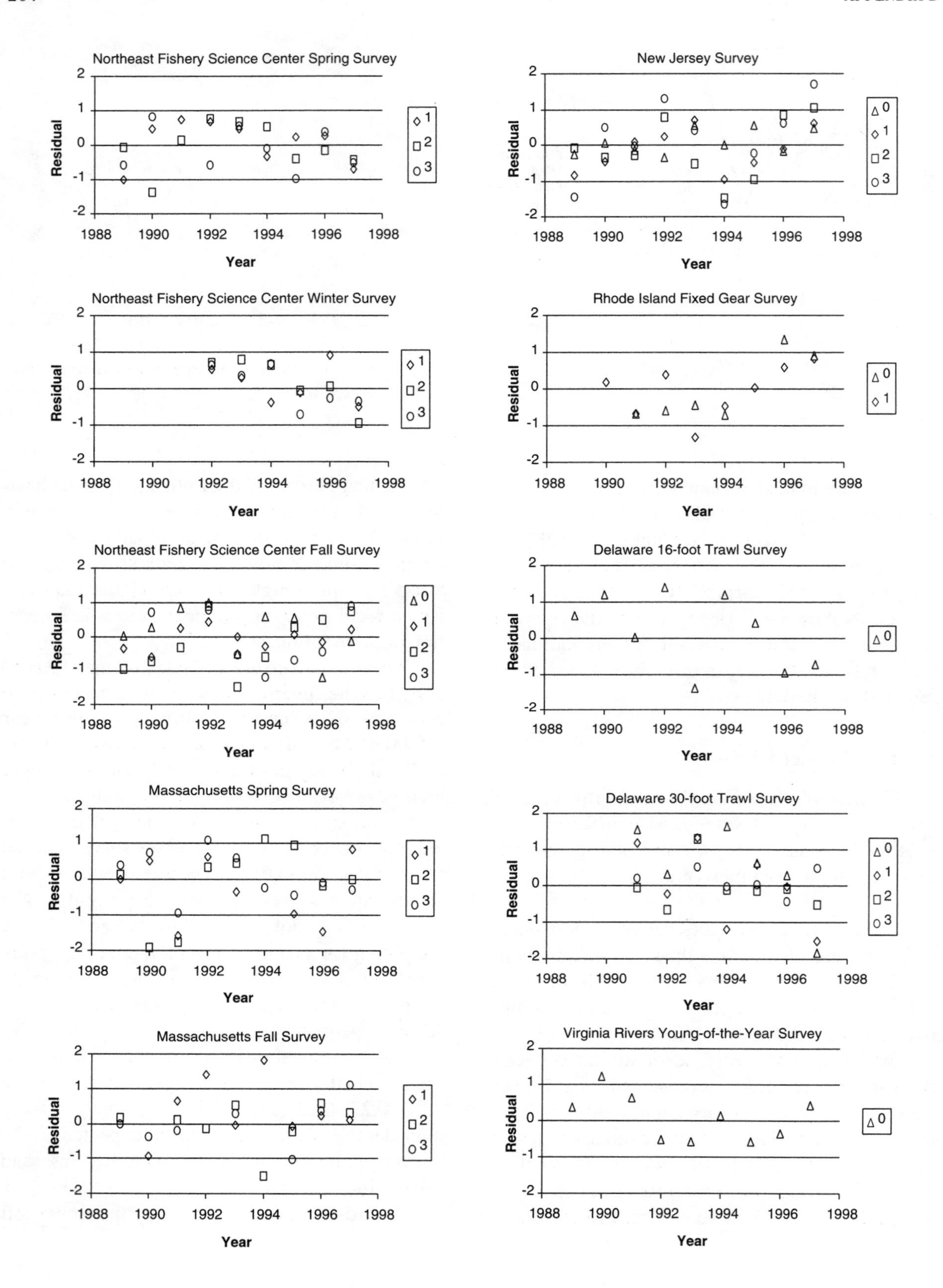
Northeast Fishery Science Center Spring Survey
New Jersey Survey
Northeast Fishery Science Center Winter Survey
Rhode Island Fixed Gear Survey
Northeast Fishery Science Center Fall Survey
Delaware 16-foot Trawl Survey
Massachusetts Spring Survey
Delaware 30-foot Trawl Survey
Massachusetts Fall Survey
Virginia Rivers Young-of-the-Year Survey
Residual
Year

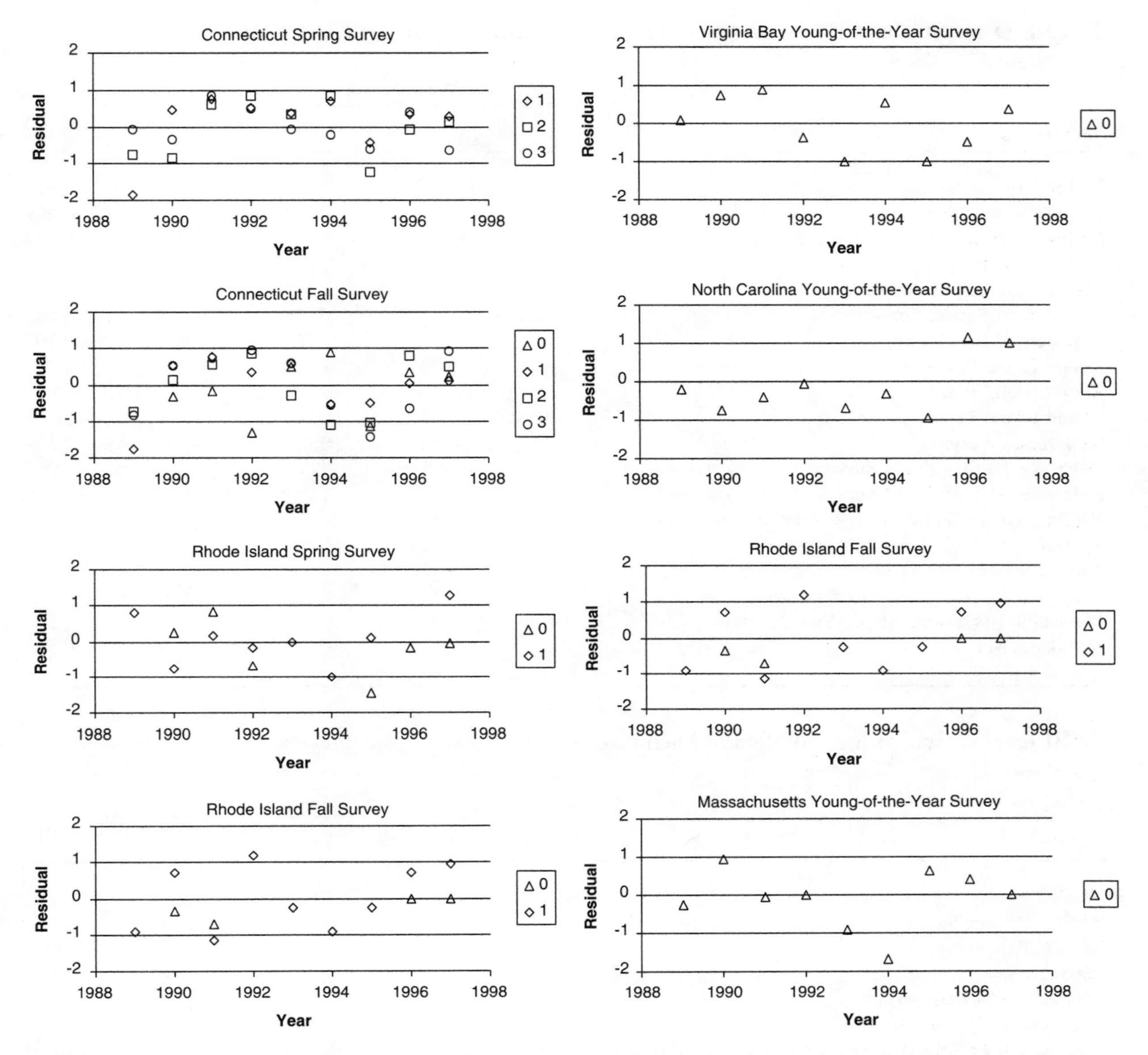

FIGURE D-6 Plots of the residuals of the natural log of catchability (Ln[Q]) of each survey from Laurec-Shepherd tuning of summer flounder data.

software package is the slope and standard error of the slope of a regression of $q(f,y,a)$ on y for each survey and age. Table D-4 shows the ratios of slope to its standard error for each available combination of survey and age. Depending on degrees of freedom, only levels of this ratio over 2 might be regarded as being significant. As might be expected from groundfish surveys, only a few of the series reach this level: only NEFSC winter survey for ages 2 and 3, the NEFSC autumn survey for age 1, and the Rhode Island fixed station survey for age 0. The significance of the NEFSC winter survey probably results because it is a short time-series with only the downswing of

TABLE D-2 Standard Errors of the Natural Log-transformed Survey Catchability

	Age			
Survey	0	1	2	3
NEFSC Spring Survey		0.555	0.52	0.616
NEFSC Winter Survey		0.625	0.82	0.569
NEFSC Fall Survey	0.735	0.468	0.719	0.609
Massachusetts Spring Survey		1.024	0.82	0.655
Massachusetts Fall Survey		1.187	0.521	0.617
Connecticut Spring Survey		0.866	0.639	0.507
Connecticut Fall Survey	0.707	0.607	0.628	0.707
Rhode Island Spring Survey	0.921	1.015		
Rhode Island Fall Survey	0.784	0.796		
Rhode Island Fixed Station Survey	0.97	0.707		
New Jersey Survey	0.427	0.626	0.834	1.131
Delaware 16-foot Trawl Survey	0.815			
Delaware 30-foot Trawl Survey	1.465	1.059	0.616	0.479
Virginia Rivers Young-of-the-Year Survey	0.532			
Virginia Bay Young-of-the-Year Survey	0.664			
North Carolina Young-of-the-Year Survey	0.732			
Maryland Young-of-the-Year Survey	0.599			
Massachusetts Young-of-the-Year Survey	0.9			
Shrinkage Mean	0.5	0.5	0.5	0.5

TABLE D-3 Weightings Applied to Each Age of Each Survey (percentages)

	Age			
Survey	0	1	2	3
NEFSC Spring Survey	0%	12%	15%	9%
NEFSC Winter Survey	0%	9%	6%	11%
NEFSC Fall Survey	6%	16%	8%	9%
Massachusetts Spring Survey	0%	3%	6%	8%
Massachusetts Fall Survey	0%	3%	15%	9%
Connecticut Spring Survey	0%	5%	10%	14%
Connecticut Fall Survey	6%	10%	10%	7%
Rhode Island Spring Survey	4%	3%	0%	0%
Rhode Island Fall Survey	5%	6%	0%	0%
Rhode Island Fixed Station Survey	3%	7%	0%	0%
New Jersey Survey	18%	9%	6%	3%
Delaware 16-foot Trawl Survey	5%	0%	0%	0%
Delaware 30-foot Trawl Survey	2%	3%	10%	15%
Virginia Rivers Young-of-the-Year Survey	11%	0%	0%	0%
Virginia Bay Young-of-the-Year Survey	7%	0%	0%	0%
North Carolina Young-of-the-Year Survey	6%	0%	0%	0%
Maryland Young-of-the-Year Survey	9%	0%	0%	0%
Massachusetts Young-of-the-Year Survey	4%	0%	0%	0%
Shrinkage Mean	13%	14%	16%	14%

TABLE D-4 Slope of the Regression of Catchability for Each Survey, Year, and Age (*q[f,y,a]*) Divided by the Standard Error of the Slope

Survey	Age 0	1	2	3
NEFSC Spring Survey		0.44	−0.83	−1.37
NEFSC Winter Survey		−1.35	−6.13	−2.34
NEFSC Fall Survey	1.29	3.02	1.72	−0.46
Massachusetts Spring Survey		0.17	0.82	0.14
Massachusetts Fall Survey		−0.49	−0.75	0.73
Connecticut Spring Survey		0.47	0.99	0.55
Connecticut Fall Survey	−0.40	−0.04	0.76	−0.14
Rhode Island Spring Survey	−1.31	−0.15		
Rhode Island Fall Survey	−0.55	−0.58		
Rhode Island Fixed Station Survey	3.33	1.50		
New Jersey Survey	1.66	1.41	0.77	1.31
Delaware 16-foot Trawl Survey	0.01			
Delaware 30-foot Trawl Survey	−0.88	−1.27	−0.16	−1.31
Virginia Rivers Young-of-the-Year Survey	−1.23			
Virginia Bay Young-of-the-Year Survey	−0.42			
North Carolina Young-of-the-Year Survey	1.90			
Maryland Young-of-the-Year Survey	1.65			
Massachusetts Young-of-the-Year Survey	1.50			

a cycle included. It would therefore seem inappropriate to adjust this series for linear trends in catchability.

The Lowestoft package allows the Laurec-Shepherd method to be run in retrospective mode (Box D-1). This shows the interpretation that the method would have placed on the current data set if the components that would have been available in past years were used to tune the VPA in those years. It thus gives some feel for the variability of the results and also can indicate biases in the analysis. Figures D-7 to D-10 show the retrospective patterns of estimates of fishing mortality rate, spawning stock biomass, total stock biomass, and recruitment at age 0 for summer flounder. The figures indicate that past interpretations might have differed by as much as double or half from more recent and presumably better estimates of the value of fishing mortality (e.g., 1992 or 1994). In recent years the results seem more consistent, as shown by more convergence among the lines that extend to later years. These results translate into somewhat reduced and opposite retrospective patterns in SSB and total stock biomass. This is in contrast to the NMFS results, which show little variation in SSB. The weightings calculated for each survey and age group in each year of the retrospective analysis are given in Table D-5.

Since the fishing mortality decreased in recent years, the advisability of using shrinkage to the mean needs to be questioned. Consequently, a fresh set of retrospective results were generated using the Laurec-Shepherd method on the 1998 tuning set for summer flounder without applying shrinkage to the mean (not shown). On the whole, these results seem more variable than those obtained using the shrinkage option, indicating that

BOX D-1
Retrospective Analysis in Stock Assessments
(NRC, 1998a)

The reliance of most stock assessment models on time-series data implies not only that each successive assessment characterizes current stock status and other parameters used for management, but also that the complete time series of past abundance estimates is updated. Retrospective analysis is the examination of the consistency among successive estimates of the same parameters obtained as new data are gathered. Either the actual results from historical assessments are used or, to isolate the effects of changes in methodology, the same method is applied repeatedly to segments of the data series to reproduce what would have been obtained annually if the newest method had been used for past assessments.

Retrospective analysis has been applied most commonly to age-structured assessments (Sinclair et al., 1991; Mohn, 1993; Parma, 1993; Anon., 1995). In such applications, the statistical variance of the abundance (or fishing mortality) estimates tends to decrease with time elapsed; estimates of the last year (those used for setting regulations) are the least reliable. In retrospective analysis, abundance estimates for the final years of each data series can vary substantially among successive updates, whereas those for the early years tend to converge to stable values. In some cases (i.e., some northwest Atlantic cod stocks, Pacific halibut, North Sea sole), early abundance estimates are consistently biased (either upward or downward) with respect to corresponding estimates obtained in later assessments. Extreme cases of consistent overestimation of stock abundance can have disastrous management consequences, as illustrated by the collapse of the Newfoundland northern cod (Hutchings and Myers, 1994; Walters and Pearse, 1996).

Retrospective biases can arise for many reasons, ranging from bias in the data (e.g., catch misreporting) to different types of model misspecification (mostly parameters that are assumed to be constant in the analysis but actually change, as well as incorrect assumptions about relative vulnerability of age classes). In traditional retrospective analyses, successive assessments use data for different periods, all starting at the same time with one year of data added to each assessment. An alternative method is to conduct successive assessments using data for a moving window of a fixed number of years (as in Parma, 1993, and Deriso et al., 1985). This method is appropriate for exploring trends in parameter estimates.

the variability seen in the data may be interpreted as observation error, bias, or a systematic trend in the underlying process (e.g., in fish catchability). Identification of the underlying process will improve predictions as the appropriate level of variation will be better represented.

ACON ADAPT

The summary statistics from the fit of log numbers at age for the 5 terminal ages are:

Age	*Param*	*CV*	*Bias (percent)*
0	9.88058	0.336	–0.0121
1	9.61357	0.281	0.0263
2	9.10554	0.272	0.0859
3	8.687	0.283	0.0981
4	6.63158	0.435	0.5283

The residuals (log[prediction/observation]) from the fit and the mean square error for each survey-age are given in Figure D-11. These patterns often have diagnostic value, which is

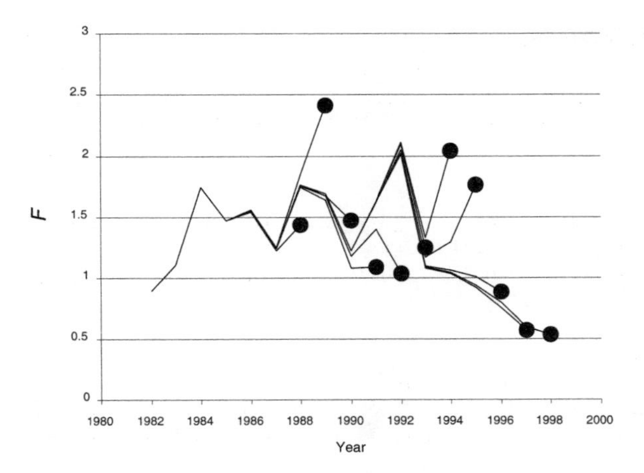

FIGURE D-7 Retrospective estimates of fishing mortality (*F*) for summer flounder using the full 1998 data set (with shrinkage to past *F*).

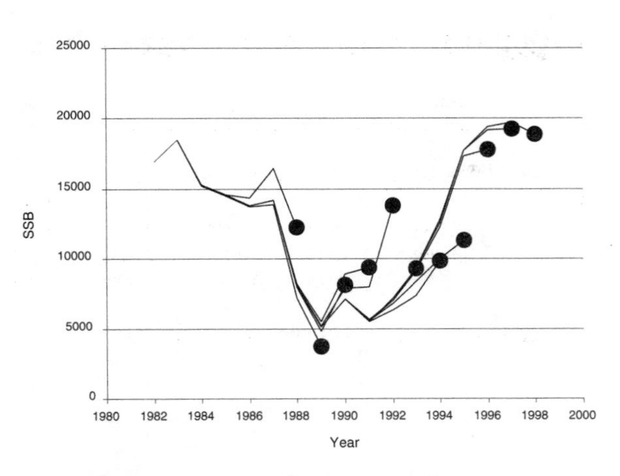

FIGURE D-8 Retrospective estimates of spawning stock biomass (SSB) for summer flounder using the full 1998 data set (with shrinkage to past *F*).

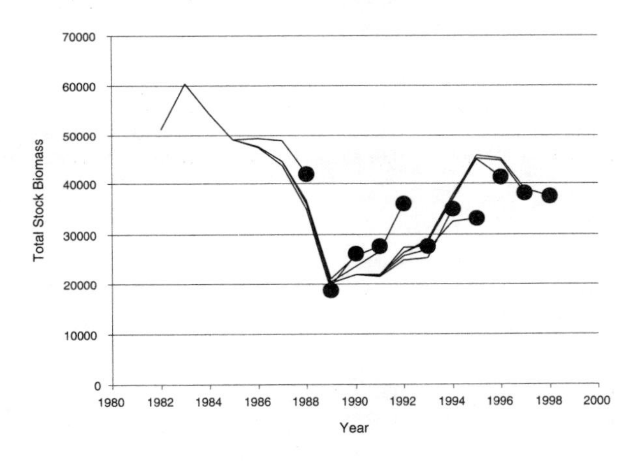

FIGURE D-9 Retrospective estimates of total stock biomass (TSB) for summer flounder using the full 1998 data set (with shrinkage to past *F*).

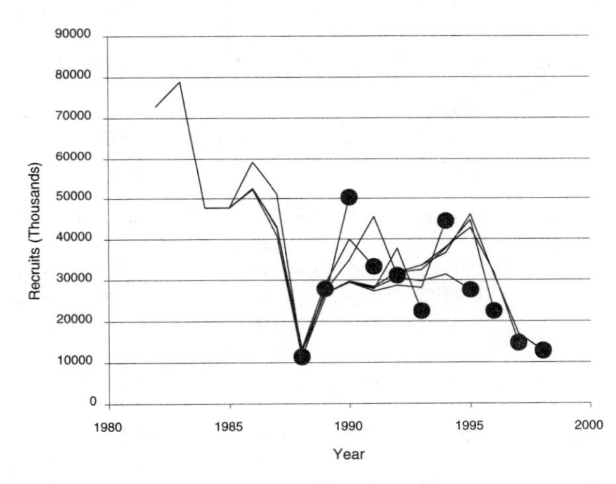

FIGURE D-10 Retrospective estimates of recruitment (at notional age 0) in millions (recruits in thousands) for summer flounder using the full 1998 data set (with shrinkage to past *F*).

somewhat confused in the present instance because of the relatively large number of indices. Vertical patterns suggest a year effect, for example, the terminal year of the NEFSC.w survey has large positive residuals for all ages. Indices with large mean square residuals may be removed. Also, a correlation analysis may be done among the residuals to identify redundant or contradictory series.

The fitted numbers for each survey and the data, after scaling for the catchabilities, are shown in Figure D-12. Figure D-13 summarizes the results of the sequential population analysis. The upper plot is the spawning stock biomass aged ahead (mortality applied) to August. The lower plot is the average (unweighted) *F* for ages 1-3.

A retrospective analysis was performed using average *F* over ages 1-3 and August biomass

TABLE D-5 Weights Applied for Each Year of Retrospective Analysis

	1988	1989	1990	1991	1992	1993	1994	1995	1996	1997	1998
Northeast Fisheries Science Center Spring Survey											
0	0	0	0	0	0	0	0	0	0	0	0
1	0.0687	0.0624	0.0592	0.0636	0.0504	0.0683	0.072	0.0739	0.0726	0.0782	0.1156
2	0.0369	0.1048	0.1207	0.0796	0.0768	0.0935	0.0971	0.1221	0.113	0.1367	0.1459
3	0.0692	0.0674	0.0696	0.0684	0.0523	0.0812	0.0885	0.0973	0.0556	0.071	0.0928
Northeast Fisheries Science Center Winter Survey											
0	0	0	0	0	0	0	0	0	0	0	0
1	0	0	0	0	0	0	0	0	0.0945	0.075	0.0912
2	0	0	0	0	0	0	0	0	0.1951	0.0882	0.0587
3	0	0	0	0	0	0	0	0	0.077	0.0792	0.1087
Northeast Fisheries Science Center Fall Survey											
	0	0.1199	0.1477	0.1381	0.0889	0.0457	0.0555	0.0512	0.0472	0.0624	0.0621
0.06											
1	0.3614	0.5409	0.5504	0.5628	0.4426	0.3331	0.3119	0.2991	0.3145	0.3264	0.1626
2	0.038	0.1406	0.1602	0.1319	0.1243	0.0911	0.1184	0.1008	0.0907	0.0934	0.0763
3	0.2244	0.2216	0.23	0.2237	0.2473	0.2109	0.1959	0.1482	0.0809	0.0862	0.0949
Massachusetts Department of Marine Fisheries Spring Survey											
0	0	0	0	0	0	0	0	0	0	0	0
1	0.0375	0.031	0.033	0.0277	0.0246	0.0412	0.0283	0.0274	0.0226	0.0251	0.034
2	0.0112	0.0314	0.0449	0.0274	0.0264	0.038	0.0366	0.038	0.0394	0.0514	0.0587
3	0.0405	0.0433	0.048	0.0446	0.0505	0.0564	0.0609	0.0685	0.0485	0.0653	0.082
Massachusetts Department of Marine Fisheries Fall Survey											
0	0	0	0								
1	0.0216	0.0172	0.0196	0.0225	0.0178	0.0267	0.0241	0.0258	0.0262	0.033	0.0253
2	0.0222	0.082	0.133	0.1386	0.0908	0.1999	0.1604	0.1405	0.1099	0.1318	0.1453
3	0.053	0.0601	0.0627	0.0726	0.0763	0.0978	0.1004	0.1187	0.0668	0.0795	0.0925
Connecticut Department of Environmental Protection Spring Survey											
0	0	0	0	0	0	0	0	0	0	0	0
1	0.0188	0.017	0.018	0.0203	0.0171	0.0268	0.0276	0.0291	0.0281	0.0343	0.0475
2	0.0525	0.1358	0.1942	0.1347	0.105	0.0949	0.075	0.088	0.0786	0.0923	0.0966
3	0.1624	0.1934	0.214	0.2199	0.1966	0.189	0.1862	0.2259	0.1163	0.1368	0.1369
Connecticut Department of Environmental Protection Fall Survey											
0	0	0	0.0889	0.0939	0.0392	0.0454	0.0611	0.0542	0.0569	0.0674	0.0648
1	0.1574	0.054	0.0537	0.0541	0.0444	0.0663	0.067	0.0668	0.0597	0.0683	0.0967
2	0.792	0.3854	0.1651	0.3176	0.2123	0.1646	0.1884	0.1538	0.0919	0.0966	0.1
3	0.2244	0.2076	0.1801	0.1733	0.1565	0.1252	0.1351	0.1207	0.057	0.0624	0.0704
Rhode Island Department of Fish and Wildlife Spring Survey											
0	0.0211	0.019	0.022	0.0222	0.0155	0.0182	0.0196	0.0216	0.0264	0.0325	0.0382
1	0.0308	0.0265	0.0283	0.0332	0.0269	0.0439	0.0475	0.0485	0.0279	0.0282	0.0346
2	0	0	0	0	0	0	0	0	0	0	0
3	0	0	0	0	0	0	0	0	0	0	0

TABLE D-5 Continued

	1988	1989	1990	1991	1992	1993	1994	1995	1996	1997	1998
Rhode Island Department of Fish and Wildlife Fall Survey											
0	0.0359	0.0324	0.0333	0.0308	0.018	0.025	0.0257	0.028	0.0401	0.0458	0.0527
1	0.0816	0.0665	0.0664	0.0455	0.039	0.0582	0.0611	0.0597	0.0494	0.0503	0.0562
2	0	0	0	0	0	0	0	0	0	0	0
3	0	0	0	0	0	0	0	0	0	0	0
Rhode Island Department of Fish and Wildlife Fixed-station Survey											
0	0	0	0	0	0	0	0	0	0.0283	0.0338	0.0344
1	0	0	0	0	0	0	0.0884	0.0813	0.0585	0.0518	0.0713
2	0	0	0	0	0	0	0	0	0	0	0
3	0	0	0	0	0	0	0	0	0	0	0
New Jersey Department of Fish and Wildlife Survey											
0	0	0	0	0	0.2307	0.0739	0.1222	0.079	0.1536	0.1417	0.1777
1	0	0	0	0	0.2136	0.158	0.1014	0.1101	0.0969	0.0882	0.0909
2	0	0	0	0	0.2345	0.1479	0.1349	0.1058	0.0604	0.0531	0.0567
3	0	0	0	0	0.0324	0.031	0.034	0.0327	0.0217	0.0239	0.0275
Delaware Department of Fish and Wildlife 16-Foot Trawl Survey											
0	0.1819	0.1225	0.0987	0.0823	0.0351	0.0439	0.0363	0.0357	0.0465	0.046	0.0488
1	0	0	0	0	0	0	0	0	0	0	0
2	0	0	0	0	0	0	0	0	0	0	0
3	0	0	0	0	0	0	0	0	0	0	0
Delaware Department of Fish and Wildlife 30-Foot Trawl Survey											
0	0	0	0	0	0	0	0	0.0112	0.0164	0.014	0.0151
1	0	0	0	0	0	0	0	0.0301	0.0297	0.022	0.0318
2	0	0	0	0	0	0	0	0.0715	0.0749	0.0951	0.104
3	0	0	0	0	0	0	0	0	0.3614	0.2607	0.1534
Virginia Rivers Young-of-the-Year Survey											
0	0.126	0.1787	0.1649	0.1421	0.0779	0.0815	0.0899	0.0989	0.1099	0.1106	0.1145
Virginia Institute of Marine Sciences Survey, including Chesapeake Bay Young-of-the-Year											
0	0	0	0	0	0.2	0.0786	0.083	0.0723	0.0736	0.0717	0.0735
North Carolina Young-of-the-Year Survey											
0	0	0	0	0.1599	0.1042	0.281	0.2438	0.2946	0.0692	0.0666	0.0605
Maryland Young-of-the-Year Survey											
0	0.1267	0.1185	0.1125	0.1019	0.0714	0.1029	0.0926	0.0849	0.1238	0.1164	0.0903
Massachusetts Young-of-the-Year Survey											
0	0.0715	0.0792	0.0748	0.0671	0.0392	0.0508	0.0403	0.0402	0.0489	0.0536	0.04
Shrinkage Mean											
0	0.317	0.302	0.2668	0.2108	0.1229	0.1432	0.1342	0.1322	0.1439	0.1378	0.1296
1	0.2221	0.1845	0.1714	0.1703	0.1234	0.1775	0.1708	0.1482	0.1194	0.1191	0.1425
2	0.0472	0.12	0.182	0.1702	0.1298	0.17	0.1892	0.1794	0.1463	0.1613	0.1578
3	0.2262	0.2067	0.1956	0.1976	0.188	0.2084	0.1991	0.1879	0.1149	0.1351	0.1408

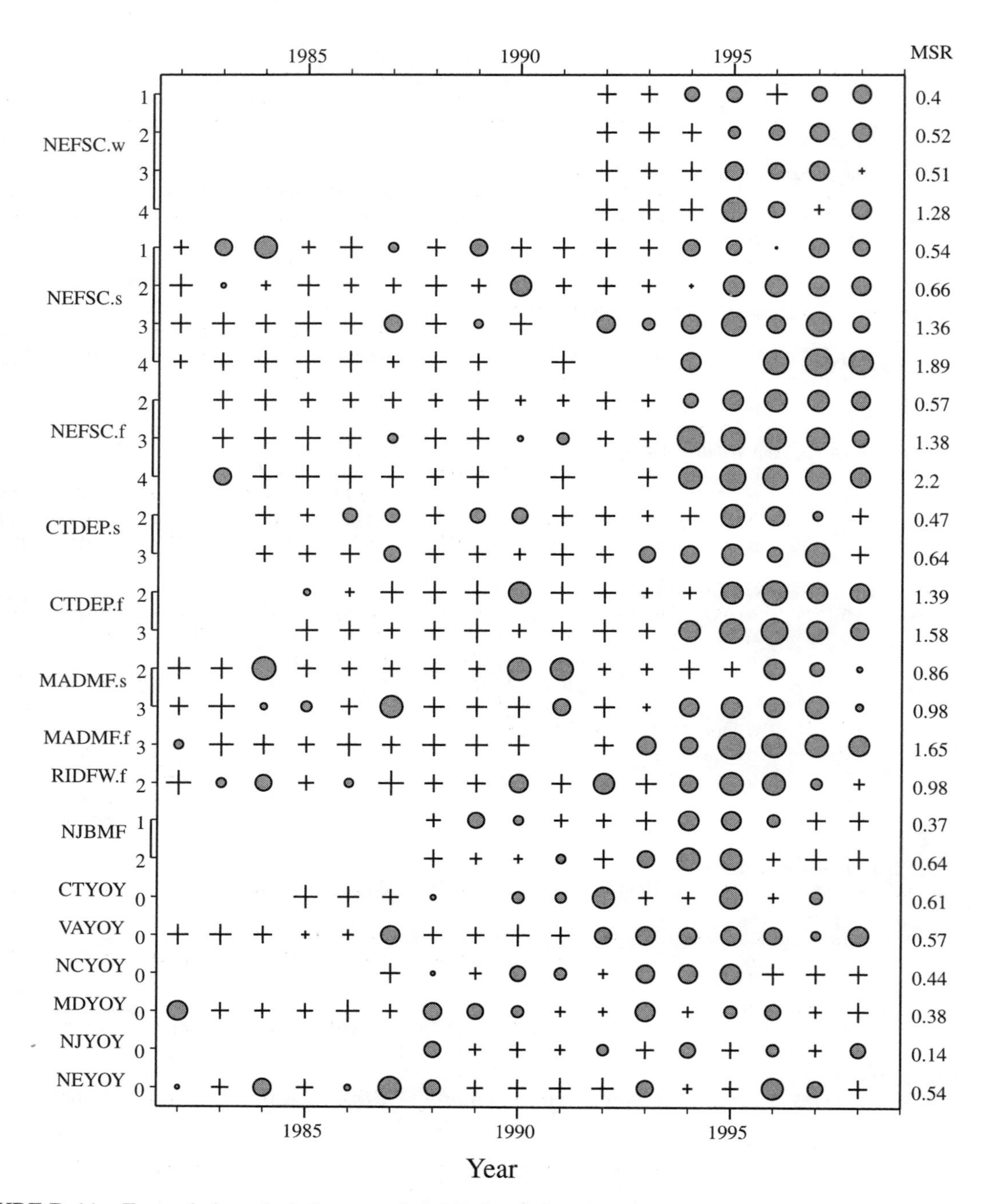

FIGURE D-11 Expanded symbol diagram of residuals of abundance measures from sequential population analysis. Circles and crosses represent negative and positive residuals respectively, with the size of the symbol being proportional to the size of the residual. The left-hand column denotes the survey and ages while the right-hand column shows the mean square residual (MSR) for each survey-age.

NOTE: CTDEP = Connecticut Department of Environmental Protection; CTYOY = Connecticut Young-of-the-Year Survey; MADMF = Massachusetts Department of Marine Fisheries; MDYOY = Maryland Young-of-the-Year Survey; NCYOY = North Carolina Young-of-the-Year Survey; NEFSC = Northeast Fisheries Science Center; NEYOY = New England Young-of-the-Year Survey; NJBMF = New Jersey Bureau of Marine Fisheries; NJYOY = New Jersey Young-of-the-Year Survey; RIDFW = Rhode Island Department of Fish and Wildlife; VAYOY = Virginia Young-of-the-Year Survey. f = fall survey; s = spring survey; w = winter survey.

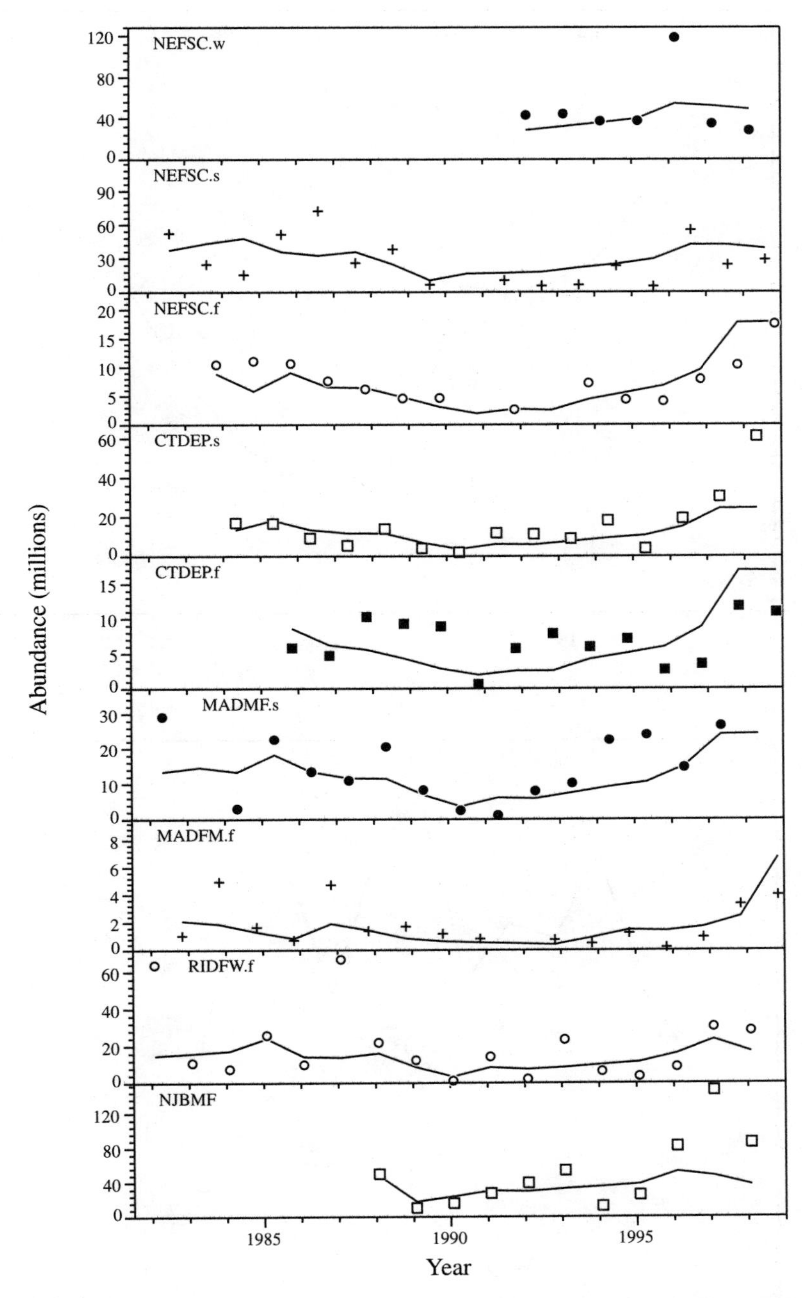

FIGURE D-12 Predicted population size and data from each survey. The data are summed over the appropriate ages, which are shown in Figure D-11.

FIGURE D-13 Sequential population analysis (SPA) estimates of F and biomass from ADAPT base run using data shown in Figure D-1. The biomass has been aged ahead to August and weighted by a maturity ogive (a continuous cumulative frequency curve).

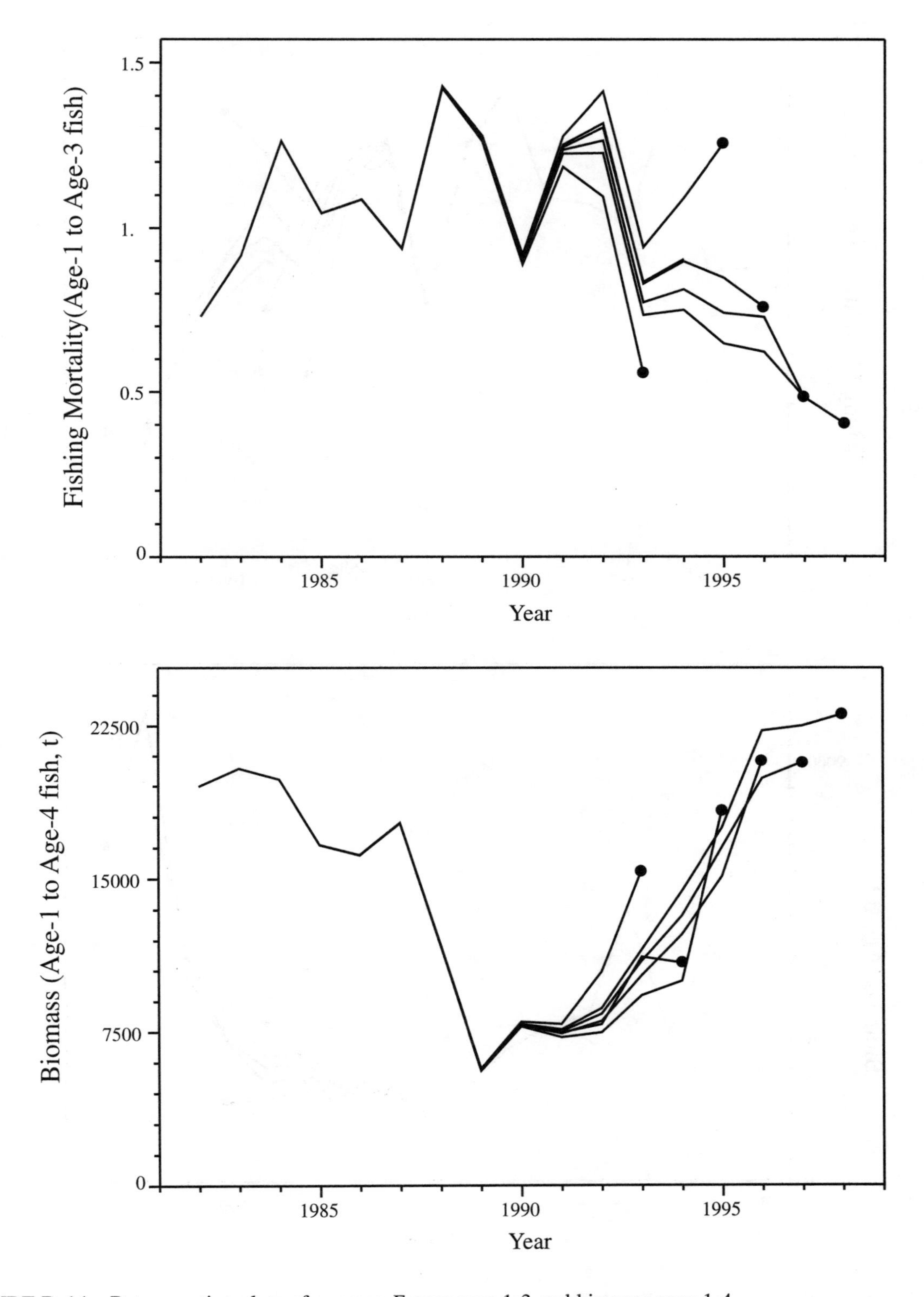

FIGURE D-14 Retrospective plots of average *F* over ages 1-3 and biomass ages 1-4.

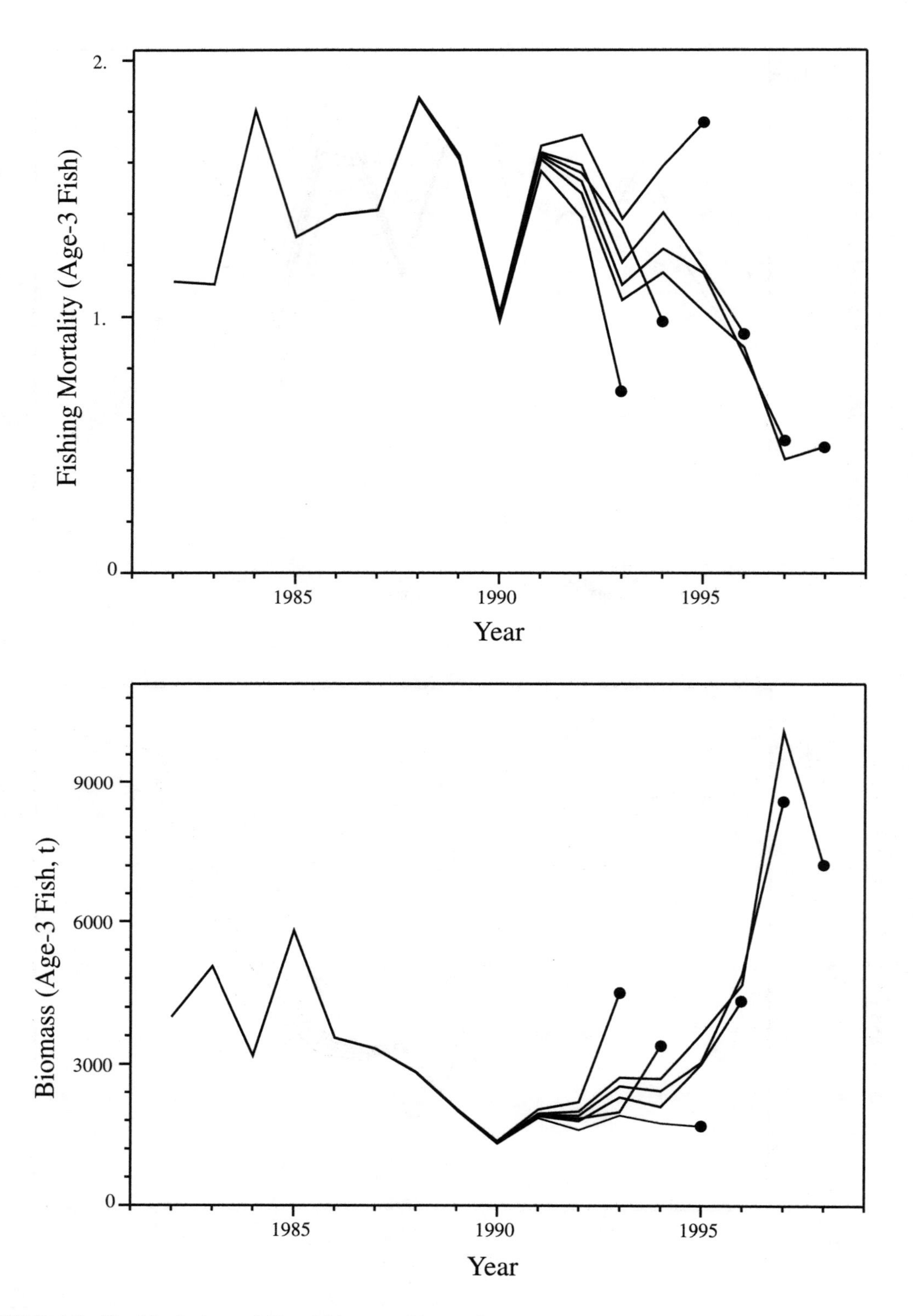

FIGURE D-15 Residual plots of F and biomass for age 3.

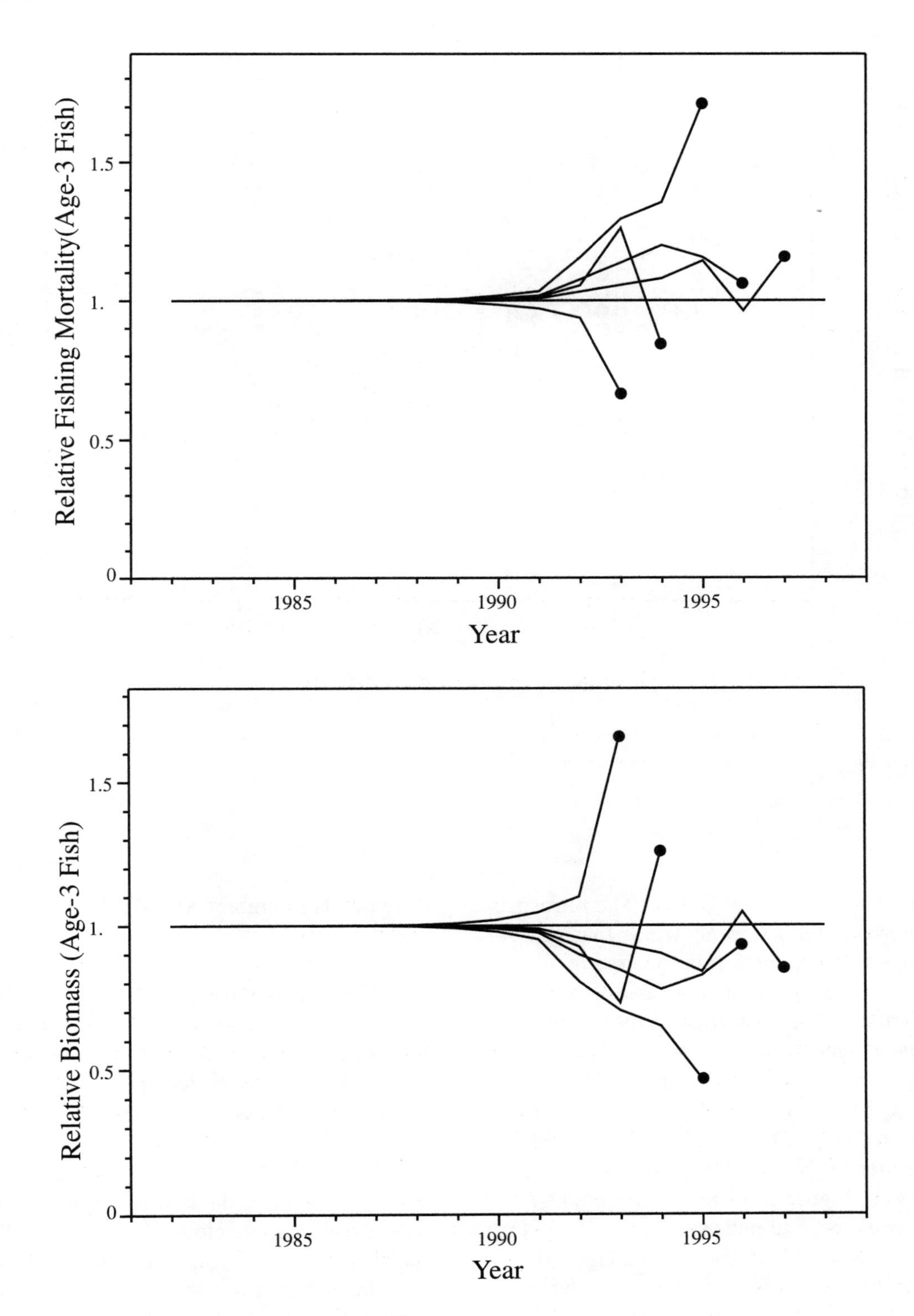

FIGURE D-16 Relative retrospective patterns for *F* and biomass at age 3. The estimates in Figure D-15 have been divided by the longest time series to obtain the values shown above.

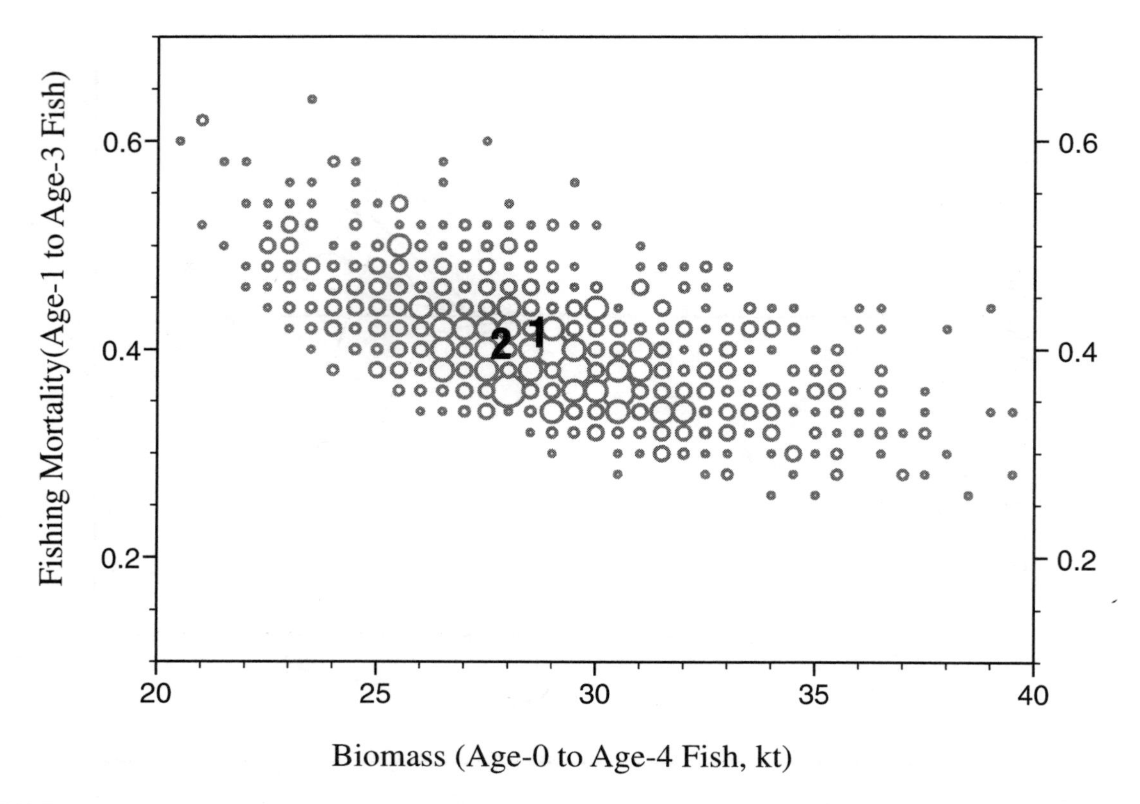

FIGURE D-17 Expanded symbol diagram of terminal F and biomass from 1,000 bootstrap trials. "1" marks the point estimate and "2" marks the bias-corrected estimate.

over ages 1-4 (see Figure D-14). The retrospective pattern for the Fs seems worse than that for the biomass, in that the former diverge more substantially. The apparent discrepancy was analyzed further. The first effect removed was the different weighting implied when F values are averaged but biomass values are summed. This was done by looking at the patterns for age 3 alone (see Figure D-15). Here, the F still has more scatter of the terminal points of retrospective curves. Figure D-16 recasts Figure D-15 as relative retrospective patterns. Each line in Figure D-15 was scaled relative to the longest series. Here the two patterns look similar, though inverted. This demonstration showed that the apparent discrepancy was due to the different weighting and the non-linear relationship between F and numbers at age and not an error in computation.

The performance of the model was further tested by a bootstrap analysis. The residuals from the base run were re-sampled to produce 1,000 bootstrapped replicate surveys. The catch data were then fit to the replicates. The terminal average F and biomass were aggregated and plotted in Figure D-17. The cloud of symbols is well behaved in the sense that it forms a continuous mass. In some instances, this type of analysis will reveal separate clouds, which can reflect instability in the analysis. The number "1" denotes the point estimate and the number "2" is the bias-corrected estimate. The bias is estimated by taking the difference between the point estimate and the mean of the bootstrap estimates, which is then

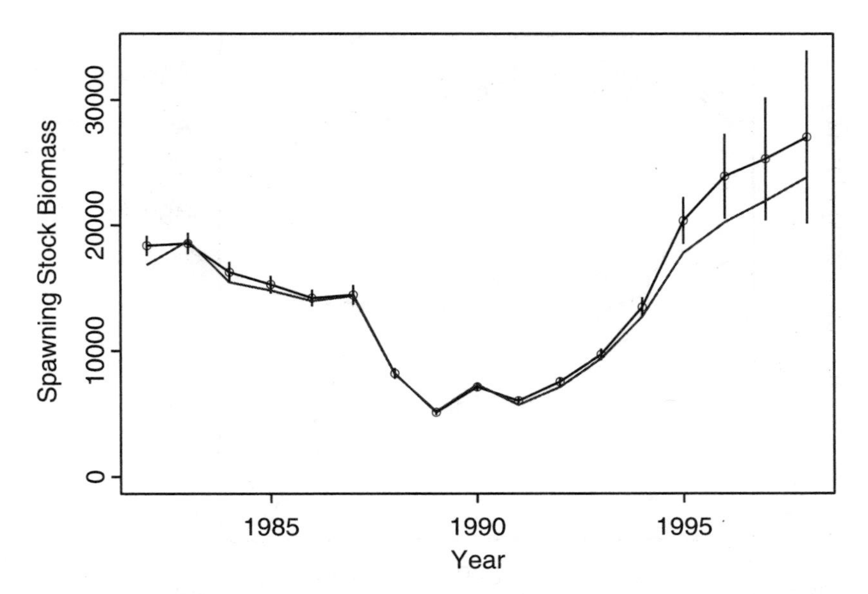

FIGURE D-18 Comparisons of CAGEAN AD Model Builder results with varying selectivity at age (with error bars) and NMFS ADAPT results (same as in Figure D-4).

applied to the point estimate to produce the bias-corrected values.

The analysis of summer flounder has more surveys than are usually included, and was performed in some haste. Although we consider this ADAPT analysis of the summer flounder data to be preliminary, we believe that the results are useful for NMFS as a basis for a more complete evaluation of their ADAPT approach. If a more complete investigation and comparison with other analyses were deemed desirable, considerable time should be set aside by NMFS for such an exercise, which would help quality control.

AD Model Builder CAGEAN

The most significant difference between the conventional CAGEAN-type analysis and the conventional VPA approach in terms of the summer flounder assessment is the assumption of separability between the year and age effects included in the CAGEAN procedure. This separability assumption carries with it the implication of a constant selectivity at age over time as employed here. Changes observed in this fishery might cause selectivity to vary with changes in regulations (e.g., mesh size), making an assumption of constant selectivity inappropriate. To address this issue a modification was made to the CAGEAN model to allow for some variation in selectivity with time. This was done by allowing deviations to occur in the log selectivity at age with a penalty applied to the sum of the squared deviations in the log-likelihood formulation used to specify the optimization surface. It was only when deviations were allowed for age classes 0 and 1 that any substantial change in the estimates result. When the CAGEAN model was used with varying selectivity at age 0 and age 1, the spawning stock biomass estimates more closely follow that derived under the 1999 NMFS assessment (Figure D-18). Selectivity estimates, as portrayed in fishing mortality estimates at age shown in Figure D-19, show a sharp drop in recent years, which the model assumes corresponds to changes in gear and targeting practices in the commercial and recreational fisheries. If such an assumption

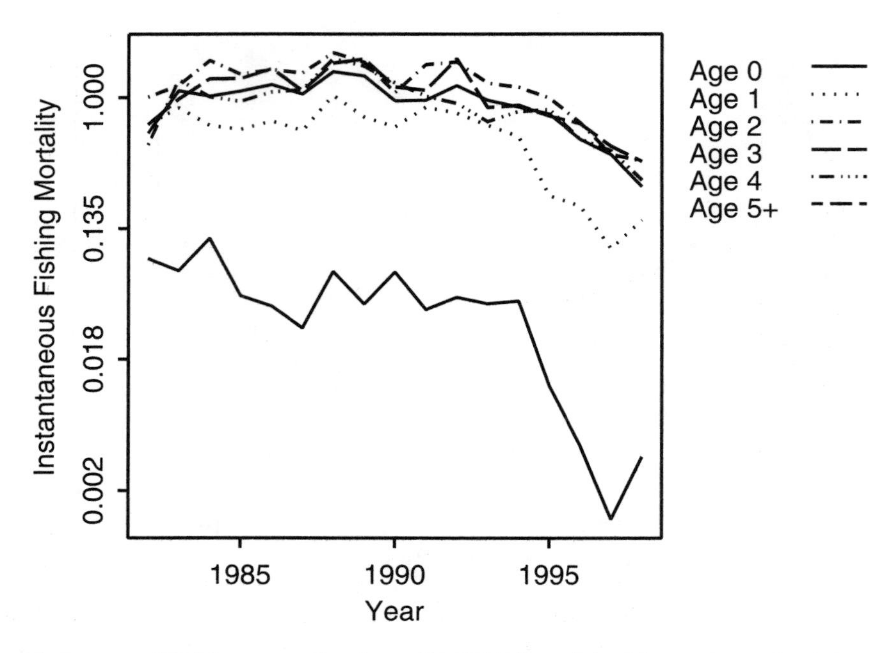

FIGURE D-19 Estimates of fishing mortality from CAGEAN model with varying selectivity at age.

is erroneous, other factors, such as a drop in recruitment, could explain the trends.

The influence of this change in selectivity under this model formulation is similar and perhaps analogous to responses in model estimates to the shrinkage assumption in F employed by the other assessment methods. All these results point to the rather tenuous nature of the estimates available for the most recent years.

SUMMARY

Three analyses of summer flounder data by three different individuals using three different stock assessment models yielded the same general decadal trends indicated by the 1999 National Marine Fisheries Service assessment (see Figures D-4 and D-5). Namely, these analyses showed that the spawning stock biomass has recovered substantially from a trough in the early 1990s and that fishing mortality dropped substantially during the same period. The committee believes both changes are probably due to strict management measures implemented in 1992.

However, the models yielded some differences in estimates. Because all the models used the same data, differences in the estimates arose from differences in the structural assumptions of each modeling method. Model structure introduces implicit assumptions, and some models are built to be easier to use by specifying assumptions in the standard computer software rather than allowing stock assessment scientists to vary the assumptions through input values. These assumptions often go unstated, but in this case conflicting signals in the data dramatically influenced model results because of the assumptions used. Our analysis does not necessarily indicate that the NMFS results are incorrect or that any of our specific results are correct. However, the analysis does show that the assumptions are influencing the results and that their influence should be acknowledged explicitly and used to help understand the dynamics of fish populations.

The committee was able to narrow the likely causes of these differences and believes they are related to the structure imposed on fishing mortality at age in the various models. This structure is implemented in the assumption of separable selectivity and full recruitment fishing mortality in the CAGEAN model and by the level set for the "shrinkage" control assumption in the VPA and ADAPT models. The separable selectivity assumption in the CAGEAN model forces selectivity at age to remain fixed over time. The shrinkage factor, used in the ADAPT and VPA models, controls how quickly fishing mortality (F) is allowed to change from year to year.

The CAGEAN model shows the most distinct drop in spawning stock biomass in recent years because it does not have the flexibility to deal with the decrease in younger fish seen in both CPUE and catch-at-age data. The behaviors of the VPA and ADAPT models appear to be related to the degree of variability in fishing mortality allowed through the shrinkage applied to fishing mortality at age. Although the controls imposed by these models take place through different mechanisms, both types of controls decrease the degree to which the estimates of fishing mortality at age can depart from what was estimated in the past. We believe that some degree of shrinkage is appropriate in all assessment models, because if the F values are not constrained they may drift, especially for younger age classes in the most recent years. Alternatively, if analysts believe that F values should not be constrained, the analysts should justify why fishing mortality at age for these younger age classes is expected to change (e.g., through changes in recruitment or in the bycatch or discard mortality rates). Such changes may not be obvious in outputs showing only annual fishing mortalities averaged over age classes.

Models differ in how they deal with conflicting data, such as seen in the survey CPUE data (Figure D-1c). The relative abundance indices may have peaked with the 1995 cohort. The trends indicated by the two observations available for the 1998 cohort (shown in Figure D-1c as age 0 and age 1) go in opposite directions; the survey CPUE of year-0 summer flounder is increasing, whereas the survey CPUE for year-1 fish is decreasing. This creates a conflict in the data that may lead to equally likely, but conflicting, trends in model estimates. To explore and address this issue, a modification was made to the CAGEAN model to allow for some variation in selectivity over time.[2] When the CAGEAN model was used with varying selectivity at age 0 and age 1, the spawning stock biomass estimates more closely followed those derived under the 1999 NMFS assessment (Figure D-18). This indicates that the selectivity assumption caused at least some of the differences in the outputs of the NMFS ADAPT and CAGEAN models. Selectivity estimates, as portrayed in fishing mortality estimates at age (shown in Figure D-19), show a sharp drop in recent years, which the model assumes corresponds to changes in selectivities in commercial and recreational fisheries. This assumption, if false, could mask more serious alternative processes that also could provide an explanation for the trends, such as a drop in recruitment in recent years.

Estimated fishing mortality rates varied greatly among the models (Figure D-5). In the last year of the series, each method produced almost the same estimate of fishing mortality, yet all are above the fisheries management target level for F of 0.24.

Stock assessment scientists responsible for summer flounder should investigate how differences among model results arise and whether such differences indicate changes needed in the models or assumptions used. NMFS should try to test the effects of shrinkage and should investigate the conflicting trends in the youngest year classes. As stated in the 1998 NRC report *Improving Fish Stock Assessments* (NRC, 1998a):

> Because there are often problems with the data used in assessments, a variety of different assessment models should be applied to the same data. . . . The different views pro-

[2] This was done by allowing deviations to occur in the log selectivity-at-age with a penalty applied to the sum of the squared deviations in the log-likelihood formulation used to specify the optimization surface.

> vided by different models should improve the quality of assessment results. (p. 113)

This advice is borne out by the current committee's reassessment of summer flounder data.

In the summer flounder case, the NMFS assessment could be improved by analyzing the same data using different models. The differences obtained should help analysts learn about problems in the data, problems in using the ADAPT model with these data, or problems with the assumptions used in the NMFS ADAPT model (e.g., related to shrinkage and changes in selectivity over time).